Interfaces in High-T_c Superconducting Systems

Subhash L. Shindé David A. Rudman
Editors

Interfaces in High-T_c Superconducting Systems

With 129 Figures

Springer-Verlag
New York Berlin Heidelberg London Paris
Tokyo Hong Kong Barcelona Budapest

Subhash L. Shindé
IBM Corporation
B-300-40 E
Hopewell Junction, NY 12533
USA

David A. Rudman
Division 814.03 NIST
325 Broadway
Boulder, CO 80303-3328
USA

Library of Congress Cataloging-in-Publication Data
Interfaces in high-T_c superconducting systems/Subhash L. Shindé, David A. Rudman.
p. cm.
Includes bibliographical references and index.
ISBN 0-387-94000-6
1. High temperature superconductors—Surfaces—Congresses.
2. Semiconductors—Junctions—Congresses. 3. Thin films—Surfaces—Congresses. I. Shindé, Subhash L. II. Rudman, David Albert.
QC611.98.H54I56 1993
537.6'236—dc20 92-21214

Printed on acid-free paper.

Production managed by Francine McNeill; manufacturing supervised by Vincent Scelta.
Typeset by Asco Trade Typesetting Ltd., Hong Kong.
Printed and bound by Edwards Brothers, Inc., Ann Arbor, MI.
Printed in the United States of America.

9 8 7 6 5 4 3 2 1

ISBN 0-387-94000-6 Springer-Verlag New York Berlin Heidelberg
ISBN 3-540-94000-6 Springer-Verlag Berlin Heidelberg New York

Preface

Over the past six years, the field of high-temperature superconductivity has advanced at a very rapid rate. Continued growth of this field demands more focused studies of various issues vital to the understanding and application of high-temperature superconductors. One key area of study is the structure and properties of interfaces involving these complex materials. Interfaces in superconducting systems include naturally occurring microstructural features, such as grain boundaries, twin boundaries, and interphase boundaries, as well as artificially fabricated boundaries, such as substrate–superconductor interfaces and heterostructures, as well as interfaces in composites. These interfaces critically affect the superconducting properties, and their understanding is crucial to the development of both bulk and thin-film applications of high-temperature superconductors.

This volume attempts to cover a part of this fascinating field. The initial chapters describe the nucleation and growth of thin films, including the important aspects of defect generation and their distribution. Defects in thin films and bulk materials (especially grain boundaries) are covered next with a discussion of the relationship between the defect morphology and the superconducting properties. The authors have used techniques like transmission electron microscopy (TEM) and scanning tunneling microscopy (STM) for this purpose. The substrate–superconductor and metal contact–superconductor interfaces are described in the following chapters. It is believed that the interaction of the first few monolayers of metal atoms with the superconductor surfaces dominates the nature of the contact. X-ray photoelectron spectroscopy has been used for these studies. Finally, superconducting properties of single and multiple grain-boundary junctions are discussed. This involves detailed electrical characterization. Despite the continued developments in the above areas since the conception of this book, it is hoped that the concepts outlined here by noted scientists will endure.

I would like to thank my colleagues T.K. Worthington, T. Shaw, T.R. Dinger, P. Choudhari, A. Schrott, and F. Himpsel for all the discussions and for critical reading of the chapters. I would also like to thank T. von Foerster and C. Daly of Springer-Verlag for the timely publication of this volume. Finally, very special thanks go to all of the authors, whose time and effort made this a worthwhile endeavor.

Hopewell Junction, New York

SUBHASH L. SHINDÉ

Contents

1
Observations on the Growth of $YBa_2Cu_3O_{7-\delta}$ Thin Films by Transmission Electron Microscopy

M. GRANT NORTON and C. BARRY CARTER

Abstract

A major research effort in high-temperature oxide superconductors is the growth of high-quality thin films. The physical properties of polycrystalline thin films are controlled by their microstructure which is influenced by the early stages of film growth and the establishment of epitaxy. In this article, the nucleation and heteroepitactic growth of $YBa_2Cu_3O_{7-\delta}$ thin films is reviewed. In particular, the role of transmission electron microscopy in these studies is highlighted.

1.1 Introduction

Since the discovery of superconductivity in the oxide systems La–Ba–Cu–O [1] and Y–Ba–Cu–O [2] there has been a surge of interest in the growth of thin films of both these materials and also the newer superconducting oxides which have higher transition temperatures. The majority of the thin films produced are polycrystalline and it is common that such films are epitactic—the term epitactic will be defined below—thus the microstructure of the film is important in determining a number of physical properties. The microstructure of the film is determined, at least in part, by the nucleation and early stages of film growth. It is therefore of great importance that the initial stages of film growth are understood. The substrate exerts an important influence during the early stages of film growth through the establishment of epitaxy. Important factors at this stage include the lattice mismatch between

Department of Materials Science and Engineering, Cornell University, Ithaca, NY 14853.

[1] M.G. Norton's present address is the Department of Mechanical and Materials Engineering, Washington State University, Pullman, WA 99164.

[2] C.B. Carter's present address is the Department of Chemical Engineering and Materials Science, University of Minnesota, Minneapolis, MN 55455.

the film and substrate and the bonding of the deposit atoms with the substrate. Epitaxy may also be influenced by the substrate temperature during deposition and other parameters associated with the deposition process.

There are a number of techniques that have been used to study the thin-film growth of oxide superconductors. These techniques include: reflection high-energy electron diffraction (RHEED) [e.g., Ref. 3], x-ray diffraction (XRD) [e.g., Ref. 4], transmission electron microscopy (TEM) [e.g., Refs. 5 and 6], Rutherford backscattering spectrometry (RBS) [e.g., Ref. 7], and more recently scanning tunneling microscopy (STM) [e.g., Refs. 8 and 9].

TEM has contributed greatly to the understanding of epitaxy [10]. The high spatial resolution of the electron microscope is well-suited for studies of the early stages of thin-film growth—very-thin films. Weak-beam imaging and high-resolution electron microscopy (HREM) can provide information on the nanometer and subnanometer scale. By the use of electron diffraction and x-ray analysis in the electron microscope, chemical as well as structural information may be obtained. Two important factors in TEM studies of thin-film growth are the need for well-defined substrate surfaces and the elimination of artifacts which can be introduced during specimen preparation for TEM analysis.

In this chapter, a novel specimen preparation technique for the examination of thin-film and near-surface effects by TEM will be discussed. This technique circumvents the problems indicated above and has successfully been used in the study of a number of different effects, these include: metal epitaxy on ceramic substrates [e.g., Ref. 11], the heteroepitactic growth of oxide thin films [e.g., Refs. 12–14], the study of solid-state reactions [e.g., Ref. 15], the study of vapor phase corrosion [e.g., Ref. 16], and the effect of ion implantation of ceramics [e.g., Ref. 17].

Before the details of the specimen preparation technique are discussed and different examples are used to illustrate the importance of this technique in the examination of the growth of $YBa_2Cu_3O_{7-\delta}$ thin films; it is necessary to discuss some general issues involved in the growth of thin films and in particular the growth of $YBa_2Cu_3O_{7-\delta}$ thin films.

1.1.1 *Types of Epitactic Growth*

The term *epitaxy* is derived from the Greek words $\varepsilon\pi\iota$ (over) and $\tau\alpha\xi\iota\sigma$ (arrangement) and is defined as "The growth of crystals on a crystalline substrate that determines their orientation; the orientation of crystals so grown" [18]. The adjectival form of epitaxy that will be used throughout this text is epitactic [19]. The occurrence of epitaxy has been recognized for well over a century [20, 21], many of the early examples of epitaxy concerned crystal growth from solution. It was thought that a small misfit between the deposit crystal lattice and the substrate was necessary for the occurrence of epitaxy. Although many of the early examples of the occurrence of epitaxy, in particu-

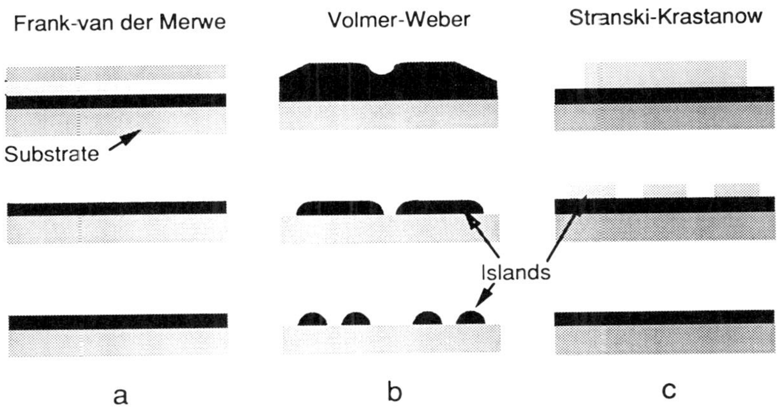

FIGURE 1.1. Three forms of epitactic growth: (a) layer-by-layer growth; (b) formation and growth from discrete nuclei; and (c) initial formation of a tightly bound layer followed by nucleation and growth (redrawn after M.J. Stowell [26]).

lar from solution, supported this concept [22] it was however, demonstrated that good epitaxy can occur even for very large lattice misfits [e.g., Ref. 23].

Because the mode of growth of an epitactic thin film is an important influence on the final microstructure of the film and on the types and amount of defects formed, it is relevant to review, very briefly, some of the prominent features of epitactic growth mechanisms. For a more detailed account of epitactic growth the reader is referred to the book edited by Matthews [24]. Three forms of epitactic growth are shown schematically in Figure 1.1; these are (1) layer-by-layer growth (Frank–Van der Merwe mode), (2) formation and growth from discrete nuclei (Volmer–Weber mode), and (3) initial formation of a tightly bound layer (or layers) followed by nucleation and growth (Stranski–Krastanow mode). Layer-by-layer growth will only occur when there is a strong interaction between the film atoms and the substrate atoms. Even in the case where there is strong interphase atomic bonding, layer-by-layer growth is only effective up to a limited film thickness for systems where there is a non-zero misfit [25]. The lattice mismatch between the film and substrate leads to a strain energy that serves as the driving force for cluster generation in many cases [25]. Therefore, for systems where there is a non-zero misfit, the most likely growth modes are either the formation of clusters on a bare substrate (Volmer–Weber mode) or on a few layers of uniform film (Stranski–Krastanow mode).

In all cases, it is possible for the initial deposit to be elastically strained to match the substrate; this is referred to as pseudomorphic growth. The dominant condition for the occurrence of this growth mode is that the strain necessary to ensure coherency at the film/substrate interface is small. Ini-

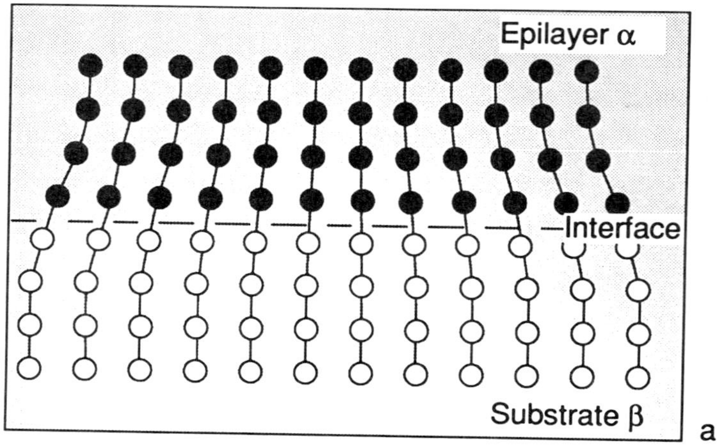

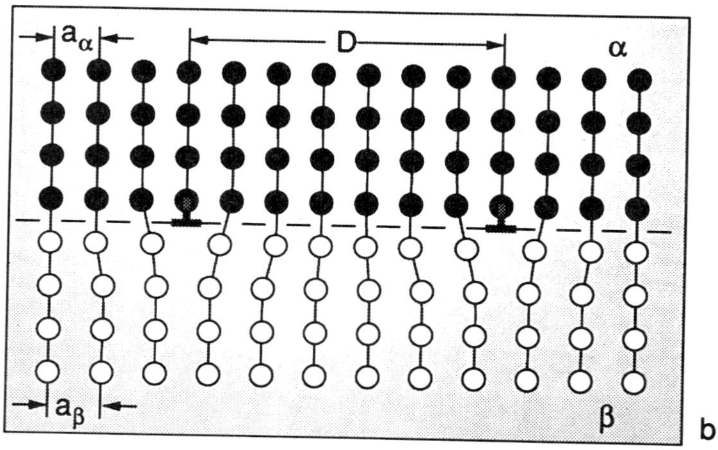

FIGURE 1.2. (a) A coherent interface with slight mismatch leads to coherency strains in the adjoining lattices (redrawn after D.A. Porter and K.E. Easterling, *Phase Transformations in Metals and Alloys* (Van Nostrand Reinhold, New York 1981)). (b) A semicoherent interface. The misfit parallel to the interface is accommodated by a series of edge dislocations (redrawn after D.A. Porter and K.E. Easterling, *Phase Transformations in Metals and Alloys* (Van Nostrand Reinhold, New York, 1981)).

tially the deposit will be elastically strained as shown schematically in Figure 1.2(a), but as the film thickens misfit dislocations will be introduced since the strain energy becomes so large that it is favorable to introduce misfit dislocations at the interface, so losing coherency [26]. The situation where misfit dislocations have been introduced at the interface is shown schematically in Figure 1.2(b).

1.1.2 *Importance of Superconducting Thin Films*

The high critical currents achieved in thin films of the $YBa_2Cu_3O_{7-\delta}$ superconductor [e.g., Ref. 27] highlight the importance of thin-film technology in this materials system. The formation of high performance weak links and tunnel junctions [e.g., Refs. 28 and 29]—important in the realization of superconductor applications, for example, Josephson elements in superconducting quantum interference devices (SQUIDS) and radiation sensor applications—further illustrate the important role that thin films have in the development of high-temperature superconductors.

1.1.3 *Deposition Techniques and Substrates*

Superconducting thin films of $YBa_2Cu_3O_{7-\delta}$ have been fabricated using a number of different techniques including electron-beam evaporation [e.g., Ref. 30], sputtering [e.g., Ref. 31], chemical vapor deposition [e.g., Ref. 32], and pulsed-laser ablation [e.g., Ref. 33]. Films have been deposited onto a number of different substrates, including single crystals of MgO [e.g., Ref. 34], $SrTiO_3$ [e.g., Ref. 35], and $LaAlO_3$ [e.g., 36] and polycrystalline substrates of MgO [37] and yttria-stabilized zirconia (YSZ) [38]. The choice of substrate has mainly been determined by two criteria:

1. The epitactic films should have a high transition temperature and a high critical current.
2. The substrate should be suitable for device or technological applications.

These two criteria are not both satisfied for many of the systems which have been studied. For example, $YBa_2Cu_3O_{7-\delta}$ thin films deposited on $SrTiO_3$ (where the lattice mismatch is $\sim 2\%$) have been shown to have a very high degree of crystalline perfection, as determined by ion-channeling in Rutherford backscattering spectrometry (RBS) studies [39]. However, $SrTiO_3$ has a very high dielectric constant (> 300 at 20°C), which precludes its usefulness in many microwave device applications. In some cases, the microstructure produced on a substrate which has a large lattice mismatch, for example, MgO (lattice mismatch $\sim 9\%$) may in itself be useful. Thin films of $YBa_2Cu_3O_{7-\delta}$ produced on MgO contain many naturally occurring weak links—high-angle grain boundaries. It is well-known that grain boundaries in $YBa_2Cu_3O_{7-\delta}$ thin films have superconducting weak-link properties which can, at least approximately, be described by the Josephson equations, and that these properties are dependent on the orientation of the two grains [40, 41]. By studying naturally occurring grain boundaries in polycrystalline superconducting thin films, insight into the nature of the Josephson-like properties of high-temperature superconducting weak links can be obtained [28].

1.2 A Technique for the Examination of Film Nucleation and the Early Stages of Growth

Ex situ and in situ deposition studies using TEM have provided some of the first direct information on the processes which occur during film growth [42–44]. The technique discussed here uses the rationale of these earlier experiments; wherein an electron-transparent thin foil acts as the substrate material. However, new specimen preparation techniques have been introduced since these earlier studies, most notably ion-milling, which are used to advantage in this present work; thus, extending the range of possible materials which can be used as substrates. Depositions can be performed directly onto these thin-foil substrates, which can be transferred directly from the deposition chamber to the electron microscope, thereby avoiding any possibility of defect introduction during post-deposition specimen preparation. It has been demonstrated that specimen preparation methods for TEM analysis (e.g., polishing, dimpling, and ion-milling) can introduce defects into $YBa_2Cu_3O_{7-\delta}$ materials [e.g., Ref. 45].

The details of preparing these thin-foil substrates is as follows: The thin-foil specimens are initially prepared using standard methods for the preparation of specimens for TEM analysis; 3-mm disks are cored from a larger single-crystal substrate. These disks are mechanically polished and dimpled to produce a specimen thickness of $\sim 20\ \mu m$. The polished specimens are then ion-milled to perforation using 5-kV Ar^+ ions. After ion-milling to perforation, the thin foils are chemically cleaned and then annealed. The clean-

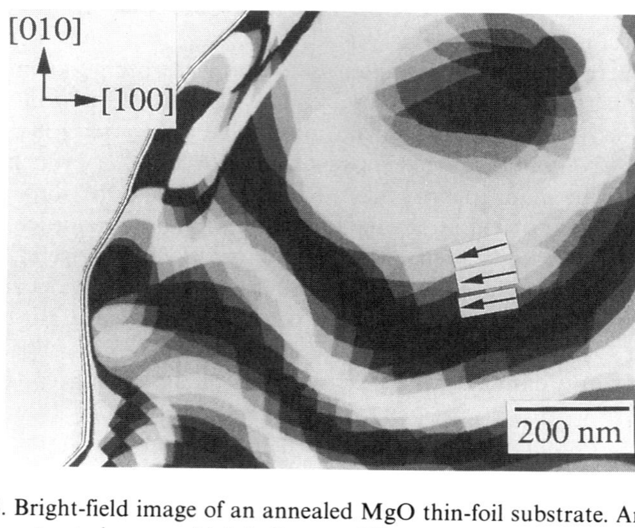

FIGURE 1.3. Bright-field image of an annealed MgO thin-foil substrate. Arrows show a series of contrast changes which indicate a different thickness, therefore a series of steps on the surface.

ing process is a multistep process whereby any contamination resulting from ion-milling is removed [46]. The cleaning process was optimized to produce a clean surface. The high-temperature annealing produces a surface consisting of a series of well-defined terraces and steps. The annealing temperature for MgO is 1350°C for 10 min and for $SrTiO_3$ the annealing temperature is 1100°C for 10 min [46]. An example of such a surface on (001)-oriented MgO is shown in the bright-field image in Figure 1.3. The surface steps give rise to a series of discrete contrast changes. Each level of contrast indicates a step terrace bound by step planes. These relatively wide terraces and long, well-defined steps provide an ideal surface for examination of the early stages of thin-film growth. Such surface structures can be formed on many oxide [e.g., Ref. 46] and non-oxide ceramics [47].

1.3 Nucleation

Film deposition directly onto the thin-foil substrates has given valuable information on both the growth mechanism of $YBa_2Cu_3O_{7-\delta}$ films and the importance of the substrate surface on film growth. Figure 1.4 shows a bright-field image of an MgO thin-foil substrate recorded close to the [001] zone axis showing islands of $YBa_2Cu_3O_{7-\delta}$ which have formed on the (001) surface during the early stages of growth [5, 48]. Film deposition was by pulsed-laser ablation; the amount of material deposited onto the individual

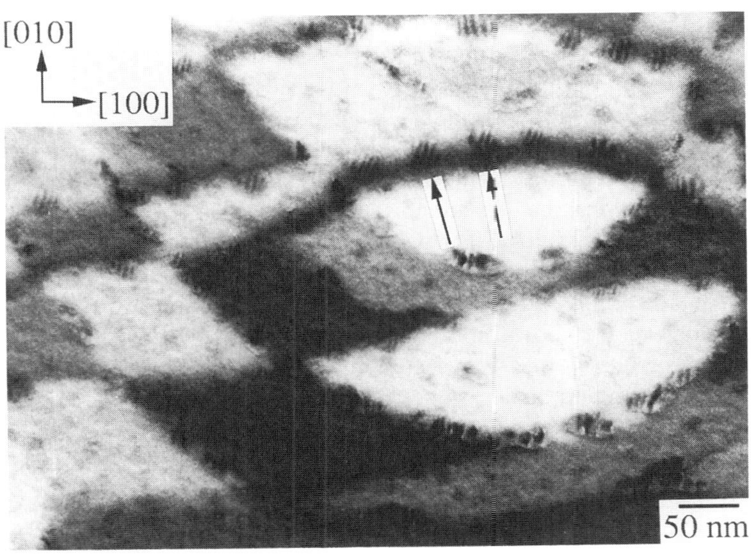

FIGURE 1.4. The early stages of film nucleation and growth The arrows indicate deposit islands aligned along the surface steps.

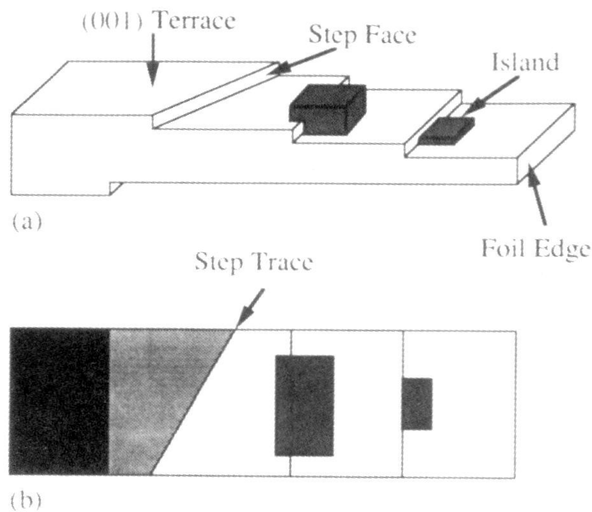

FIGURE 1.5. (a) Schematic showing the preferential nucleation of islands at step edges formed on an electron-transparent thin foil; (b) schematic representation of a bright-field image of (a), viewed in plan geometry. The contrast changes in the bright-field image can be seen to arise from different thicknesses of the thin-foil substrate. (Courtesy L.A. Tietz.)

substrates was controlled by controlling the number of laser pulses. The preferential nucleation of the islands at the surface steps is illustrated by arrows in Figure 1.4. The formation of islands at the step edges is shown schematically in Figure 1.5. Also illustrated in Figure 1.5 is how the contrast changes, observed in the bright-field images in Figures 1.3 and 1.4, result from the formation of steps on the substrate surface. Selected-area diffraction (SAD) patterns recorded from such samples showed that the [001] zone axes of the substrate and the individual islands were aligned. It has been proposed [5, 48] that the alignment of the $YBa_2Cu_3O_{7-\delta}$ nuclei with the surface steps increases the perfection of the alignment observed between the different grains during film growth. This preannealing is thus important for the formation of high-quality ultrathin films [49]. The deposition of films onto a substrate consisting of a series of steps is an example of graphoepitaxy, the alignment during epitactic growth being controlled by the surface topography [5, 48].

1.3.1 *Importance of the Substrate Surface*

The observation of growth at steps shows that the topography of the substrate surface is important in influencing the early stages of film growth. Two further examples of the importance of the substrate surface are illustrated below. The first example shows the importance of the substrate preparation procedure [49–52]. The second example shows how by formation of a well-defined step structure on a bulk substrate the formation of oriented films can be obtained [37].

The surface quality of $SrTiO_3$ substrates was examined by RBS ion-channeling studies [52], in this study it was shown that the surface quality of polished substrates could be improved by high-temperature annealing. For $SrTiO_3$ substrates the annealing was performed at 950°C for 1 h in oxygen. At an annealing temperature of 1100°C for 10 min it was shown by TEM that reconstruction of the (001) surface of $SrTiO_3$ had occurred and surface steps had formed [46]. For MgO substrates, annealing at >1100°C was found to improve the surface quality of the substrate [49, 51].

An examination of the microstructure and properties of $YBa_2Cu_3O_{7-\delta}$ films deposited by pulsed-laser ablation onto single-crystal MgO substrates prepared in different ways illustrates the importance of the surface preparation [50]. The substrates were prepared in three ways: chemically polished; mechanically polished; mechanically polished and then annealed at 1100 to 1200°C for 12–24 h. The $YBa_2Cu_3O_{7-\delta}$ films were deposited under identical conditions using pulsed-laser ablation. The microstructure and crystalline quality of the deposited films was very different depending on the substrate pretreatment. Films grown on chemically polished substrates contain a large number of grains which are rotated about the [001] zone axis. These rotations give rise to tilt boundaries with angular misorientations of 45°, 27°, 16°, 9°, and ~1°. All of these angles can be accounted for by preferred epitactic relations between the film and the substrate surface [e.g., Ref. 53]. For films deposited on mechanically polished substrates, the x-ray pole-figure analysis indicated a 10%–20% density of rotated grains occurring at a misorientation angle of ~45°. For $YBa_2Cu_3O_{7-\delta}$ films grown on annealed MgO the pole-figure analysis presents no evidence of any high-angle tilt boundaries—the *a* and *b* axes of the film were aligned with the substrate crystal axes. X-ray rocking-curve measurements yield further information on the crystalline quality of films deposited on the different substrates, these result are shown in Table 1.1 alongside ion-channeling measurements from the same

TABLE 1.1. A summary of the results of the structural and superconductivity measurements made on films on various MgO substrate surfaces. All the films were deposited by pulsed-laser ablation using the same deposition parameters.

Substrate preparation	RBS analysis χ min	Rocking-curve analysis on (005) peak (FWHM)	$T_c(R = 0)$ (K)	J_c at 77 K (10^6 A/cm^2)
Chemically polished	62.0%	1.02°	82.5–84.0	0.0032–0.0250
Mechanically polished	18.7%	0.72°	83.0–86.0	0.5000–3.3000
Annealed	18.0%	0.30°	87.5–88.5	3.4000–6.0000

films. Also shown in Table 1.1 are transition temperatures and critical currents determined for the individual films.

The presence of steps on a surface is important in a number of different situations. For example, they provide active sites for catalysis and influence surface diffusion processes [54]. They also influence properties which are affected by point defects such as segregation and also provide sources and sinks for the vacancies which produce the surface space charge in ionic materials [55]. The presence of steps on the surface of a thin-foil substrate was shown to provide preferential sites for island nucleation [5, 48]. The ideas developed from those previous studies were carried over to bulk substrates; where the growth of $YBa_2Cu_3O_{7-\delta}$ thin films on vicinal MgO was investigated [37]. A vicinal surface may be defined as one "approximating, resembling, or taking the place of a fundamental crystalline form or face" [56], that is, a vicinal surface can be thought of in terms of terraces parallel to low-index, low-energy planes and steps [54]. For MgO, which has the rocksalt structure, the surface with the lowest energy is (001), which is also the cleavage plane. Figure 1.6 shows a schematic representation of a vicinal surface formed on MgO. The surface consists of (001) terraces, at an angle θ to the original substrate surface, which are bound by steps. The steps have height h and are themselves parallel to (100) planes.

There are three very significant differences between the microstructure of $YBa_2Cu_3O_{7-\delta}$ films formed on vicinal MgO and those formed on (001)-oriented MgO. For the films grown on vicinal surfaces the c axis of the $YBa_2Cu_3O_{7-\delta}$ tends not to be parallel to the [001] axis of the substrate, instead the [001] zone axis of the film was aligned parallel to the substrate normal. X-ray rocking-curve measurements show that, for small vicinal angles, there was a large mosaic spread indicating that some grains are misoriented towards the (001) direction of the substrate [37, 57]. However, no evidence was found for grains oriented parallel to the substrate [001] direction. This observation indicates that growth of $YBa_2Cu_3O_{7-\delta}$ thin films on vicinal surfaces is a form of graphoepitaxy. It is not conventional heteroepitaxy, where the lattice parameter and the crystal structure of the substrate

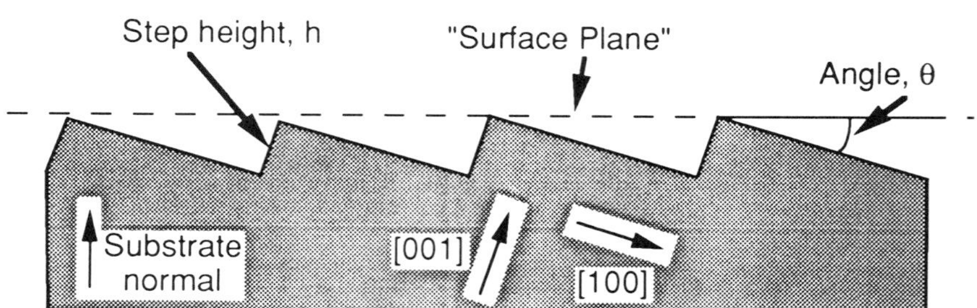

FIGURE 1.6. Schematic representation of a stepped surface. The step height is h and the angle of deviation from the low-energy (001) surface is θ.

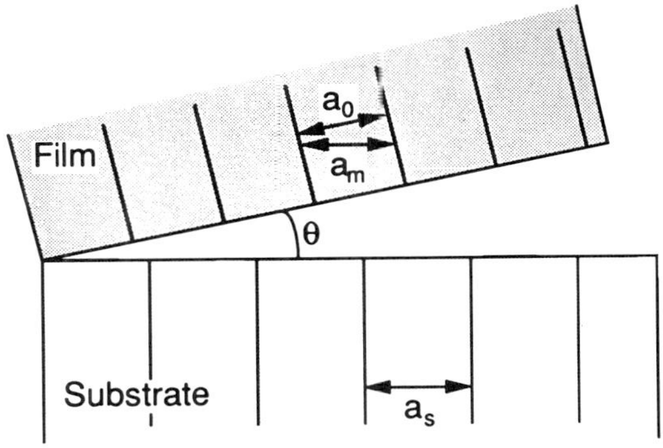

FIGURE 1.7. Schematic showing the general process of film growth proposed for $YBa_2Cu_3O_{7-\delta}$ films on vicinal surfaces (not drawn to scale).

TABLE 1.2. Variation in lattice misfit with angle of rotation of the tilt boundary.

Angle of rotation, deg	Misfit, % [calculated from Eq. (2)]
0	10.21
2.5	10.21
5	9.92
10	8.78
15	6.58
20	3.43

controls the microstructure. The result is that the misfit at the interface is accommodated, in part, by a tilt boundary which is superimposed on the interphase interface. Figure 1.7 illustrates the general process of film growth proposed here. Such a mechanism is common in both other oxide systems [58] and has recently been reported for metals [59]. The calculated misfit for different angles of rotation is shown in Table 1.2. In the situation where there is no lattice rotation, the misfit f can be calculated from the equation

$$f = a_s - a_o/a_o \tag{1}$$

if the stress-free lattice parameters of the substrate and overgrowth are a_s and a_o, respectively, and the thickness of the film is less than the thickness of the substrate [24]. For the situation where the lattice planes of the overgrowth are rotated, the modified lattice parameter for the film becomes a_m, where $a_m = (a_o/\cos\theta)$. If a_m is substituted for a_o in Eq. (1), then the modified lattice misfit f' is given by Eq. (2):

$$f' = (a_s - a_m)/a_m. \tag{2}$$

From Table 1.2 it can be seen that the lattice misfit decreases with increasing angle of rotation of the tilt boundary. It has been observed that the FWHM decreases for films grown on vicinal MgO where the vicinal angle is $>10°$ [37]. A possible reason for this observation is that the mismatch is significantly reduced at these high angles thereby favoring, on energetic grounds, growth parallel to the substrate normal.

No comparable study has been performed on the growth of $YBa_2Cu_3O_{7-\delta}$ thin films on a more closely lattice-matched substrate, for example, $SrTiO_3$ or $LaAlO_3$. However, growth of $YBa_2Cu_3O_{7-\delta}$ on a rough $SrTiO_3$ substrate showed that the *c* axis of the film was aligned with the $SrTiO_3$ (001) crystallographic direction on areas inclined a few degrees from the (001) direction [60]. This cited study was not a comprehensive examination of film growth on misoriented surfaces, but may provide an insight into film growth on vicinal surfaces where there is not a large lattice mismatch. For the systems $YBa_2Cu_3O_{7-\delta}/SrTiO_3$ and $YBa_2Cu_3O_{7-\delta}/LaAlO_3$ where the lattice misfit is 2.3% and −1.1%, respectively, there may be no reason based on energetic considerations for formation of the tilt boundary at the interface.

X-ray pole-figure analysis indicated that there were no high-angle grain boundaries present in $YBa_2Cu_3O_{7-\delta}$ films grown on MgO where the vicinal angle was 5°. The film microstructure as examined by TEM was very different from that observed for $YBa_2Cu_3O_{7-\delta}$ films grown on (001)-oriented MgO. $YBa_2Cu_3O_{7-\delta}$ films grown on (001)-oriented MgO typically have a

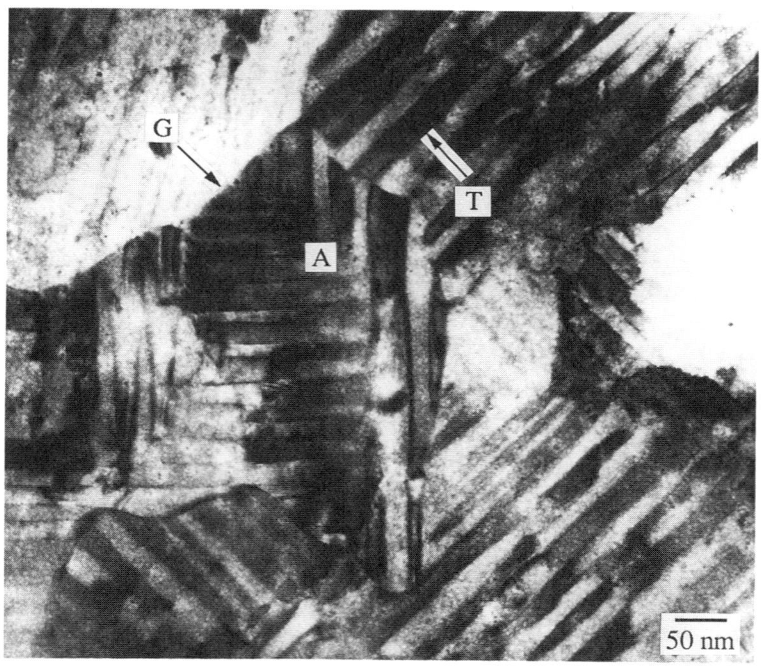

FIGURE 1.8. Bright-field image of a $YBa_2Cu_3O_{7-\delta}$ thin film formed on (001)-oriented MgO. The presence of twins (T) and grain boundaries (G) in the film are indicated.

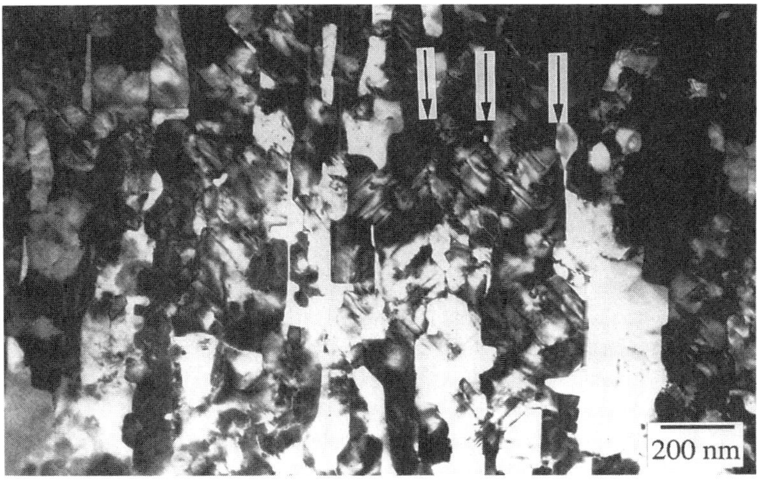

FIGURE 1.9. Bright-field image of a $YBa_2Cu_3O_{7-\delta}$ thin-film on 5° vicinal MgO. The predominant direction of the grain boundaries is indicated by arrows.

mosaic-type microstructure [e.g., Refs. 5, 61], as illustrated in Figure 1.8, where the *c* axis oriented grains are equiaxed with a diameter up to ~1 μm. The mosaic microstructure is common for $YBa_2Cu_3O_{7-\delta}$ films grown on substrates where there is a large lattice mismatch [e.g., Ref. 62]. Figure 1.9 shows a bright-field image of a thin film of $YBa_2Cu_3O_{7-\delta}$ grown on vicinal MgO, where the vicinal angle was 5°. The microstructure consists of *c* axis oriented grains, which have a much larger aspect ratio than that observed in Figure 1.8. The predominant direction of the grain boundaries is shown by arrows in Figure 1.9. This microstructure can be described by a growth mechanism where the step edges act as nucleation sites. The alignment of the islands with the step edges prevents in-plane rotational misalignment, which would occur if the islands were not constrained; enabling formation of these long grains [63]. Thus, growth on vicinal surfaces is another illustration of graphoepitaxy. This process enables growth of highly oriented microstructures on a substrate which is not lattice-matched.

1.4 The Early Stages of Film Growth: Ultrathin $YBa_2Cu_3O_{7-\delta}$ Films

The early stages of film growth (ultrathin films) have been studied primarily on two substrate materials—MgO and $SrTiO_3$. The specimen preparation technique for TEM analysis, as described above, has been used to study the early stages of growth of $YBa_2Cu_3O_{7-\delta}$ films by pulsed-laser ablation on

(001) orientations of MgO, $SrTiO_3$, and $LaAlO_3$ substrates [64]. Other techniques that have been used to determine the nature of film growth have included RHEED [e.g., Ref. 3], XRD [e.g., Ref. 4], and STM [8, 9]. Growth of $YBa_2Cu_3O_{7-\delta}$ on $SrTiO_3$ has been proposed to occur by both layer-by-layer and island growth mechanisms. These differences may be due, in part, to different deposition techniques or to different deposition conditions such as the substrate temperature. For films deposited by sputtering both the STM and XRD studies indicate film growth by an island mechanism. However, $YBa_2Cu_3O_{7-\delta}$ films deposited by activated reactive evaporation were reported to be formed in a layer-by-layer fashion. This growth mechanism has also been proposed [51] to explain the good superconducting properties obtained with ultrathin (~3-nm) films of $YBa_2Cu_3O_{7-\delta}$ grown on $SrTiO_3$ by inverted cylindrical magnetron sputtering. Consideration of the atomic bonding and lattice misfit would suggest that the growth of $YBa_2Cu_3O_{7-\delta}$ thin films on $SrTiO_3$ would occur by an island mechanism.

Figure 1.10 shows a thin (~12-nm) film of $YBa_2Cu_3O_{7-\delta}$ which has been deposited by pulsed-laser ablation on a thin-foil substrate of (001)-oriented $SrTiO_3$. The film appears to be continuous and smooth. No moiré fringes or dislocations are apparent in the image suggesting that film growth is pseudomorphic; at least at thicknesses <12nm. This observation is supported by RBS ion-channeling studies on $YBa_2Cu_3O_{7-\delta}$ thin films on $SrTiO_3$ [65], which showed that the films were slightly strained. The strain was directly observed in a 9-nm-thick film by channeling at an angle of 45° from the surface normal, but was not found in films with thicknesses >13 nm.

The growth of $YBa_2Cu_3O_{7-\delta}$ films on MgO was found to occur by an island mechanism [e.g., Refs. 5, 6, 8, 13, 48, 66] as can be seen in Fig-

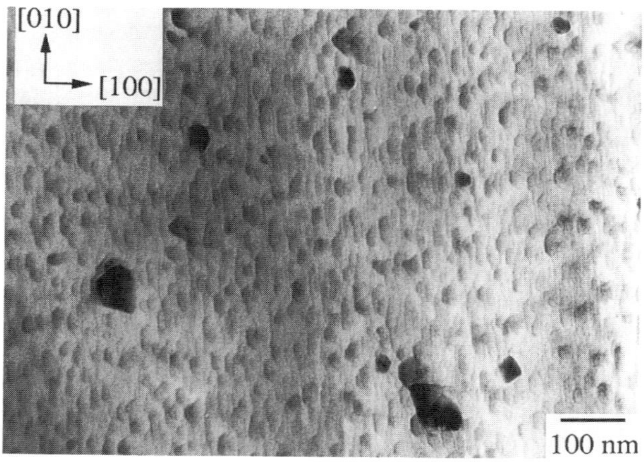

FIGURE 1.10. Bright-field image showing a thin $YBa_2Cu_3O_{7-\delta}$ film on (001)-oriented $SrTiO_3$.

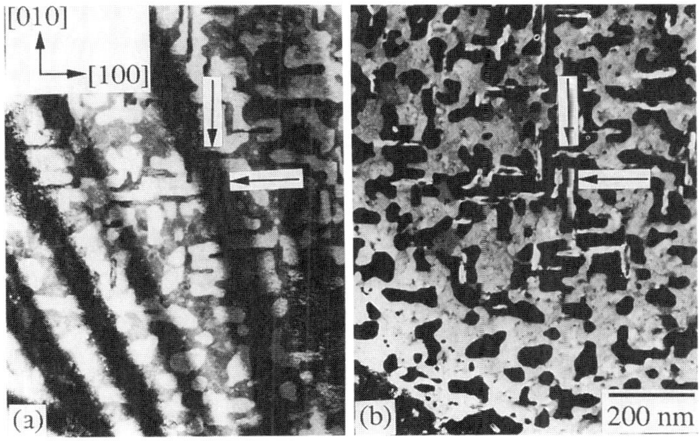

FIGURE 1.11. (a) Bright-field image of $YBa_2Cu_3O_{7-\delta}$ thin film on (001)-oriented MgO, and (b) dark-field image from the same area. The dark-field image was formed using a (110) reflection from the film, thus the film appears bright, while the substrate areas are dark.

ures 1.11(a) and 1.11(b). The bright-field and dark-field images show a thin ($\sim$12 nm) $YBa_2Cu_3O_{7-\delta}$ film which has been deposited directly onto a thin-foil substrate of (001)-oriented MgO. It can be seen from the bright-field image, Figure 1.11(a), that the film is not continuous at this stage of growth, but it is interconnected. Different grain morphologies are visible; a contiguous area which comprises the majority of the film and needle-shaped grains, identified by arrows, which are aligned in orthogonal directions along the [010] and [100] crystal axes of the MgO. The dark-field image shown in Figure 1.11(b) was recorded using a {110} reflection from the $YBa_2Cu_3O_{7-\delta}$ crystal lattice, so the bright areas correspond to epitactic areas of the film, while the dark areas are either bare substrate or grains in a different orientation or having a different composition. A SAD pattern recorded from the above area is shown in Figure 1.12(a) and schematically in Figure 1.12(b). From the SAD pattern the film appears to be oriented exclusively with the *c* axis of the $YBa_2Cu_3O_{7-\delta}$ perpendicular to the (001) surface of the MgO, no indication of the presence of any other grain orientations is visible. (By using much longer exposure times for the diffraction patterns very faint reflections from these other grains were visible [67].) The different grain morphologies can be explained if the grains grow fastest in directions normal to the *c* axis [66].

The microstructure observed for $YBa_2Cu_3O_{7-\delta}$ thin films grown on MgO by pulsed-laser ablation is qualitatively similar to that observed for ultra-thin films grown on MgO by single-target off-axis magnetron sputtering [6], where it was further demonstrated that $YBa_2Cu_3O_{7-\delta}$ films grow on MgO substrates by an island mechanism. Cross-section TEM studies of these films

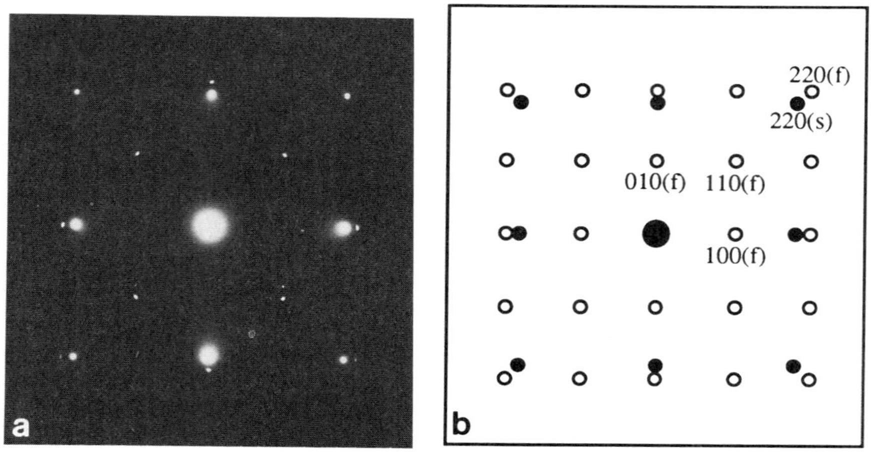

FIGURE 1.12. (a) SAD pattern showing alignment of the [001] zone axes of $YBa_2Cu_3O_{7-\delta}$ and MgO; (b) corresponding schematic.

confirmed that the areas between the islands were indeed bare substrate, confirming the earlier postulate [13]. Taken together these studies indicate that $YBa_2Cu_3O_{7-\delta}$ film growth on MgO occurs by a Volmer–Weber growth mode, as shown schematically in Figure 1.1(b). The island growth mechanism of $YBa_2Cu_3O_{7-\delta}$ on MgO has recently been confirmed by STM studies [8].

1.5 Moiré Fringes and Their Application to the Study of Film Growth

The periodic structure of a crystal can be revealed by means of moiré patterns from overlapping crystals [68]. Moiré fringe patterns are formed in electron microscope images as a result of interference between diffracted beams from two overlapping crystals and, in the bright-field case the direct beam. If the two crystals being considered are the deposited film and the substrate, then information can be obtained about the orientation relationship between the two. For a detailed discussion on the formation of moiré fringe patterns in electron microscope images the reader is referred to the book by Hirsch et al. [68]. For the purposes of this article it is sufficient to consider the formation of moiré fringes in a purely geometrical manner. Figure 1.13 shows an analog representing the overlapping of two line gratings. In Figure 1.13(a), the two overlapping gratings are parallel, but have different spacings—the situation that could be envisaged in the growth of a thin film on a substrate where the lattice parameters of the two crystals are different—producing a structure with a double periodicity. Beats are formed between

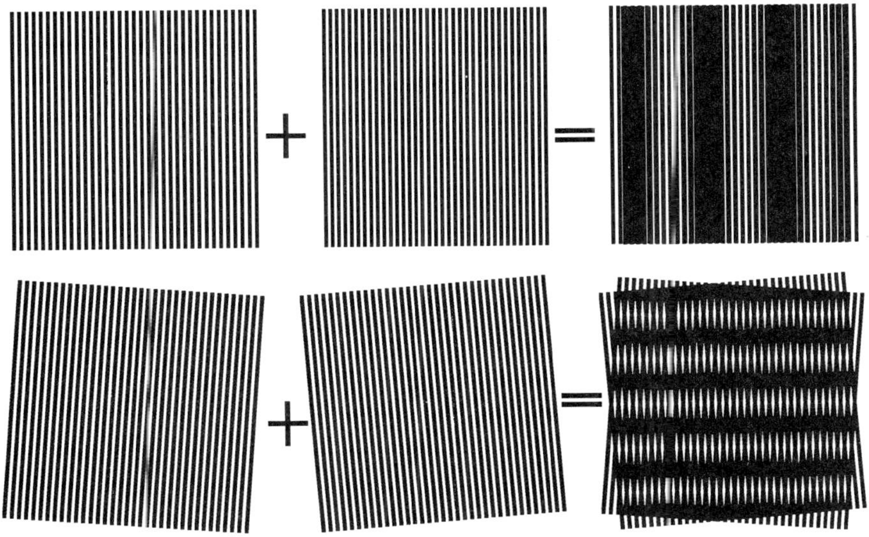

FIGURE 1.13. An optical analog demonstrating the formation of moiré fringe patterns by two overlapping gratings: (a) parallel moiré patterns; (b) rotation moiré patterns (redrawn after P.B. Hirsch et al. [68]).

TABLE 1.3. (200) lattice spacings and calculated moiré fringe spacings.

d_1($YBa_2Cu_3O_{7-\delta}$)/nm	d_2(substrate)/nm	D(moiré fringe spacing)/nm
0.191	0.211 (MgO)	2.05
0.191	0.195 ($SrTiO_3$)	8.88
0.191	0.189 ($LaAlO_3$)	18.05

these two periodicities, with a spacing D given by Eq. (3):

$$D = d_1 d_2/|(d_1 - d_2)|. \tag{3}$$

This beat pattern, or moiré fringe pattern, represents the periodicity with which the two gratings go in and out of register, and D is often sufficiently large to be resolvable at low magnification in the electron microscope; even though the individual lattice spacings may be below the resolution limit. For example, the calculated (200) moiré fringe spacings for some $YBa_2Cu_3O_{7-\delta}$/substrate combinations are shown in Table 1.3.

The type of pattern shown in Figure 1.13(a) is called a parallel moiré fringe pattern. If the overlapped gratings have the same spacing d, but are rotated with respect to each other by a small angle α, the moiré pattern observed will be that shown in Figure 1.13(b)—the rotation moiré pattern. The fringe spac-

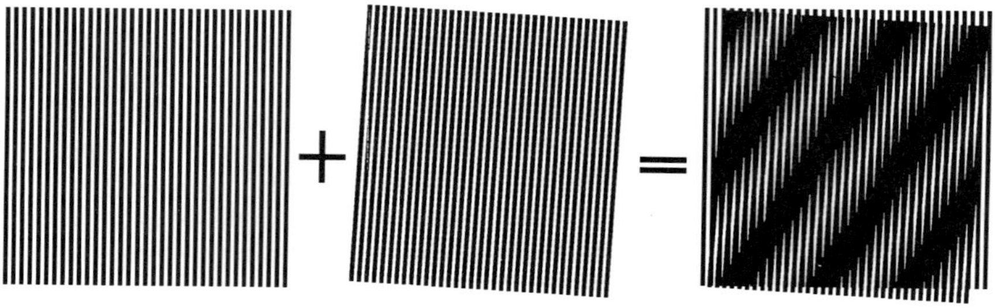

FIGURE 1.14. Optical analog showing that a rotation of one grating leads to a magnified rotation of parallel moiré patterns (redrawn after P.B. Hirsch et al. [68]).

ing for the rotation moiré pattern is given by Eq. (4):

$$D = d/(2\sin(1/2)\alpha). \tag{4}$$

It is usual to define a moiré magnification M, by $M = D/d$. For rotation moiré patterns $M = 1/\alpha$ and for parallel moiré patterns $M = d/(d_1 - d_2)$, where $d \sim d_1 \sim d_2$. Mixed moiré patterns can be formed, due to the overlapping of gratings which have different spacings and are also rotated. The spacing in this case is given by eq. (5):

$$D \sim d_1 d_2/((d_1 - d_2)^2 + d_1 d_2 \alpha^2)^{1/2}. \tag{5}$$

The effect of small rotations on parallel moiré fringes shows that the fringe rotation is magnified by the moiré magnification [68]. This is shown in the optical analog in Figure 1.14.

Several workers have used moiré fringes to investigate the imperfection structure of deposited films [e.g., Refs. 43, 69–72]. These studies have mainly been concerned with deposited metal films on substrates such as mica and molybdenum disulfide. With the development of the sample preparation technique described earlier, the use of moiré fringe patterns in the examination of the heteroepitactic growth of thin films has been applied to many more systems; for example, Fe_2O_3/Al_2O_3, $BaTiO_3/MgO$, and $YBa_2Cu_3O_{7-\delta}/MgO$.

1.5.1 *Defect Introduction During Film Growth*

Examination of the $YBa_2Cu_3O_{7-\delta}$ islands shown in Figures 1.11(a), and 1.11(b), indicated that a number of the islands were rotationally misaligned about the [001] zone axis [50]. This misalignment is attributed to the large lattice mismatch between MgO and $YBa_2Cu_3O_{7-\delta}$ [13]. Examination of the islands at higher magnification shows the presence of moiré fringes in the image of the islands as illustrated in Figure 1.15. The presence of moiré fringes indicates a mismatch in the lattice spacings of the two crystals. Therefore, the film/substrate interface cannot be coherent; part of the strain being relieved by the introduction of misfit dislocations. A schematic representa-

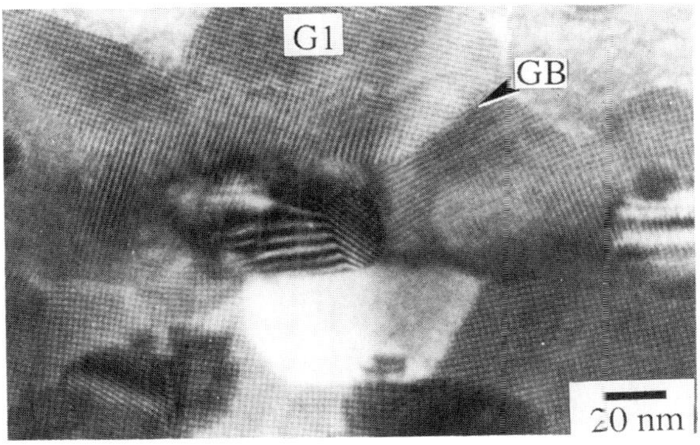

FIGURE 1.15. Bright-field image showing moiré fringe pattern in the image of $YBa_2Cu_3O_{7-\delta}$ islands on an (001)-oriented MgO thin-foil substrate.

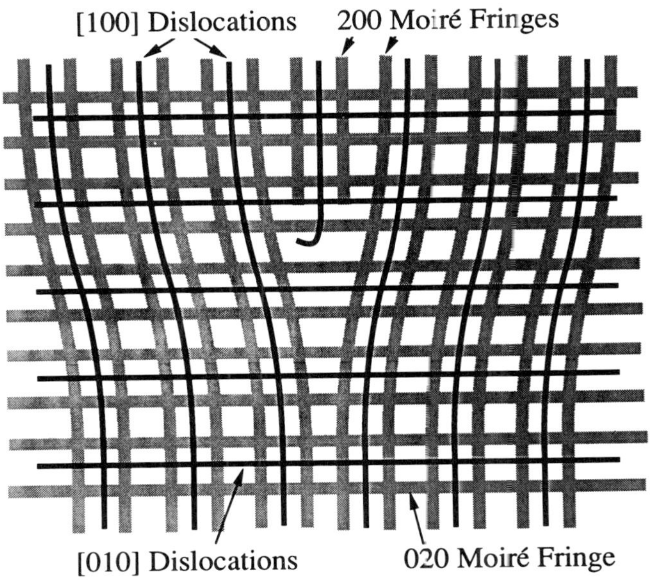

FIGURE 1.16. A schematic representation showing the relationship between the moiré fringe pattern and the proposed array of misfit dislocations.

tion of the relationship between the moiré fringe pattern and the proposed array of misfit dislocations is shown in Figure 1.16, where one of the misfit dislocations has become a threading dislocation (see the discussion below). The spacings of the [200] and [020] moiré fringes are half those of the corresponding misfit dislocations. In an epitactically aligned *c*-axis oriented grain the misfit will be accommodated by [100] and [010] misfit dislocations.

(a)

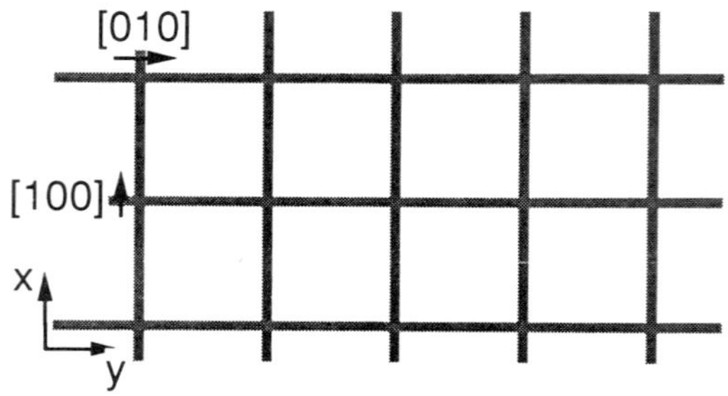

(b)

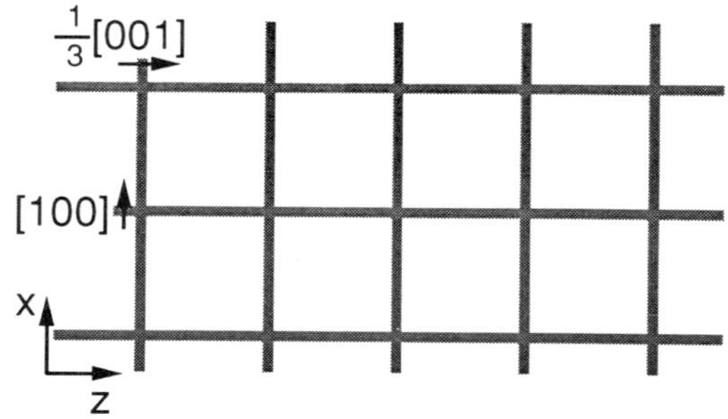

FIGURE 1.17. Mechanism for the accommodation of the lattice misfit in a *c*-axis oriented grain (a) and an *a*- (or *b*-) axis oriented grain (b).

In an *a*- (or *b*-) axis oriented grain one misfit dislocation will be [001] or a combination of 1/3[001] dislocations. These two situations are shown schematically in Figure 1.17.

The moiré fringe spacing was measured from Figure 1.15 to be 2.20 nm. In the general situation the parallel moiré spacing can be calculated from formula (3). In this case, $d_1 = d_{200}$ ($YBa_2Cu_3O_{7-\delta}$) or d_{020} ($YBa_2Cu_3O_{7-\delta}$) and $d_2 = d_{200}$ (MgO) giving predicted spacings for the moiré fringes of 2.06 (D_{200}) and 2.56 nm (D_{020}). Deviations from these values can be caused either by elastic strain in the film or by rotation of the film away from exact alignment with the substrate or by a combination of the two factors. The distorted nature of the moiré fringes in Figure 1.15, indicates that not all the strain at the interface has been relieved. The presence of misfit dislocations in thin $YBa_2Cu_3O_{7-\delta}$ films on MgO was detected by RBS ion-channeling studies [7]. Direct evidence for the formation of a periodic array of interfacial dislocations in $YBa_2Cu_3O_{7-\delta}$ thin films on MgO was made by cross-section high-resolution electron microscopy (HREM) studies [61].

Individual grains with different rotations about [001] are visible in the area of film shown in Figure 1.15. As discussed earlier the rotation of the moiré fringes is larger, by an order of magnitude in this case [13], than the actual lattice rotation of the film with respect to the substrate. Thus, for grain 1 (G1) in Figure 1.15, the rotation of the moiré fringes is 15°, but the actual

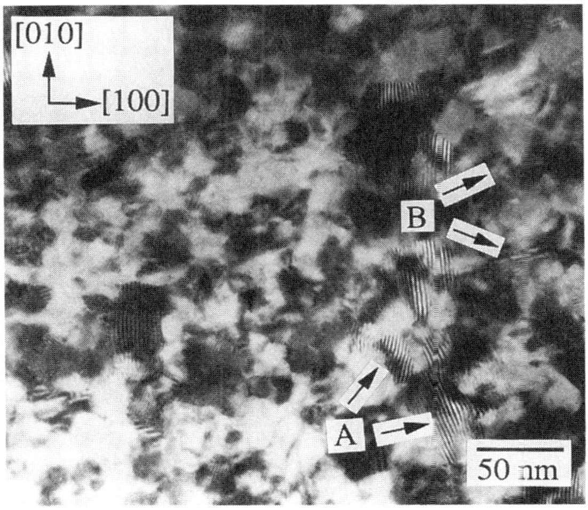

FIGURE 1.18. Bright-field image showing moiré fringes which have different spacings.

rotation of the lattice fringes is 1.2° [13]. Likewise, the actual relative rotation of two adjoining grains is much smaller than the apparent rotation of the moiré fringes at the grain boundary. Thus, the grain boundary (GB) indicated in Figure 1.15 is actually a low-angle grain boundary. Low-angle tilt boundaries with rotations about [001] are characteristic of the microstructure of these films [e.g., Ref. 5].

Of particular interest are high-angle grain boundaries in $YBa_2Cu_3O_{7-\delta}$ thin films [e.g., Ref. 73]. It has already been observed, by the use of selected-area electron diffraction, that 45° rotated domains are formed during the early stages of film growth—at the film/substrate interface [67]. These domains may also be identified through examination of moiré fringe patterns. Figure 1.18 shows a bright-field image of a thin ($\sim$12 nm) $YBa_2Cu_3O_{7-\delta}$ film on (001)-oriented MgO. Two sets of moiré fringes are visible in the image of the islands. One set, labeled A in Figure 1.18, have a fringe spacing $\sim$2.20 nm—consistent with previously observed values and in agreement with the predicted value for the parallel $\{200\}$ moiré fringe spacing. The second set of moiré fringes, labeled B in Figure 1.18, have a much finer fringe spacing, $\sim$1 nm. The calculated moiré fringe spacing for 45° rotated domains is $\sim$0.9 nm [74].

Moiré patterns can be used to identify microstructural defects in epitactic deposits. In particular, the presence of threading dislocations generated during film growth will be revealed in the moiré fringe pattern. If a threading edge dislocation is present in the layer it will be visible as a terminating fringe as can be seen, for example, at A in Figure 1.19. A number of possible mechanisms for introducing imperfections during growth have been discussed by

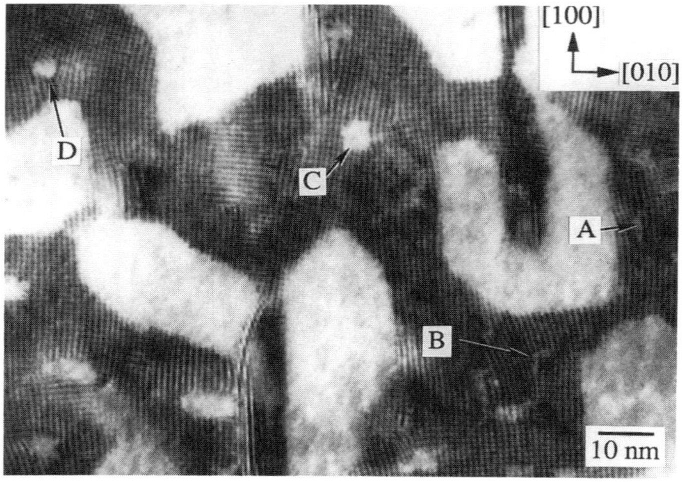

FIGURE 1.19. Thin $YBa_2Cu_3O_{7-\delta}$ film showing moiré fringes in the images of the islands. The labeled areas correspond to regions where terminating moiré fringes are visible, these are described in the text.

Pashley [69]. The accommodation of rotational displacements between agglomerating islands that are close to the exact epitactic orientation can be seen in area B in Figure 1.19. This situation is shown schematically in Figure 1.20. At a threading [100] dislocation, two [200] moiré fringes will terminate—this appears to be the most common situation [6, 75]. However, single terminating fringes are also found, indicating the presence of partial dislocations [6, 75].

The formation of epitactic films through the coalescence of misoriented islands results in a mosaic structure as shown in Figure 1.8. In some areas of the interconnected film, holes remain as the film thickens. So-called incipient dislocations are often observed to be associated with such holes as shown in area C in Figure 1.19 when two moiré fringes terminate on one side of the hole it implies that one dislocation would have to be formed if the hole were to be eliminated and this may be energetically unfavorable [76]. A similar area where terminating moiré fringes are present can be seen at D in Figure 1.19; where the moiré fringes curve as they leave the hole. Again, if further film growth occurred, it might be difficult to fill this hole. A tubular void would therefore remain in the film.

These TEM studies have shown how different defects can be introduced into a growing film—threading dislocations (accommodation of the rotational misfit between agglomerating islands), grain boundaries (island coalescence and grain growth) and voids. Other defects have also been observed in $YBa_2Cu_3O_{7-\delta}$ thin films, these include stacking faults [62, 77, 78], twin boundaries [e.g., Ref. 79], and more recently screw dislocations [8, 9]. These and other defects will be considered in more detail elsewhere in this book.

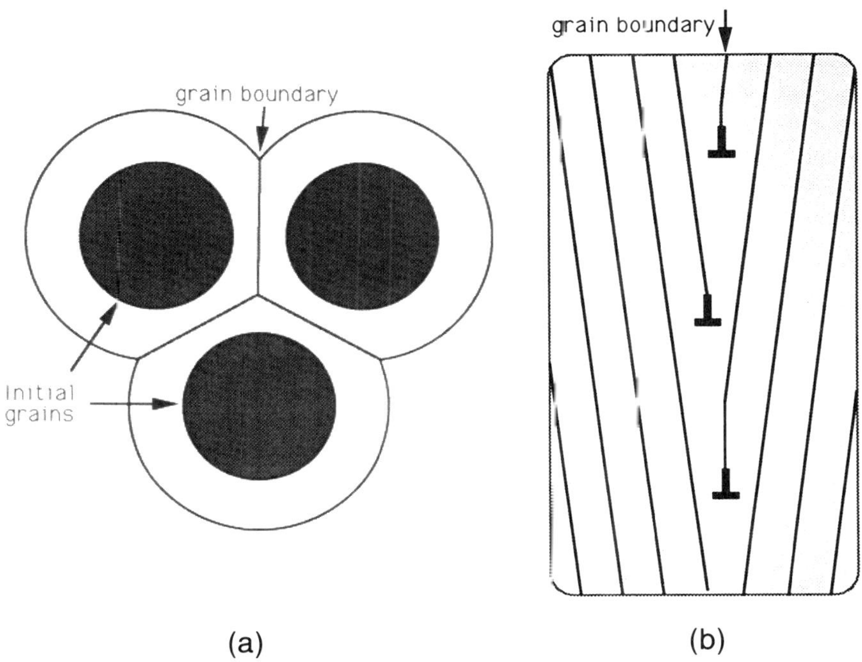

FIGURE 1.20. Accommodation of rotational misfit: (a) between three small islands and (b) between two large islands.

1.6 Concluding Remarks

This chapter has reviewed the nucleation and early stages of growth of $YBa_2Cu_3O_{7-\delta}$ thin films, particularly on nonlattice matched substrates. These studies have principally utilized the transmission electron microscope. In conclusion, the following general remarks can be made about the nucleation and heteroepitactic growth of $YBa_2Cu_3O_{7-\delta}$ thin films.

Film growth occurs, on both lattice matched and nonlattice matched substrates by an island mechanism. Steps on the substrate surface act as preferential sites for island nucleation. This effect may be expected, thereby, to influence the microstructure of the final film. Indeed, the condition (topography) of the substrate surface has been demonstrated, in a number of different studies, to affect not only the microstructure of the thin film but also its properties. The clearest example of this effect was the growth of $YBa_2Cu_3O_{7-\delta}$ thin films on vicinal MgO substrates. The early stages of film growth are influenced not only by the type and nature of the substrate and film material, but also by the substrate temperature during deposition and the growth rate. Further film deposition causes the initial nuclei to grow and eventually the islands coalesce to form an interconnected network structure. A number of defects are introduced into the film during the early stages of growth, for example, during island coalescence. These defects appear to be formed due to the rotational misalignment of the individual islands. At this stage of growth it was observed that some of the islands still contain holes (incipient dislocations).

These studies have produced a great deal of understanding of the nucleation and heteroepitactic growth mechanism of $YBa_2Cu_3O_{7-\delta}$ thin films. Continuations of these studies include increased use of scanning tunneling microscopy (discussed elsewhere in this book) as a means of examining film growth mechanisms; not only of $YBa_2Cu_3O_{7-\delta}$ thin films but also other types of high-temperature superconducting oxides. Other important areas are in determining the effect that the deposition conditions have on film growth; particularly in optimizing film growth for the desired material properties. This latter consideration also involves controlling the substrate topography (e.g., the use of lithographic techniques to pattern the substrate surface).

Acknowledgments. The support of the Office of Naval Research (N00014-89-J-1692), the Consortium for Superconducting Electronics (MDA 972-90-C-0021) and the Materials Science Center at Cornell is gratefully acknowledged. The authors would like to thank their colleagues for many helpful discussions and technical assistance, especially Dr. Lisa Tietz, Dr. Stuart McKernan, Brian Moeckly, and Professor Robert Buhrman.

References

1. J.G. Bednorz and K.A. Muller, Z. Phys. B **64**, 189 (1986).
2. M.K. Wu, J.R. Ashburn, C.J. Torng, P.H. Hor, R.L. Meng, L. Gao, Z.J. Huang, Y.Q. Wang, and C.W. Chu, Phys. Rev. Lett. **58**, 908 (1987).
3. Y. Bando, T. Terashima, K. Iijima, K. Yamamoto, K. Hirata, K. Hayashi, K. Kamigaki, and H. Terauchi, Proc. XII Intl. Cong. for Electron Microsc. **2** (1990).
4. J.Q. Zheng, X.K. Wang, M.C. Shih, S.J. Lee, S. Williams, J. So, P. Dutta, J.B. Ketterson, and R.P.H. Chang, Proc. 2nd. Intl. Conf. on Electronic Mater. **85** (1990).
5. M.G. Norton and C.B. Carter, J. Cryst. Growth **110**, 645 (1991).
6. S.K. Streiffer, B.M. Lairson, C.B. Eom, B.M. Clemens, T.H. Geballe, and J.C. Bravman, Phys. Rev. B **43**, 13007 (1991).
7. J. Geerk, G. Linker, and O. Meyer, Mater. Sci. Rep. **4**, 193 (1989).
8. M. Hawley, I.D. Raistrick, J.G. Beery, and R.J. Houlton, Science **251**, 1587 (1991).
9. C. Gerber, D. Anselmetti, J.G. Bednorz, J. Mannhart, and D.G. Schlom, Nature **350**, 279 (1991).
10. H. Poppa, in *Epitaxial Growth Part A*, edited by J.W. Matthews (Academic Press, New York, 1975), p. 215.
11. K.J. Morrissey and C.B. Carter, Mater. Res. Soc. Symp. Proc. **41**, 137 (1985).
12. L.A. Tietz, S.R. Summerfelt, G.R. English, and C.B. Carter, Appl. Phys. Lett. **55**, 1202 (1989).

13. M.G. Norton, L.A. Tietz, S.R. Summerfelt, and C.B. Carter, Appl. Phys. Lett. **55**, 2348 (1989).
14. M.G. Norton and C.B. Carter, J. Mater. Res. **5**, 2762 (1990).
15. D.W. Susnitzky and C.B. Carter, Proc. 45th Annual Meeting of the Electron Microscopy Society of America, 1987, p. 154.
16. Y.K. Simpson and C.B. Carter, Philos. Mag. Lett. **53**, L1 (1986).
17. M.G. Norton, E.L. Fleischer, W. Hertl, C.B. Carter, J.W. Mayer, and E. Johnson, Phys. Rev. B **43**, 9291 (1991).
18. *The Oxford English Dictionary*, 2nd. ed., prepared by J.A. Simpson and E.S.C. Weiner (Clarendon Press, Oxford, Oxford University Press, New York, 1989).
19. S.W. Bailey, V.A. Frank-Kamenetskii, S. Goldsztaub, A. Kato, A. Pabst, H. Schultz, H.F.W. Taylor, M. Fleischer, and A.J.C. Wilson, Acta. Crystallogr. A **33**, 681 (1977).
20. M.L. Frankhenheim, Ann. Phys. **37**, 516 (1836).
21. H. Siefert, in *Structure and Properties of Solid Surfaces*, edited by R. Gomer and C.R. Smith (University of Chicago Press, Chicago, 1953), p. 218.
22. L. Royer, Bull. Soc. Franc. Mineral **51**, 7 (1928).
23. L.G. Schultz, Acta. Crystallogr. **4**, 483 (1951).
24. *Epitaxial Growth Part B*, edited by J.W. Matthews (Academic Press, New York, 1975).
25. M.H. Grabow and G.H. Gilmer, Surf. Sci. **194**, 333 (1988).
26. M.J. Stowell, in *Epitaxial Growth Part B*, edited by J.W. Matthews (Academic Press, New York, 1975).
27. A. Inam, M.S. Hegde, X.D. Wu, T. Venkatesan, P. England, P.F. Miceli, E.W. Chase, C.C. Chang, J.M. Tarascon, and J.B. Wachtman, Appl. Phys. Lett. **53**, 908 (1988).
28. D.K. Lathrop, B.H. Moeckly, S.E. Russek, and R.A. Buhrman, Appl. Phys. Lett. **58**, 1095 (1991).
29. R.W. Simon, J.F. Burch, K.P. Daly, W.D. Dozier, R. Hu, A.E. Lee, J.A. Luine, H.M. Manasevit, C.E. Platt, S.M. Schwarzbek, D. St.John, M.S. Wire, and M.J. Zani, in *Science and Technology of Thin Film Superconductors 2*, edited by R.D. McConnell and R. Noufi (Plenum Press, New York, 1990), p. 549.
30. M. Naito, R.H. Hammond, B. Oh, M.R. Hahn, J.W.P. Hsu, P. Rosenthal, A.F. Marshall, M.R. Beasley, T.H. Geballe, and A. Kapitulnik, J. Mater. Res. **2**, 713 (1987)
31. X.X. Xi, G. Linker, O. Meyer, E. Nold, B. Obst, F. Ratzel, R. Smithey, B. Strehlau, F. Weschenfelder, and J. Geerk, Z. Phys. B **74**, 13 (1989).
32. P.H. Dickinson, T.H. Geballe, A. Sanjurjo, D. Hidenbrand, G. Craig, M. Zisk, J. Collman, S.A. Banning, and R.E. Sievers, J. Appl. Phys. **66**, 444 (1989).
33. D. Dijkkamp, T. Venkatesan, X.D. Wu, S.A. Shaheen, N. Jisrawi, Y.H. MinLee, W.L. Mclean, and M. Croft, Appl. Phys. Lett. **51**, 619 (1987).
34. S.E. Russek, B.H. Moeckly, R.A. Buhrman, J.T. McWhirter, A.J. Sievers, M.G. Norton, L.A. Tietz, and C.B. Carter, in *High Temperature Superconductors: Fundamental Properties and Novel Materials Processing*, edited by J. Narayan, C.W. Chu, L.F. Schneemeyer, and D.K. Christen, Mater. Res. Soc. Symp. Proc. **169**, 455 (1990).
35. G. Koren, A. Gupta, E.A Giess, A. Sergmuller, and R.B Laibowitz, Appl. Phys. Lett. **54**, 1054 (1989).
36. R.W. Simon, C.E. Platt, A.E. Lee, G.S. Lee, K.P. Daly, M.S. Wire, J.A. Luine, and M. Urbanik, Appl. Phys. Lett. **53**, 2677 (1988).

37. B.H. Moeckly, S.E. Russek, D.K. Lathrop, R.A. Buhrman, M.G. Norton, and C.B. Carter, Appl. Phys. Lett. **57**, 2951 (1990).
38. D.P. Norton, D.H. Lowndes, J.D. Budai, D.K. Christen, E.C. Jones, K.W. Lay, and J.E. Tkaczyk, Appl. Phys. Lett. **57**, 1164 (1990).
39. X.X. Xi, J. Geerk, G. Linker, Q. Li, and O. Meyer, Appl. Phys. Lett. **54**, 310 (1989).
40. S.E. Russek, D.K. Lathrop, B.H. Moeckly, R.A. Buhrman, D.H. Shin, and J. Silcox, Appl. Phys. Lett. **57**, 155 (1990).
41. D. Dimos, P. Chaudari, J. Mannhart, and F.K. LeGoues, Phys. Rev. Lett. **61**, 219 (1988).
42. G.A. Bassett, Proc. European Regional Conference on Electron Microscopy, Delft, **1**, 270 (1960).
43. D.W. Pashley, M.J. Stowell, M.H. Jacobs, and T.J. Law, Philos. Mag. **10**, 127 (1964).
44. G. Honjo and K. Yagi, in *Current Topics in Materials Science*, edited by Kaldis (North-Holland Publishing Company, Amsterdam, 1980), Vol. 6, p. 195.
45. M. Sarikaya, B.L. Thiel, I.A. Aksay, W.J. Weber, and W.S. Frydrych, J. Mater. Res. **2**, 736 (1987).
46. M.G. Norton, S.R. Summerfelt, and C.B. Carter, Appl. Phys. Lett. **56**, 2246 (1990).
47. M.G. Norton, S. McKernan, and C.B. Carter, in *Materials Reliability Issues in Microelectronics*, Mater. Res. Soc. Symp. Proc. **225**, 349 (1991).
48. M.G. Norton, L.A. Tietz, S.R. Summerfelt, and C.B. Carter, in *High Temperature Superconductors: Fundamental Properties and Novel Materials Processing*, edited by J. Narayan, C.W. Chu, L.F. Schneemeyer, and D.K. Christen, Mater. Res. Soc. Symp. Proc. **169**, 509 (1990).
49. B.H. Moeckly, S.E. Russek, D.K. Lathrop, R.A. Buhrman, J. Li, and J.W. Mayer, Appl. Phys. Lett. **57**, 1687 (1990).
50. B.H. Moeckly, D.K. Lathrop, S.E. Russek, R.A. Buhrman, M.G. Norton, and C.B. Carter, IEEE Trans. Magn. **27**, 1017 (1991).
51. O. Meyer, J. Geerk, Q. Li, G. Linker, and X.X. Xi, Nucl. Instrum. Methods Phys. Res. B **45**, 483 (1990).
52. O. Meyer, F. Weschenfelder, X.X. Xi, G.C. Xiong, G. Linker, and J. Geerk, Nucl. Instrum. Methods Phys. Res. B **35**, 292 (1988).
53. L.A. Tietz, C.B. Carter, D.K. Lathrop, S.E. Russek, and R.A. Buhrman, in *High Temperature Superconductors*, edited by M.B. Brodsky, R.C. Dynes, K. Kitazawa, and H.L. Tuller, Mater. Res. Soc. Symp. Proc. **99**, 715 (1988).
54. P.W. Tasker and D.M. Duffy, Surf. Sci. **137**, 91 (1984).
55. R.B. Poeppel and J.M. Blakely, Surf. Sci. **15**, 507 (1969).
56. *The American Heritage Dictionary*, 2nd College Edition (Houghton Mifflin, Boston, 1985).
57. S.K. Streiffer, B.M. Lairson, and J.C. Bravman, Appl. Phys. Lett. **57**, 2501 (1990).
58. C.B. Carter and H. Schmalzried, Philos. Mag. A **52**, 207 (1985).
59. R. Du and C.P. Flynn, J. Phys. Condensed Matter **2**, 1335 (1990).
60. T.E. Mitchell, S. Basu, M.A. Nastasi, and T. Roy, in *High Resolution Electron Microscopy of Defects in Materials*, edited by R. Sinclair, U. Dahmen, and D.J. Smith, Mater. Res. Soc. Symp. Proc. **183**, 357 (1990).
61. R. Ramesh, D. Hwang, T.S. Ravi, A. Inam, J.B. Barner, L. Nazar, S.-W. Chan, C.Y. Chen, B. Dutta, T. Venkatesan, and X.D. Wu, Appl. Phys. Lett. **56**, 2243 (1990).
62. L.A. Tietz, C.B. Carter, D.K. Lathrop, S.E. Russek, R.A. Buhrman, and J.R. Michael, J. Mater. Res. **4**, 1072 (1989).

63. M.G. Norton, B.H. Moeckly, C.B. Carter, and R.A. Buhrman, J. Cryst. Growth **114**, 258 (1991).
64. M.G. Norton and C.B. Carter, in *Laser Ablation for Materials Synthesis*, edited by D.C. Paine and J.C. Bravman, Mater. Res. Soc. Symp. Proc. **191**, 165 (1990).
65. X.X. Xi, J. Geerk, G. Linker, Q. Li, and O. Meyer, Appl. Phys. Lett. **54**, 2367 (1989).
66. M.G. Norton, S. McKernan, and C.B. Carter, Philos. Mag. Lett. **62**, 77 (1990).
67. M.G. Norton and C.B. Carter, Physica C **172**, 47 (1990).
68. P.B. Hirsch, A. Howie, R. Nicholson, D.W. Pashley, and M.J. Whelan, *Electron Microscopy of Thin Crystals* (Krieger, Malabar, FL, 1975), pp. 353–387.
69. D.W. Pashley, in *Thin-Films* (Amer. Soc. Metals, Metals Park, OH, 1964), p. 59.
70. G.A. Bassett, J.W. Menter, and D.W. Pashley, Proc. R. Soc. London A **246**, 345 (1958).
71. J. W. Menter, Adv. Phys. **7**, 300 (1958).
72. J.W. Matthews, Surf. Sci. **31**, 241 (1972).
73. S. McKernan, M.G. Norton, and C.B. Carter, J. Mater. Res **7**, 1052 (1992).
74. M.G. Norton and C.B. Carter, Physica C **182**, 30 (1991).
75. M.G. Norton and C.B. Carter, Scanning Microscopy **6**, 385 (1992).
76. F.R.N. Nabarro, *Theory of Crystal Dislocations* (Dover, New York, 1987), p. 303.
77. R. Ramesh, D.M. Hwang, J.B. Barner, L. Nazar, T.S. Ravi, A. Inam, B. Dutta, X.D. Wu, and T. Venkatesan, J. Mater. Res. **5**, 704 (1990).
78. R. Ramesh, D.M. Hwang, T. Venkatesan, T.S. Ravi, L. Nazar, A. Inam, X.D. Wu, B. Dutta, G. Thomas, A.F. Marshall, and T.H. Geballe, Science **247**, 57 (1989).
79. Y. Zhu, M. Suenaga, and Y. Xu, Philos. Mag. Lett. **60**, 51 (1989).

2 Scanning Tunneling and Atomic Force Microscope Studies of Thin Sputtered Films of $YBa_2Cu_3O_7$

IAN D. RAISTRICK and MARILYN HAWLEY

2.1 Introduction

Although the fundamental attributes of superconductors (e.g., T_c, coherence length, penetration depth) are determined by the atomic and electronic structure of the material, many practically important properties, including critical current density and high-frequency conductivity, are dominated by the interaction between these properties and the microstructure. This is particularly true for the high-temperature superconducting (HTS) cuprates, because their coherence length, which is the scale over which the Ginzberg–Landau order parameter can vary, may be as short as a few Angstroms in some directions. Thus, perturbations of the structure associated with point defects, dislocations, grain boundaries, twins, etc., can lead to spatial fluctuations in the superconducting properties, with profound implications for magnetic flux entry and motion, intergrain coupling energy, etc. Additionally, these materials have layered structures, which lend an almost two-dimensional character to the superconductivity, and intergranular misorientations are expected to be very significant.

From a practical point of view, thin-film forms of the HTS materials show the most desirable properties obtained to date. Critical current densities greater than 10^6 A/cm^2 at 77 K in zero field are obtained in many laboratories in c-axis-oriented films of $YBa_2Cu_3O_7$, and surface resistances (at 4.2 K and 10 GHz) of a few tens of milliwatts have been demonstrated [1]. Compared with bulk material, the superior properties of thin films are usually ascribed to (1) an absence of weak-link behavior due to preferred crystallographic orientation and (2) excellent flux pinning. A considerable body of work has now shown that large-angle boundaries between grains, rotated with respect to each other around the c axis, have low J_c's [2] and may exhibit resistively shunted Josephson (RSJ) behavior [3]. It has also been shown recently that low surface resistance is correlated with an absence

Los Alamos National Laboratory, Los Alamos, NM 87545.

of such large angle, in-plane grain boundaries, in c-axis-oriented films of $YBa_2Cu_3O_7$ [1]. It remains true, however, that (1) the nature of the defects responsible for flux pinning in thin films (especially when the magnetic field is parallel to the c axis) is unknown, and (2) the residual surface resistance of even the best films is many orders of magnitude greater than expected from BCS theory. It is possible that the residual R_s is associated with regions of depressed order parameter associated with microstructural defects [4–6]. It is, therefore, obviously important that the relationship between microstructure and properties be well understood.

Much also remains to be learned about the growth mechanism of thin films. The microstructure is, of course, a product of the nucleation and growth processes. A partial list of deposition variables that influence nucleation and growth for a single type of deposition process (such as sputtering) includes substrate choice (lattice mismatch, chemical considerations), substrate orientation and surface preparation, deposition temperature, total pressure and oxygen partial pressure, deposition rate, film and target composition, and cooling protocol. Where these variables lie on a multidimensional phase plot determines not only the general film morphology, for example, surface roughness, outgrowths, and the presence of voids, but also the relative substrate-film crystallographic orientations (e.g., c- or a-axis normal growth [7]), in-plane rotational order, and whether the growth occurs by a layer-by-layer process or by competing mechanisms, such as three-dimensional island growth. Direct information about the growth process is usually best obtained by in situ measurements (e.g., x-ray diffraction [8] or reflection high-energy electron diffraction (RHEED) [9]) or by interrupted growth ex situ techniques (i.e., investigation of surface microstructure).

Many laboratories, of course, are exploring the microstructure of thin films of $YBa_2Cu_3O_7$ using a variety of experimental probes, including electron microscopies [10], ion-beam channeling [11, 12] and x-ray diffraction methods. Recently, scanning tunneling and atomic force microscopies (STM and AFM) have been added to this list of techniques [13, 14]. These microscopies extend the resolution range beyond that of the scanning electron microscope (SEM), while adding the potential capability of obtaining three-dimensional atomic or unit-cell detail without the extensive and destructive sample preparation required by transmission electron microscopy (TEM). The combination of extremely high vertical resolution and local *nonaveraged* detail has made the STM a vital tool for these particular types of studies.

In this chapter we present and review recent results of studies of the microstructures of $YBa_2Cu_3O_7$ thin films on a variety of single-crystal substrates obtained using STM and AFM, and we attempt to place these results in the wider context of our present understanding of the nucleation and growth processes and how these influence the properties of the films. Most of our own results have been obtained using off-axis magnetron sputtering as the deposition process, but relevant results obtained elsewhere on films deposited by other techniques, such as pulsed laser deposition, are also discussed.

TABLE 2.1. Lattice parameters and mismatch at ambient temperature for several common substrates.

Substrate	Plane	Dimensions (nm) of pseudocubic plane	Mismatch (%)[b] a	b	$c/3$	JCPDS-ICDD Ref. No.[d]
MgO	{100}	0.4213	10.20	8.41	8.19	4–829
	(100)	45° rotation	3.90	2.21	2.00	
$SrTiO_3$	{100}	0.3905	2.14	0.49	0.29	35–734
$NdGaO_3$	{110}	0.3864 × 0.3855	1.07,	−0.60,	−0.77,	21–972
			0.84	−0.80	−1.00	
	{001}	0.3864 (90.7°)[c]	1.07	−0.6	−0.77	
$LaAlO_3$	{012}[a]	0.3792 (90.1°)[c]	−0.81	−2.42	−2.62	31–22
$LaGaO_3$	{001}	0.3892 (90.3°)[c]	1.80	0.15	0.67	24–1102

[a] Referred to primitive rhombohedral axes.
[b] Mismatch (%) = 100 × $(d_{substrate} - d_{ybco})/d_{ybco}$; $YBa_2Cu_3O_7$ lattice parameters are a_o = 0.3823 nm, b_0 = 0.3886 nm, and c_o = 1.1681 nm ($c_o/3$ = 0.3894 nm).
[c] In-plane angle in pseudocubic setting.
[d] ICDD-PDF 1991.

2.2 Growth of *c*-Axis Oriented Films

2.2.1 *Introduction*

High-quality *c*-axis-oriented films of $YBa_2Cu_3O_7$ are now grown on several different types of single-crystal substrate materials, including various perovskites; MgO; cubic zirconia; and buffered sapphire. Several of the perovskites, such as $SrTiO_3$, $LaAlO_3$, $LaGaO_3$, and $NdGaO_3$ are not only closely related structurally to the superconductor, but also have a rather close lattice match (see Table 2.1). However, for high-frequency applications, their dielectric properties are marginal or unsatisfactory, and MgO and buffered sapphire are preferred. MgO is cubic and has the rock-salt structure. In contrast to the perovskite oxides, it has a very poor lattice match to orthorhombic $YBa_2Cu_3O_7$. We might, therefore, expect significant differences in the growth behavior and microstructure between the films deposited on these two types of substrates.

The usual in-plane orientation of *c*-axis-perpendicular films, with respect to the substrate for the {100}[1] surfaces of both perovskites and MgO, is $[100]_{ybc}//[100]_{substrate}$. For the perovskites $LaAlO_3$ and $SrTiO_3$, the mismatch is accommodated by interface dislocations, and for thin (<110 nm) films on $SrTiO_3$, the growth is pseudomorphic [15]. Ramesh [10] has shown that for $SrTiO_3$, where the mismatch puts the film in tension, the dislocations are in the film. In $LaAlO_3$, which leads to a compressive stress in the film, the

[1] Referred to pseudocubic axes.

extra half-plane is in the substrate. On MgO the interface is much more disordered and has been described as incoherent or polycrystalline. Norton et al. [16] have shown, by observation of moiré fringes in very thin films, that the superconductor is not strained into coherence with the MgO. The large mismatch is presumed to favor island nucleation (Volmer–Weber mechanism) rather than layer-by-layer growth.

Besides this commonly observed cube-on-cube epitaxy, other orientations, related by rotations about the *c* axis, are seen on MgO. These orientations have been analyzed by the Bellcore–Rutgers group [17] in terms of "near coincident site lattice theory" (NCSLT) [18]. A commonly seen orientation is that of a 45° in-plane rotation ($[110]_{ybc}//[100]_{MgO}$). In this orientation, a semicoherent growth is achieved with 3 YBCO [110] growth units fitting with 4 MgO [100] units, with a reduced mismatch. A periodic array of edge dislocations in the $YBa_2Cu_3O_7$ is seen in TEM cross sections. As mentioned above, the 45° grain boundaries associated with the rotation appear to have weak-link character and to be strongly correlated with high-frequency properties. Their elimination in *c*-axis films grown for microwave applications is therefore a topic of current interest [1], as is their controlled production by epitaxial rotations [3].

Views of the growth mechanism in the literature appear to be somewhat conflicting, and these probably reflect, in part, the results of different deposition methods and growth conditions, as well as different substrate preparation procedures. In situ RHEED studies of films grown by reactive co-evaporation, suggest that a layer-by-layer growth process is operating on both $SrTiO_3$ and MgO [9]. High-resolution TEM studies of very thin (1- to 10-nm-thick) sputter-deposited films on MgO by Streiffer et al. [19] showed an initially very high density of two-dimensional nuclei (one or two unit cells high) on the surface. Additional material increases coverage rather than thickness, through propagation of unit-cell high ledges. The Cornell group, again using TEM, however, have shown that, in their laser-ablated films, true island growth takes place [16], nucleation occurs preferentially at the steps on the MgO surface, and the observed alignment of the film with the substrate is not a true epitaxial process, but rather a "graphoepitaxy," determined by the preponderant facets of the steps in the surface [20]. The Cornell substrates probably differ significantly from those used at Stanford [19], or in the present STM work described below [13, 21], in that they were heated at high temperature after a thorough cleaning and etching process, which resulted in a very stepped surface.

2.2.2 *Substrate Characterization*

Although other factors are important in influencing the properties of films deposited on MgO (100) single-crystal faces, surface roughness and morphology are expected to be critical in determining the nucleation and growth processes. Eom et al. [22] reported the use of AFM to determine the surface

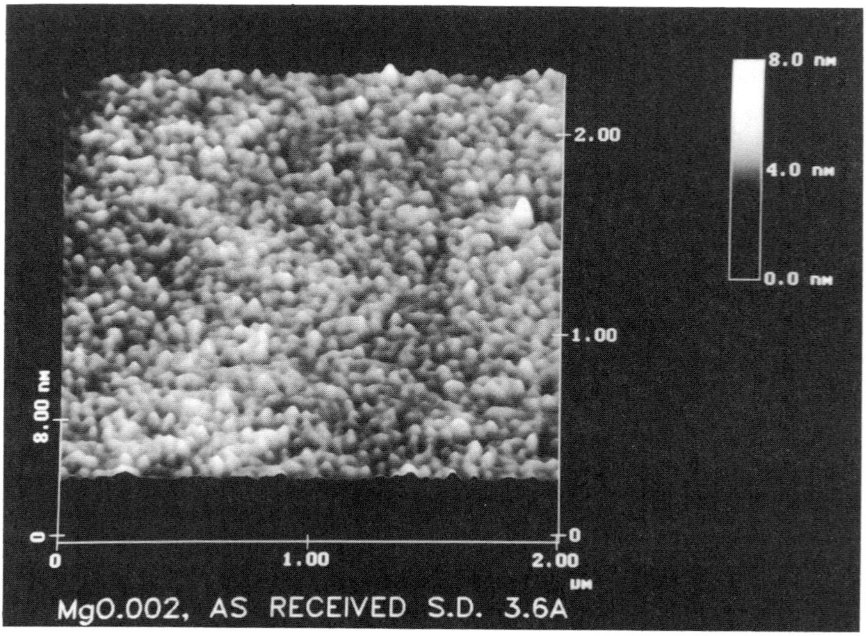

FIGURE 2.1. AFM image of the (100) surface of MgO, as received from the vendor.

roughness of their substrates, and found it to be about 0.3 to 0.4 nm on Stanford polished material.[2] In Figure 2.1 we show an AFM image of the (100) surface of a single-crystal MgO substrate as received from the vendor [23]. The AFM images were obtained using a Nanoscope II and a commercial Au-coated 100-μ triangular Si_3N_4 cantilever with a force constant of approximately 0.6 N/m. The standard deviation of the roughness over this area is about 1 unit cell (0.4 nm) and, as can be seen, no particular surface features, such as steps, are visible. Overall, the roughness has a granular appearance with an average lateral size scale of perhaps 50 nm. We also characterized a similar substrate that had been heated to 1200°C in air for 18 hr [Figures 2.2(a) and 2.2(b)]. In addition to surface steps [Figure 2.2(a)], each of which is one unit cell high, we sometimes found a significant population of small crystals on the surface [Figure 2.2(b)]. Overall, this surface roughening was disastrous for the properties of the film deposited on the surface. In Figure 2.3 we compare the x-ray phi (azimuthal) scans for films deposited on these two samples. Although both films were deposited under identical conditions, and both were fully oriented with *c*-axis perpendicular

[2] No details given.

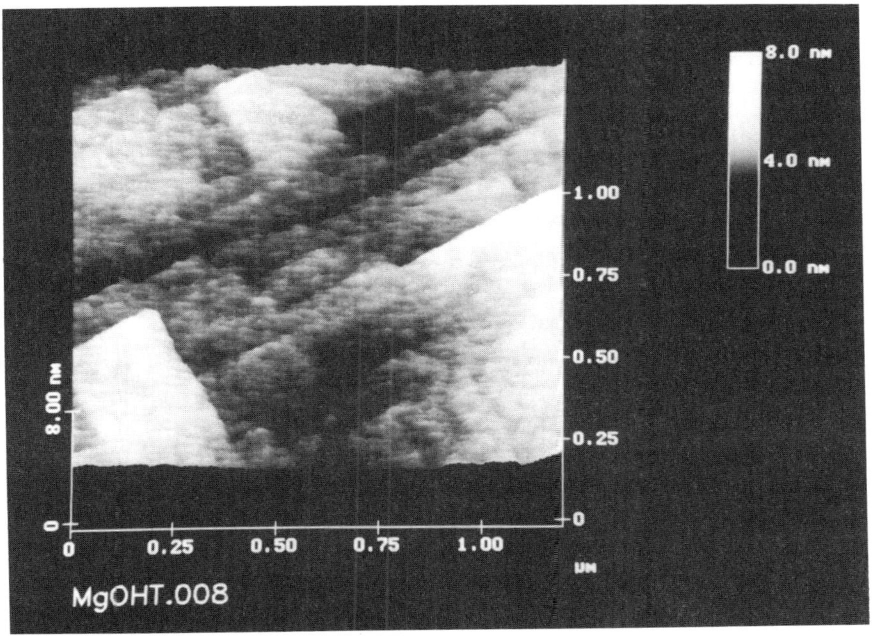

(a)

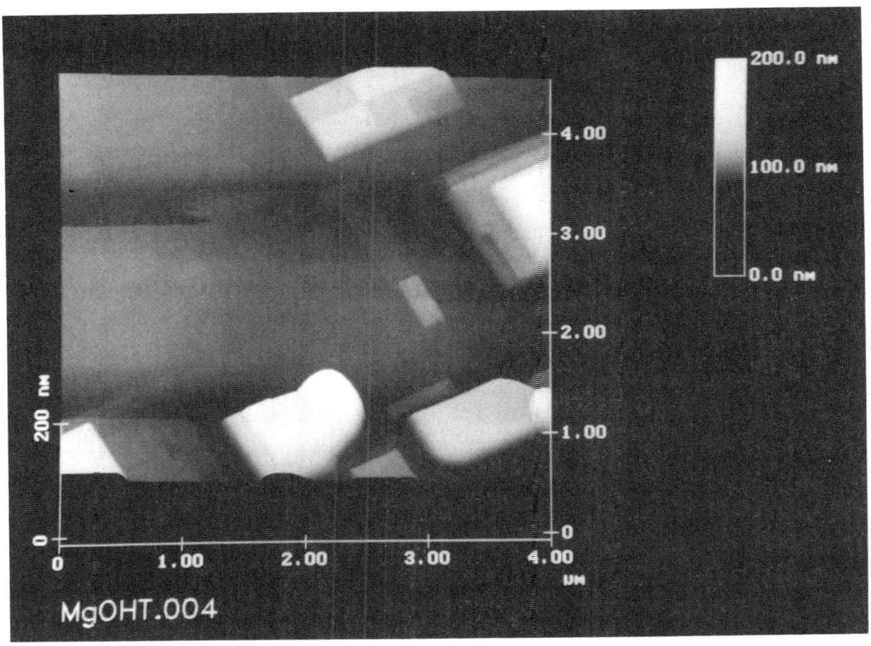

(b)

FIGURE 2.2. AFM images of MgO after heat treatment, showing (a) a large number of unit-cell-high steps, and (b) microcrystals on the surface.

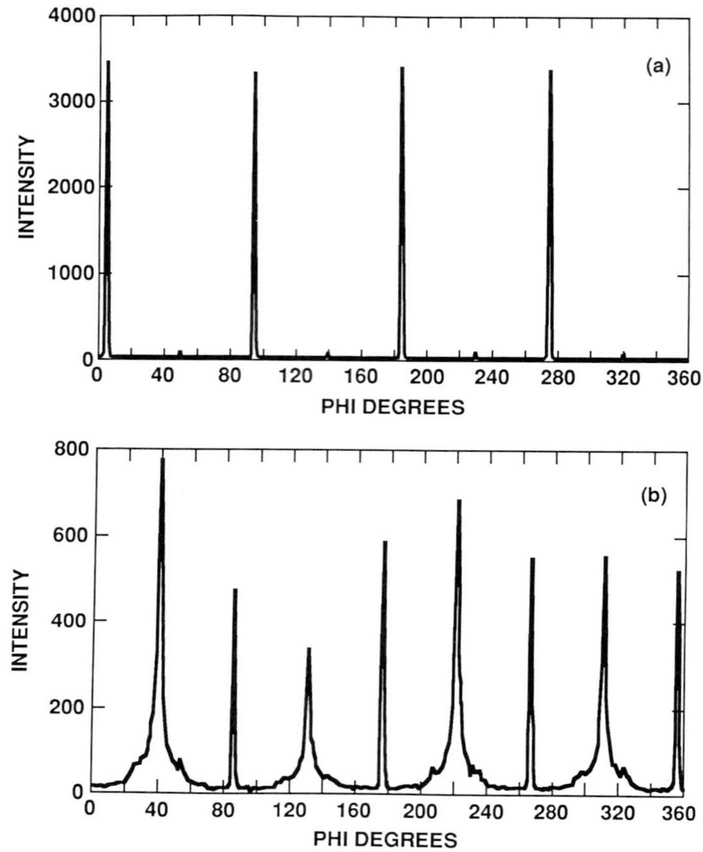

FIGURE 2.3. X-ray phi (azimuthal) scans of the {012} $YBa_2Cu_3O_7$ reflections, showing the in-plane order of (a) a typical film deposited on as-recieved MgO (100), and (b) deposited on a heat-treated sample from the same source. Note the presence of a strong 45° rotated compononent which corresponds, to $[110]_{ybc}//[100]_{MgO}$, and a wide distribution of orientations in this direction.

to the substrate, it is clear that multiple orientations, related by rotations about the *c* axis, are present. In general, we have found that annealed substrates, regardless of whether they are originally polished or cleaved, exhibit increased surface roughness, sometimes with very large (multiple unit cell) surface steps. Two further examples of polished annealed substrates are shown in Figure 2.4. In one case, faceting clearly exhibits a preferred directionality, while in the other considerable curvature is present.

Eom et al. [22] reported that after a high-temperature anneal at low oxygen partial pressures, a significant roughening of the surface took place (to about 2 nm), although no change in surface roughness was reported for films

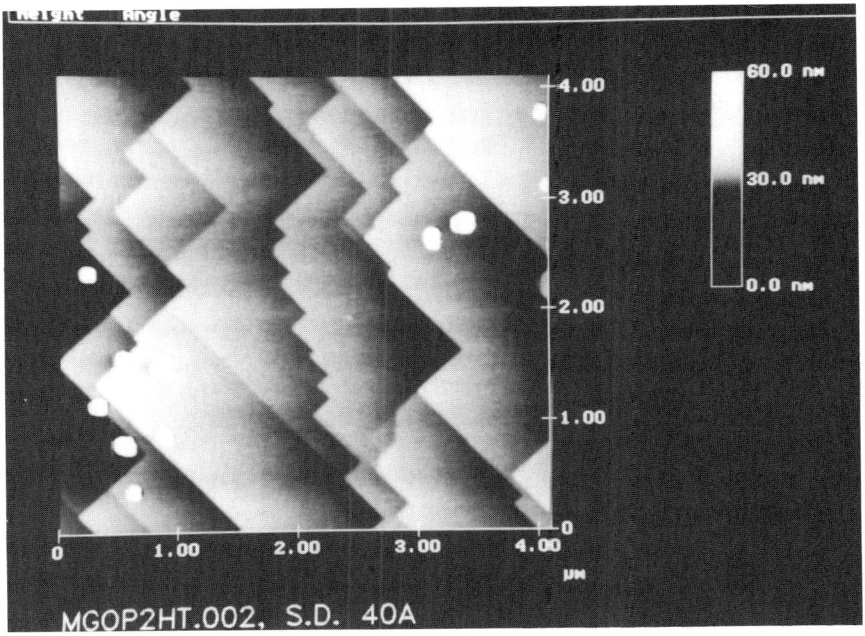

(a)

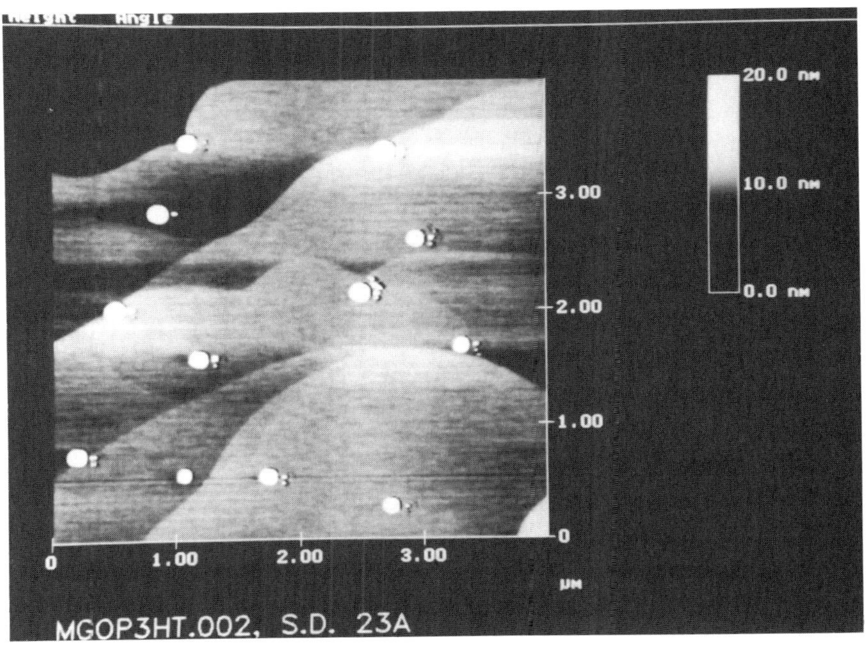

(b)

FIGURE 2.4. AFM scans of polished MgO (100) faces after heat treatment in air at 1200°C. The two samples are from different vendors The steps are more than one unit cell high, and lead to a significant increase in the standard deviation of the surface roughness.

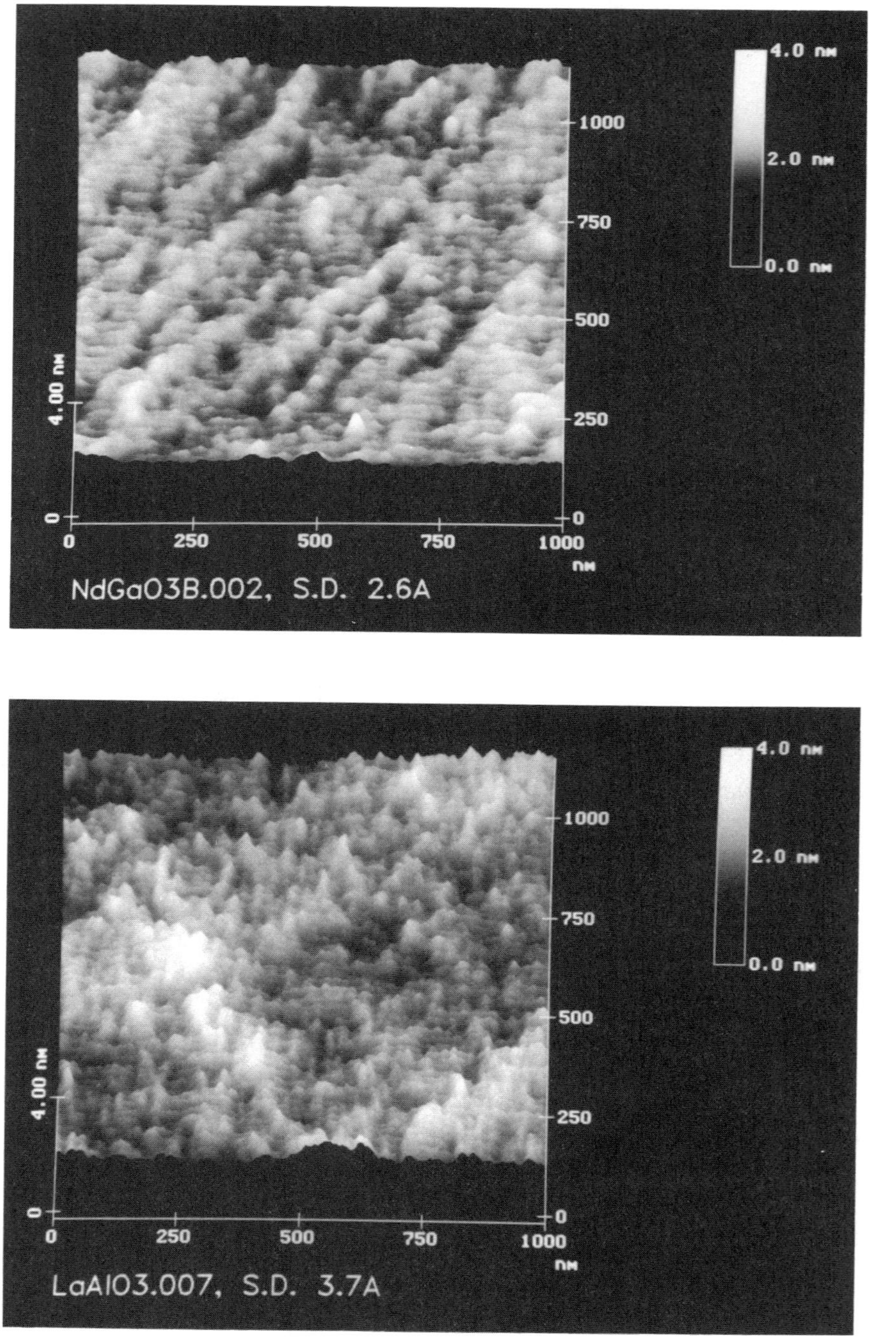

FIGURE 2.5. AFM scans of (a) $NdGaO_3$ and (b) $LaAlO_3$ polished substrates.

annealed at a pressure of 1 atm of oxygen. The Cornell group, however, report that well-defined surface steps, each of unit-cell height with the principal facets in the [100] and [110] directions, are introduced by annealing in air or oxygen [20]. There are apparent discrepancies between these various results. The surface condition of MgO depends critically on how the surface was prepared (e.g., polished, cleaved), the atmosphere to which it has been exposed (wet or dry, for example), the annealing conditions (including oxygen partial pressure) and the presence or absence of trace amounts of certain species (such as Cl ions) which are known to catalyze the reconstruction of the surface [24]. A systematic study of these effects and how they influence nucleation and growth of $YBa_2Cu_3O_7$ would be valuable.

Examples of two as-polished perovskite surfaces are shown in Figures 2.5(a) ($NdGaO_3$) and 2.5(b) ($LaAlO_3$). Again surface roughness is on the order of one or two unit cells.

2.2.3 *Microstructure and Growth Mechanism*

We have used STM and AFM to examine c-axis oriented $YBa_2Cu_3O_7$ films of different thicknesses, grown using off-axis RF magnetron sputtering [25, 22], on both MgO and perovskite substrates. The films deposited on (100) MgO were grown at 100 mtorr total pressure (60 : 40 oxygen : argon), 705 to 710°C, and an RF power of 50 W. The growth rate of the films was 1 to 1.5 nm/min. The substrates were used without further pretreatment (MgO) or with a standard solvent clean (perovskites). Good quality films can be grown on perovskite substrates over a wider range of temperatures and pressures than on MgO, and the growth conditions associated with a particular film are given below as appropriate.

X-ray diffraction indicated that the films grown on MgO were essentially completely oriented with their *c* axes perpendicular to the substrate, and had good in-plane epitaxy [$<1\%$ volume fraction misaligned: Figure 2.3(a)]. Ion-beam channeling yields were less than 10% for 160-nm-thick films. Dechanneling increases as the films become thinner. Depending on the temperature and oxygen pressure, films grown on perovskites can contain a considerable amount of *a*-axis material. In this section we consider only films which are completely *c*-axis oriented.

The superconducting transitions in these sputtered films occur at slightly reduced temperatures (85–87 K), when compared with those of bulk materials or post-annealed films (>90 K). Transitions were examined using both DC and low-frequency (10-kHz) eddy current methods and were very sharp in both cases. The films have high transport critical-current densities ($J_c > 10^7$ A/cm^2 at 64 K in zero field, $> 10^5$ A/cm^2 at 6 T, $\mathbf{B}//c$).

The STM images were acquired using the Nanoscope II and III in constant current mode with a cold-worked PtIr scanning tip. Sample bias voltages were typically 500 to 1000 mV and set-point currents ranged from 70 to 100 pA. Under these conditions stable, reproducible images were obtained

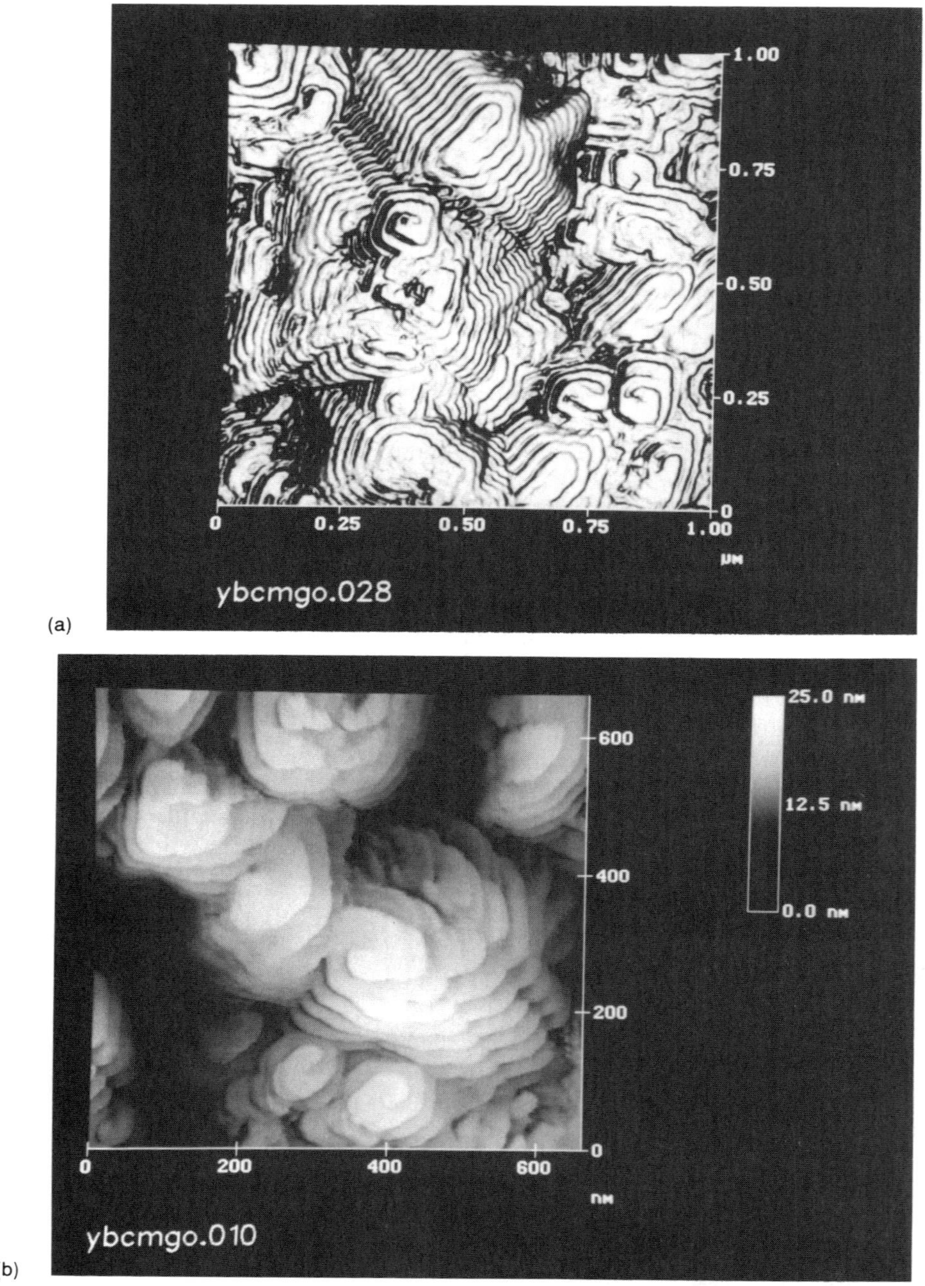

FIGURE 2.6. (a) Low magnification STM image of a film deposited on MgO by sputtering. The film is approximately 160 nm thick. The terraced surface and spiral growth pattern are clearly visible. The apparent heights of the various features are greatly exaggerated compared with the horizontal distances. The vertical height difference between terraces is one unit cell (1.17 nm). (b) Higher magnification image of a few grains of the same film.

over several scans. Either no filtering, or simple low-pass filtering to remove high-frequency noise was employed in producing the final images. The AFM images were taken for the same samples, using the conditions mentioned earlier. No special surface preparation of the films was necessary. To minimize image distortion, scan speeds were generally kept at 1 to 2 Hz at minimal current feedback gains. Further, to ensure that the data collected were representative of the particular film's surface morphology, multiple images were taken at scan widths ranging from 10 μm down to a few tens of nanometers.

The most striking features of the STM images [Figures 2.6(a) and 2.6(b): 160-nm-thick film; Figures 2.7(a) and 2.7(b): 500-nm-thick film] are the obvious granularity of the films and the spiral growth pattern of each grain. The grain diameters depend on the thickness of the film (see also Sect. 2.4 below) and lie between 100 and 500 nm, with the thicker films always exhibiting a larger surface grain size. In Figure 2.8 the approximate grain size is plotted vs the film thickness for several films deposited under identical conditions. In the thicker film, the grains are also much squarer, with well-defined faceting in the principal crystallographic directions of the substrate. The degree of polygonization depends on deposition conditions, as well as on film thickness.

As is true in any STM experiment, the relative contributions of electronic and topographical factors to the surface structure seen in the image cannot be separated. With the addition of the AFM capability, however, commonalties in the images observed with the two techniques can be confidently ascribed to topography. AFM imaging techniques faithfully duplicated most of the details of the grain structure produced by the STM (Figure 2.9). In particular, the spiral growth pattern, overall grain sizes, and step-to-step horizontal and vertical distances within a grain were reproduced.

Of particular interest is the spiral growth pattern of the grains. The vertical height difference between "terraces" is usually one unit cell, about 1.2 nm. STM and AFM line scans across similar grains are shown in Figures 2.10(a) and 2.10(b). In the AFM scans, a sequence of steps and terraces is clearly seen. Each is approximately one unit cell deep. The terraces are often tilted rather than exactly parallel to the substrate surface. The tilt angle, however, is exaggerated because the vertical scale is much different from the horizontal in Figure 2.10; it is actually about 1°. In these films, roughly equal proportions of left- and right-handed spirals are observed.

The implication of these observations is that each grain or column grows by adding atoms to a spirally expanding step edge (Figure 2.11). This is exactly the classical spiral-growth mechanism proposed by Frank [26] and developed theoretically by Burton, Cabrera, and Frank [27], and often found for crystals grown under conditions of low supersaturation, especially from the vapor phase.

The helicoid surface, clearly visible in the STM images, is produced by the intersection of a dislocation with the growth planes of the surface. The height

(a)

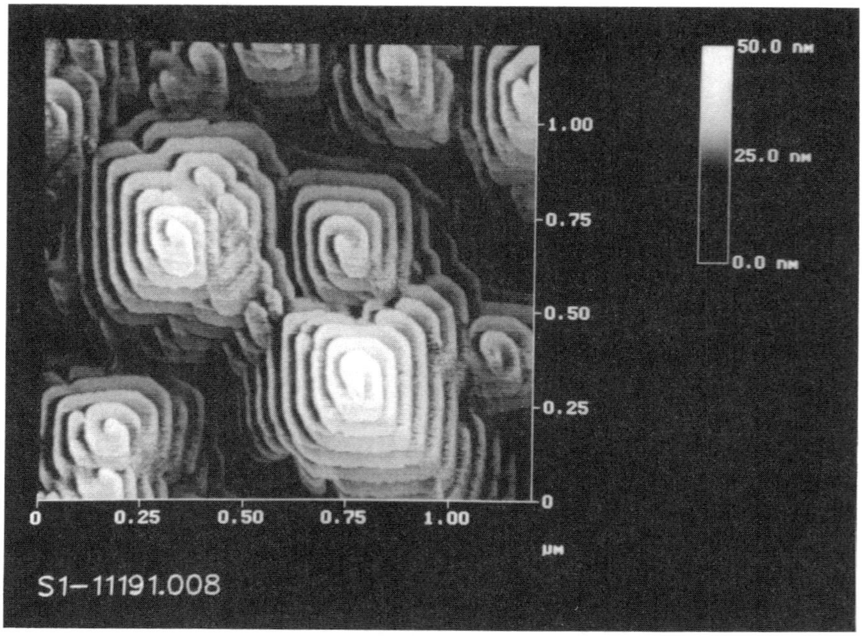

(b)

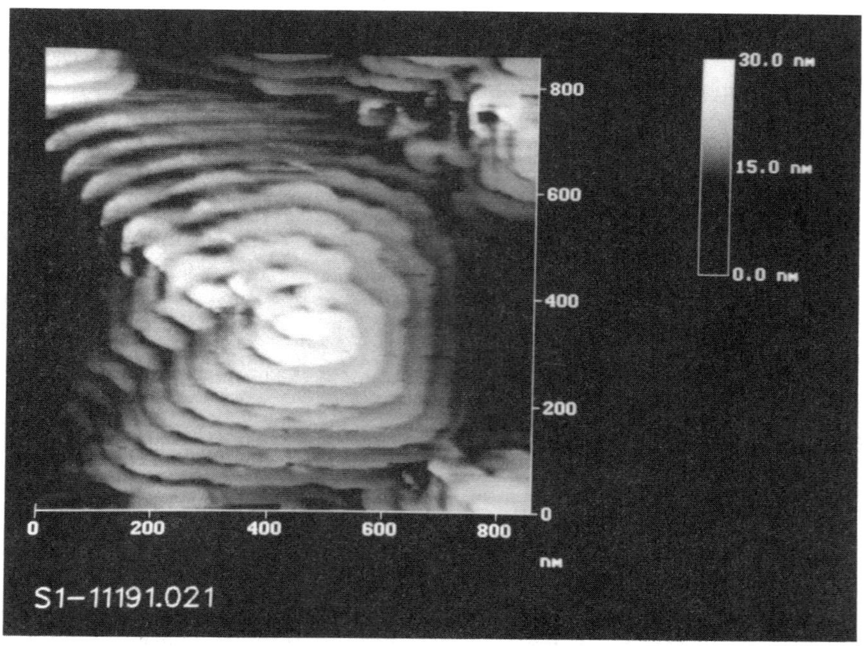

FIGURE 2.7. (a) STM image of a 500-nm-thick sputtered film on MgO, and (b) higher magnification image of a single grain of the same film. Note the interlacing of the terraces in the lower left quadrant of the grain.

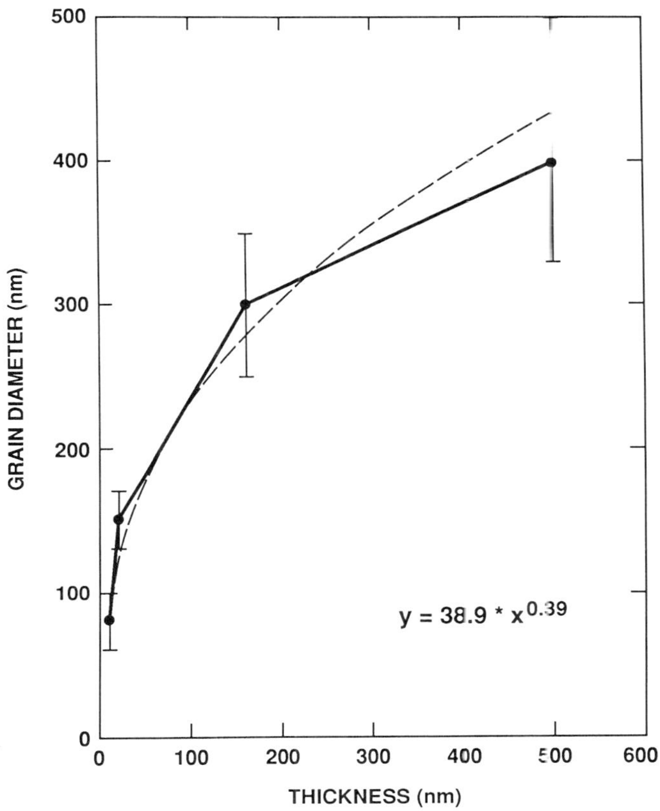

FIGURE 2.8. Plot of average surface grain size vs thickness for films prepared under indentical conditions. The dotted line is a fit of the data to the function $y = ax^b$, with the parameters shown in the figure.

of the surface step (in this case $c_0[001]$) corresponds to the component of the Burger's vector (**b**) of the dislocation that intersects the surface. The simplest case would be that of a screw dislocation with a $|\mathbf{b}|$ equal to a unit cell height and a direction parallel to [001]. Most authors have assumed that this is the situation. In principle, however, an edge or mixed dislocation with a larger **b**, inclined at some angle to the surface could produce a similar topology. To date, the dislocation has not been identified by transmission electron microscopy. The role of dislocations in growth processes is discussed more fully in a later section, as is the possible origin of these defects.

More recently, Lang et al. [28] and Krebs et al. [29] have shown that *laser-ablated* films of $YBa_2Cu_3O_7$ on MgO substrates also can grow by a dislocation mediated mechanism. This is an important result because the conditions of growth using laser ablation are typically quite different from

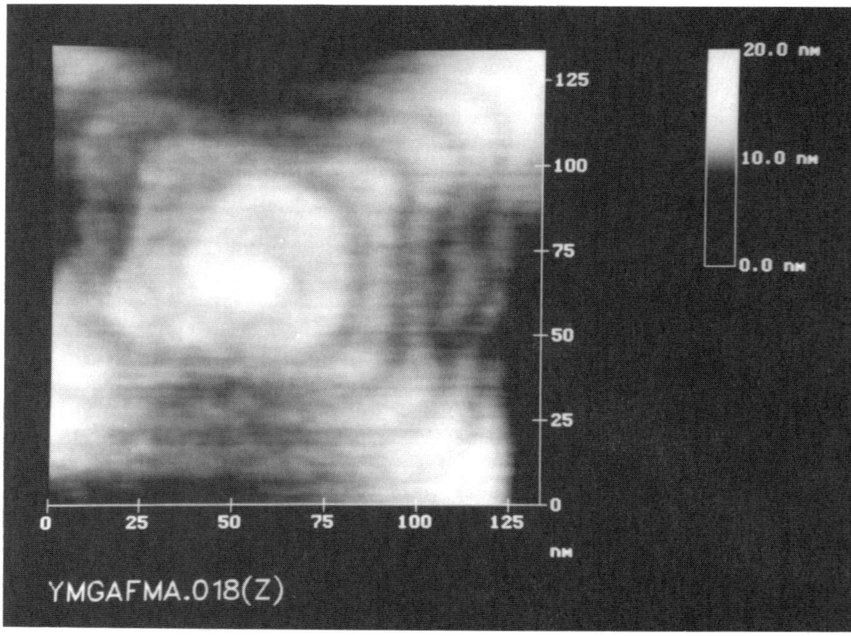

FIGURE 2.9. AFM image of a single grain from the same film imaged by STM in Figure 2.6.

those used in sputtering (higher temperatures, oxygen partial pressures and rates). The influence of the growth technique and conditions is discussed more fully below.

As mentioned earlier, films grown on MgO substrates typically contain a small fraction of material misoriented by rotations around the *c* axis. The volume fraction depends on the deposition conditions and substrate surface preparation [30], and even small amounts can have a profound effect on electrical properties such as surface resistance [1], critical current, and current-voltage behavior. The films discussed here usually contain between 0.1% and 1% of 45° misaligned material, as determined by x-ray diffraction phi scans. These rotated grains can also be seen in the STM: an example of a cluster of a few misaligned grains is shown in Figure 2.12.

Also visible [e.g., in Figure 2.7(b), lower left quadrant and in Figure 2.12] are features oriented at 45° to the principal grain and substrate directions. These features appear to correspond to the boundary between two grains which have terraces at different levels. They are therefore likely to be antiphase boundaries running parallel to the *c* axis, and are probably caused by the presence of stacking faults in the grains. The interlacing of the terraces at

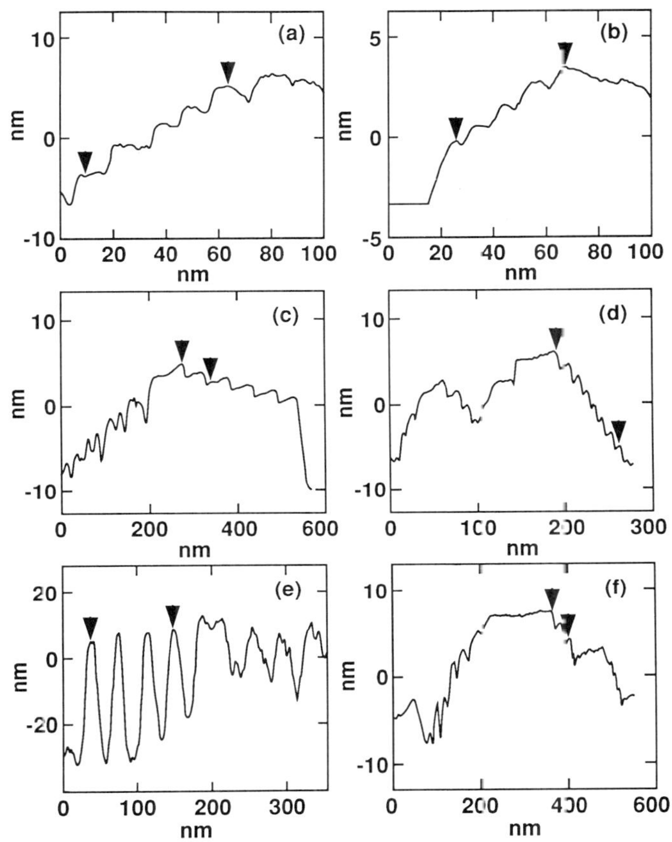

FIGURE 2.10. Line scans across grains of c-axis $YBa_2Cu_3O_7$ films taken from (a) STM and (b) AFM images of the same 160-nm-thick RF-sputtered film on MgO. Distances between markers are (a) vertical 3.74 nm; horzontal 42.55 nm; (b) vertical 4.43 nm; horizontal 53.2 nm; (c) STM image of a 500-nm-thick RF-sputtered film on MgO (vertical 2.2 nm; horizontal 59.9 nm); (d) STM image of a 160-nm-thick DC-sputtered film on MgO (vertical 10.23 nm; horizontal 69.89 nm); (e) STM image of a 200-nm-thick RF-sputtered film (vertical 3.69 nm; horizontal 203.4 nm); and (f) STM image of a 30-nm-thick laser-ablated film an $NdGaO_3$ (vertical 2.71 nm; horizontal 35.2 nm).

levels differing by (probably) $c_0/2$ can be seen on close inspection. The possible role of stacking faults in the growth process is discussed in Sect. 4. A line scan through a grain from this 500-nm-thick film is shown in Figure 2.10(c).

[We have found much larger amounts of rotationally disordered grains in films deposited on substrates heated in air at high temperatures. As noted above this is in contrast to the findings of the Cornell group. An x-ray phi scan of a film prepared on heat-treated MgO is shown in Figure 2.3(b). In

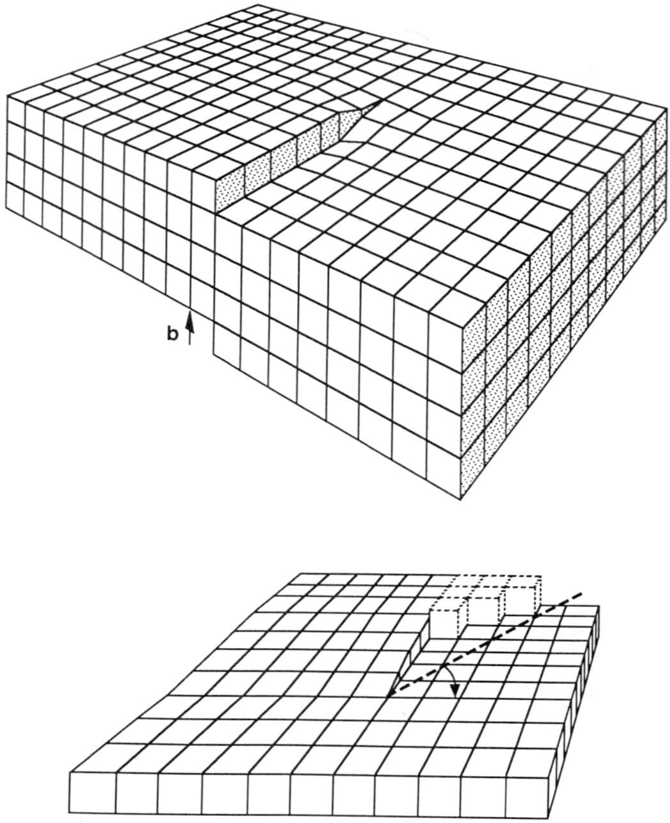

FIGURE 2.11. Line drawing schematic of spiral growth, showing the addition of growth units to a step edge.

addition to a very strong (>50%) 45° component, other smaller rotations are also present. The STM image of this film showed a much less regular growth pattern and an almost complete absence of the spiral growth morphology.]

A film prepared by DC sputtering (but under otherwise identical conditions to RF sputtered films discussed above) is shown in Figure 2.13. It exhibits extremely well-defined spiral grains with an almost circular cross-section, as viewed from above. Unit-cell high steps are evident in the line scan across one of these grains shown in Figure 2.10(d). Clearly, small variations in the growth conditions have a significant effect on the details of the film microstructure.

The spiral growth mechanism is not restricted to MgO. Results of STM studies of films deposited on {100} perovskite substrates have also been published. The IBM–Zürich group [14] have deposited films of $YBa_2Cu_3O_7$ on $SrTiO_3$ by hollow-cathode sputtering, and the STM images are almost inter-

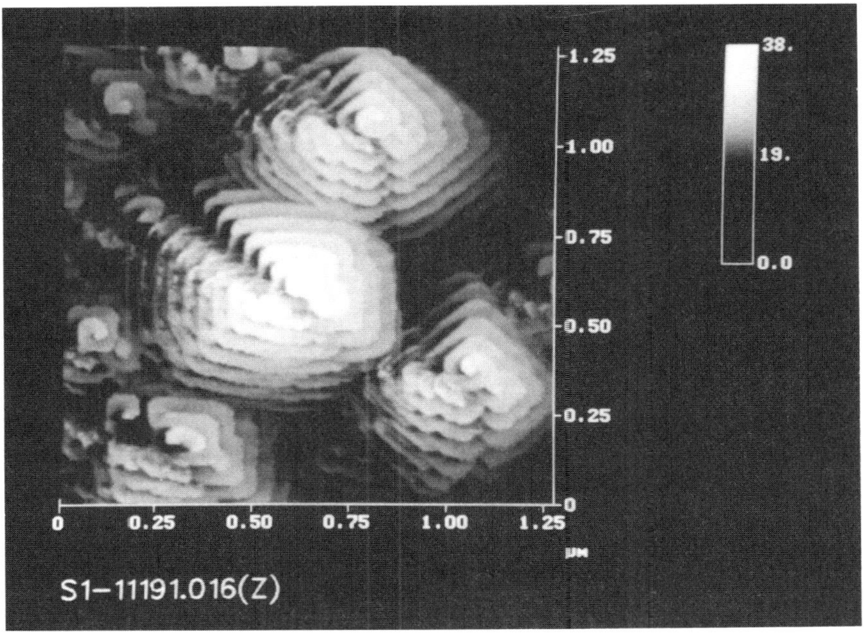

FIGURE 2.12. A small cluster of grains from a 500-nm-thick film on MgO, showing 45° rotational relative misorientations.

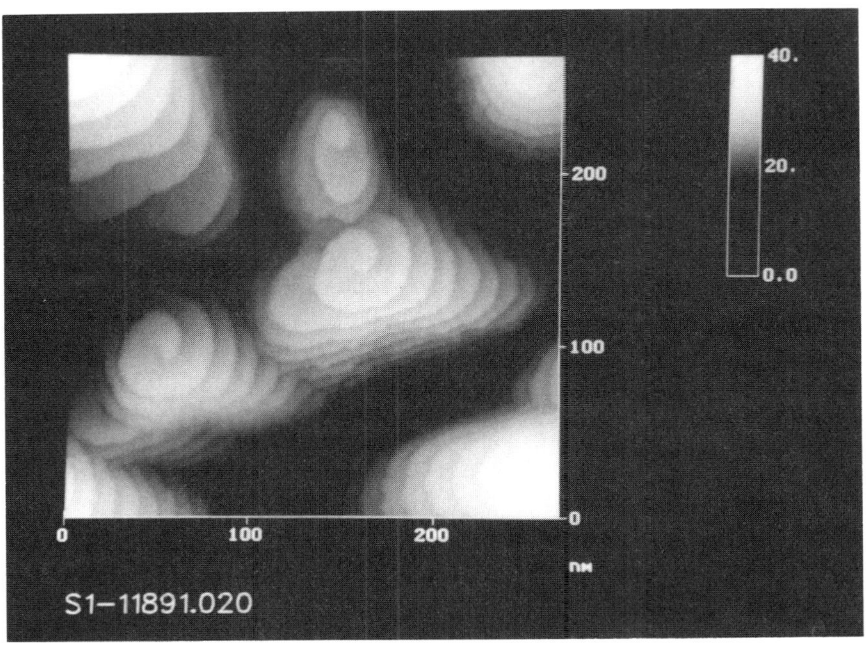

FIGURE 2.13. STM image of a film grown on (100) MgO by DC sputtering. Most of the other images were prepared by RF sputtering.

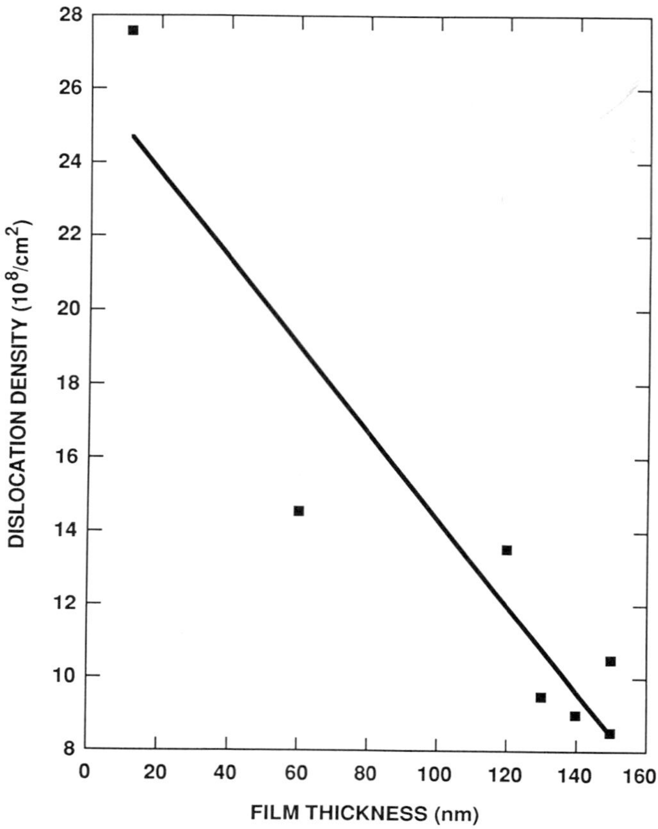

FIGURE 2.14. Plot of dislocation density vs thickness for films prepared by sputtering on $SrTiO_3$. The data are taked from Ref. [14].

changeable with those found for films of similar thickness deposited on MgO by off-axis sputtering [13].

The same authors also suggested that the density of surface dislocations was proportional to the rate of growth of the film. In fact, a much better correlation exists with the thickness, than with the deposition rate. Using the data from Ref. 14, this correlation is plotted in Figure 2.14.

It is therefore clear that, under certain deposition regimes, films can grow by a spiral-growth mechanism on perovkite substrates. It appears, however, that the well-developed spiral structure has been most evident on MgO. Recently, Norton et al. [31] have reported that laser-ablated films on $SrTiO_3$ and $LaAlO_3$ only show positive evidence of spiral growth at less-than-optimum growth temperatures, as judged by critical-current measurements (680°C vs 730°C), and Moreland [32] et al. only observed spiral, as opposed to the more general ledge, growth on MgO.

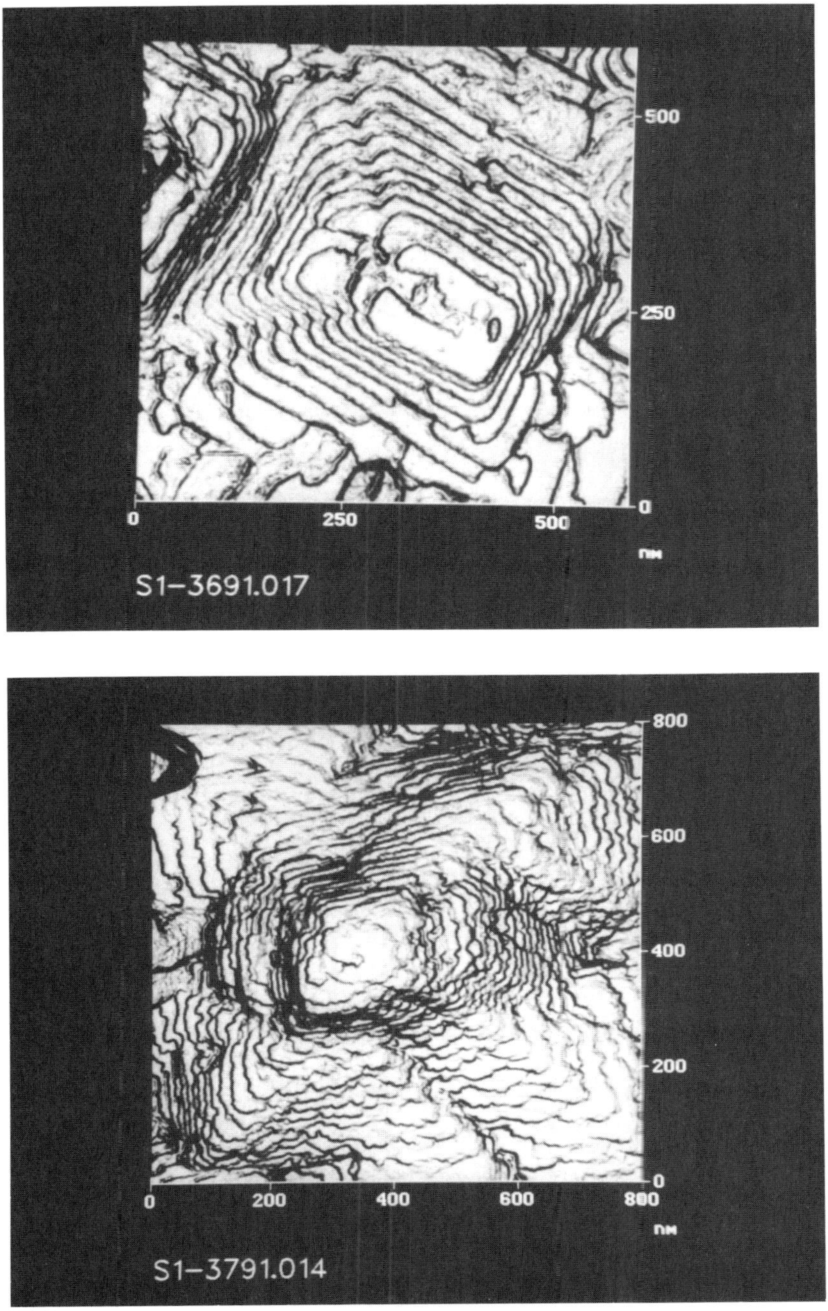

(a)

(b)

FIGURE 2.15. *c*-axis films grown by sputtering on pseudocubic (100) $NdGaO_3$. The films were deposited at the same temperature (765°C), but at different pressures: (a) 40 mtorr O_2, 160 mtorr Ar, and (b) 20 mtorr O_2, 80 mtorr Ar. The growth rate for the film shown in (b) is about twice that of the film shown in (a), which may be related to the appearance of the wavy steps in (b) (kinetic roughening).

We have found that the detailed microstructure of *c*-axis films grown on $NdGaO_3$ by off-axis sputtering depends strongly on the deposition conditions. Two examples are shown in Figure 2.15. The detailed microstructure of these films depends on the deposition conditions. The film shown in Figure 2.15(a) clearly exhibits the layered growth of a *c*-axis-oriented island structure, with step edges of unit-cell height. It is not obvious whether this structure is the result of spiral or ledge growth. Holes in the center of many of the grains suggest that a dislocation is present during growth. In the presence of the stress field associated with the spiral growth, it can become energetically favorable for the creation of a free surface, that is, a dislocation with a hollow core [33]. The film shown in Figure 2.15(b) appears to terminate in a spiral, but the terraces are clearly much more disordered. These films were grown at the same temperature (766°C) but at different gas pressures (and consequently different rates).

A common feature observed using the STM on films of $YBa_2Cu_3O_7$ is that the terraces do not always appear to be horizontal [13, 32]. An extreme example of this is shown in the image of a film deposited on (100) $SrTiO_3$ (Figure 2.16). A line scan is included in Figure 2.10(e). Whether this feature corresponds to real topography or is some combination of topography and electronic properties is still unclear. It has been reported that this dip in the

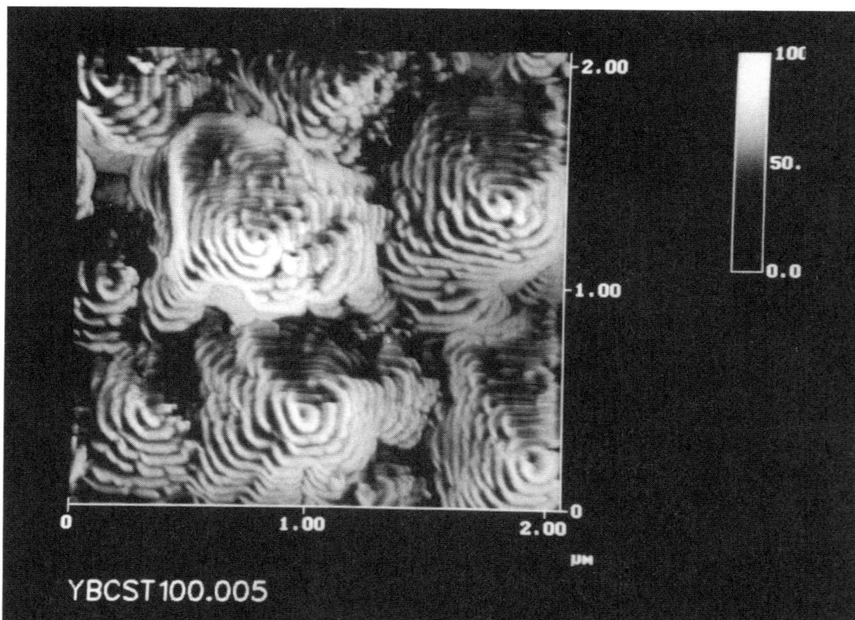

FIGURE 2.16. STM image of a film prepared by sputtering on (100) $SrTiO_3$. A line scan of this film is shown in Figure 2.10.

terrace disappears when the samples are cooled into the superconducting state, although no specific explanation has been presented [32].

2.2.4 *Early Stages of Growth*

In order to investigate the earlier stages of growth on MgO, we have used AFM and STM to examine the microstructure of some thin films prepared by off-axis sputtering. Deposition conditions were identical to those used for the thicker films discussed above, and the films were grown on substrates with a surface morphology similar to that shown in Figure 2.1. The films, which were nominally 20 nm and 10 nm thick, were grown for 18 and 9 min, respectively. Rutherford backscattering (RBS) experiments gave average thicknesses of 21.5 and 10.0 nm, based on theoretical density.

Results for the two thin films are compared here with those described earlier for a 160-nm-thick film (Figures 2.4 and 2.9). Low-resolution AFM images of the three films are shown in Figure 2.17. Image contrast corresponds to the same vertical height difference in each of the three images. Two features are striking: (1) the average grain size of films increases with increasing thickness, as noted previously, and (2) islands are clearly visible in the case of the 10-nm-thick film, and voids are present in the case of the 20-nm-thick film. Higher resolution AFM images of the three films are shown in Figure 2.18. The difference in grain size between the three films is clearly evident. In the thinnest film, regions of very fine grains and islands of larger grains may be seen. That all regions of the 10-nm-thick film are conducting was shown by broad-area scans using the STM. The thin, fine-grained regions do not, therefore, consist of bare MgO or nonconducting oxide components. STM images of the two thin films are shown at the same resolution in Figure 2.19. The coverage of the substrate with a continuous, fine-grained layer, appears to be complete very early in the growth of the films, before a thickness of 10 nm is achieved. The density of nuclei in this thin layer is comparable to the feature size on the substrate surface. Once this layer is formed, widely spaced islands, with a larger grain size, appear on the surface.

These results are in broad agreement with those of Streiffer et al. [19], who studied the microstructure of very thin films (approximately 10 to 100 A thick) using TEM. In the thinnest films, which average only one unit cell in thickness, rather than complete surface coverage, those authors found that connected islands of one, two, and occasionally three, unit cells deep nucleated at random all over the surface. As material was added to these nuclei, rather than becoming taller, they spread out until complete surface coverage was achieved at a thickness approaching 10 nm. The mechanism, therefore, is perhaps best described as two-dimensional island nucleation, presumably assisted by the large lattice mismatch between MgO and $YBa_2Cu_3O_7$, followed by attachment of new units to the step edges. This latter process is favored by (1) the layered nature of the superconductor, (2) the anisotropy of the growth rates, and, at low growth rates, (3) the presence of dislocations.

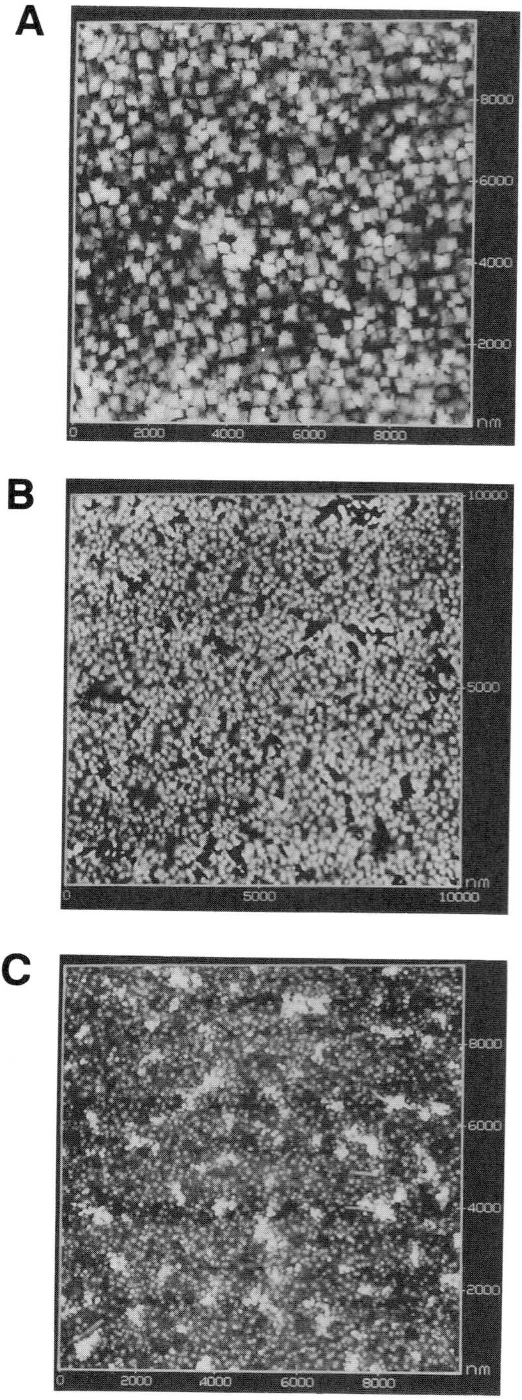

FIGURE 2.17. Low-resolution AFM images of the three films: (a) 200 nm thick; (b) 20 nm thick; and (c) 10 nm thick. Note the progression of grain sizes, and the appearance of voids and islands in (b) and (c), respectively.

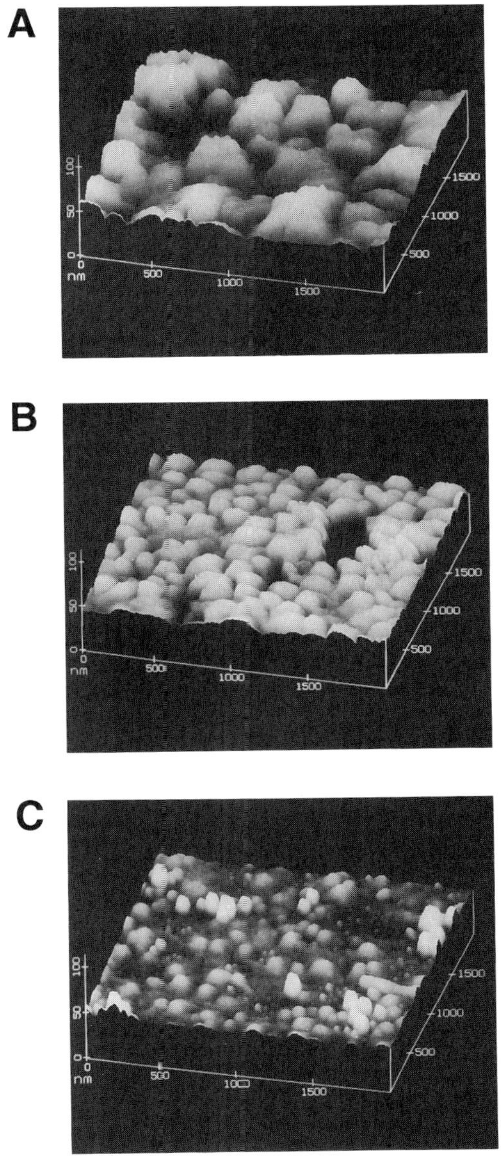

FIGURE 2.18. Profiles of high-resolution AFM scans of the three films: (a) 200 nm thick; (b) 20 nm thick; and (c) 10 nm thick.

As mentioned earlier, the Cornell group [16] using highly stepped or vicinal substrates also see island growth (Volmer–Weber) with preferential nucleation at step edges.

A consequence of this mechanism is that nuclei that are rotated with respect to one another, or nuclei that form a nonintegral number of basal plane unit cells apart will not coalesce exactly, and will lead to threading disloca-

A

B

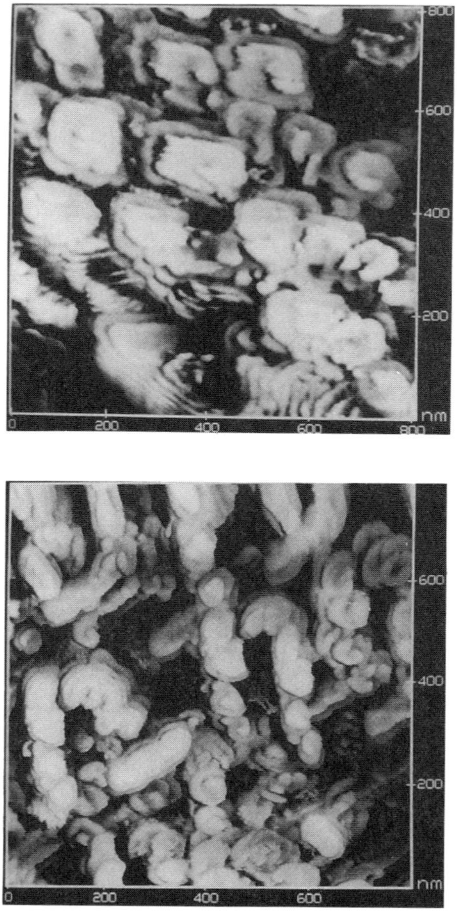

FIGURE 2.19. High-resolution STM images of two thin films: (a) 20 nm thick, and (b) 10 nm thick. Growth spirals are clearly visible, even in the thinnest films.

tions when they impinge on one another. Streiffer et al. measured a high density of such dislocations (about 10^{11} cm^{-2}) in a 7.2-nm-thick film.

It is evident from the AFM and STM images that as the thickness increases, the average grain size also increases; in other words, some of the grains in the first layer grow faster than others, and overgrow grains that are growing more slowly. If only some of the grains in the early stages contained screw dislocations, then a few grains would tend to grow rapidly and overgrow the dislocation-free grains in the first layer. This will, eventually, significantly reduce the number of growing grains. Since the grain size is probably limited by the lateral impingement of a grain upon its neighbors, this will lead, in time, to a few large grains. Alternatively, the strain associated with the dislocation itself may be a mechanism for grain coarsening. In a larger grain, the shear forces operate over a greater volume, and therefore the average stress is reduced.

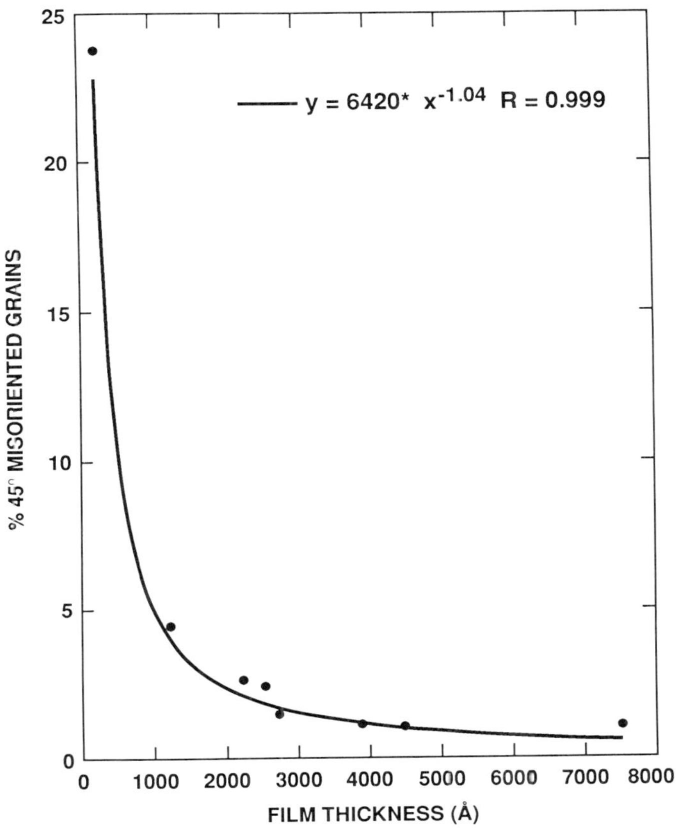

FIGURE 2.20. Volume fraction of 45°-misoriented material plotted as a function of thickness for films on an MgO (100) substrate.

The images of Figure 2.19 suggest, in a qualitative way, that the material in the first few layers is much more disordered than that in thicker films. We have measured the volume fraction of 45°-misoriented material in a series of films grown to different thicknesses, using x-ray phi scans (Figure 2.20). As may be seen, the amount of rotated material falls off quickly with increasing thickness. It is not known whether this means that the large angle grain boundaries close to the interface are annealing out during the longer depositions while new misoriented grains are being produced, or whether it is only the near-interface region that contains a significant number of rotated grains, and, once produced, they are locked in. The same ambiguity exists with respect to grain growth in general. Does the smaller grain size we see in thinner films mean that close to the substrate there is a permanent region of fine grains, or is this only a temporary situation, and during a longer deposition significant grain coarsening occurs, even close to the interface?

2.3 Growth of *a*-Axis Films

Several groups have demonstrated that, especially on perovskite substrates, a reduction in the deposition temperature causes a transition to an *a*-axis-perpendicular film growth [7, 11, 34]. Eom et al. [7] showed that on $SrTiO_3$, essentially pure *a*-axis-oriented $YBa_2Cu_3O_7$ could be obtained at about 640°C. This material had the additional attribute of having a relatively smooth surface. At temperatures between 640°C and about 740°C mixed *a*- and *c*-perpendicular orientations are obtained, with the *c*-axis-perpendicular material lying next to the substrate and *a*-axis material nucleating on the *c*-axis material at a critical thickness. The exact reasons for the switch in orientation are unknown, but it is a very general phenomenon in thicker films of $YBa_2Cu_3O_7$. Ex situ films prepared by coevaporation also show a similar transition. Eom et al. postulate that the transition is induced by a tradeoff between film–substrate interactions, which favor *a*-axis oriented material at low temperatures, and minimization of the free-surface energy, which evidently favors *c*-axis-perpendicular material. At high temperatures, surface mobility is high enough to allow minimization of the surface energy to overcome favorable film–substrate interactions.

A simpler argument based on relative growth rates can also be made. Most of the evidence suggests that the films grow much faster in the [100] and [010] directions, than in [001] (see e.g., Ref. [16]). This is presumably because new material is more easily added to the edges of the layered structure than to the top of a completed unit cell. Thus, once an *a*-axis grain has nucleated on the growing surface of *c*-axis material, it will grow faster. Eventually, the surface will become covered with *a*-axis material. In all cases in which the film has been subject to cross-sectional TEM investigation, it appears that there is a layer of *c*-axis material present at the interface. Obviously, in films containing a high percentage of *a*-axis material, the *c*-axis layer will be very thin.

In order to investigate (1) the microstructure of *a*-axis films, (2) the role of substrate in determining the microstructure of *a*-axis perpendicular material, and (3) the nature of the nucleation of *a*-axis material at the critical thickness, we have undertaken a more extensive study of films deposited on perovskite substrates.

a-axis material appears to the STM as clusters of platey grains, viewed edge-on (Figure 2.21). The film shown in Figure 2.21(a) contains 30% of *c*-axis material according to x-ray analysis but shows no evidence of *c*-oriented-grains on the surface. The film shown in Figure 2.21(b), contains almost exclusively *a*-axis material.

Striations perpendicular to the long axis are obvious in higher resolution images [Figure 2.22(a)]. Striations are presumably a consequence of a growth mechanism in which material is added to the top edges of the platelets comprising the grains. At very high resolutions [Figure 2.22(b)], the texture approaches unit-cell size and exhibits considerable granularity. For *a*-axis

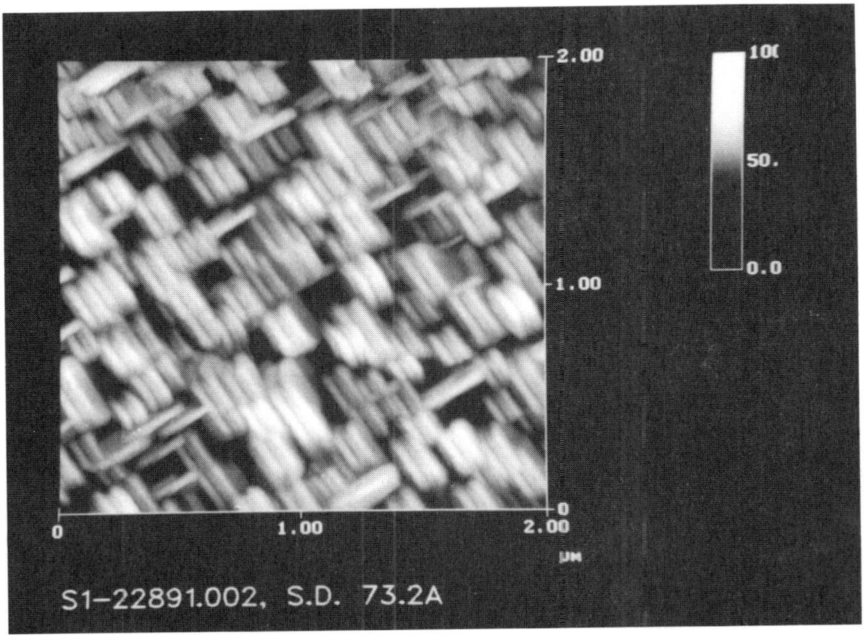

(a)

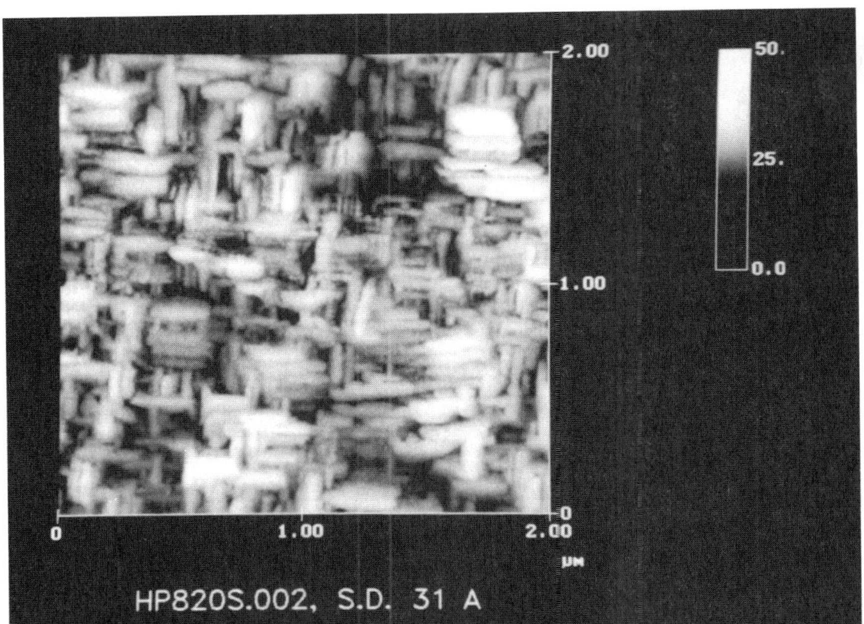

(b)

FIGURE 2.21. Comparison of growth morphlogy of *a*-axis grains on (a) pseudocubic (100) $NdGaO_3$, and (b) (100) $SrTiO_3$.

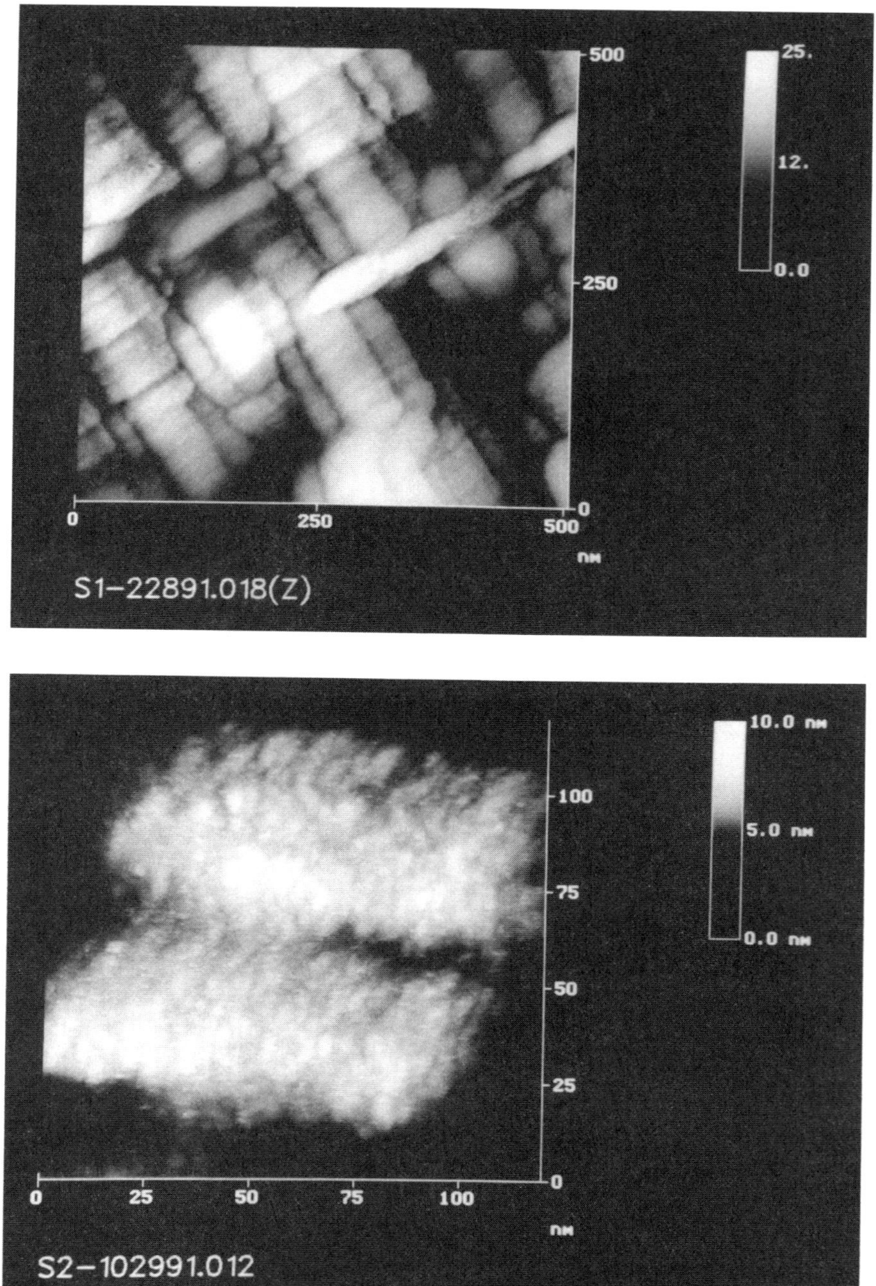

(a)

(b)

FIGURE 2.22. Higher magnification images of *a*-axis grains. The film in (a) was grown on (001) $NdGaO_3$, and the film in (b) on (100) $SrTiO_3$.

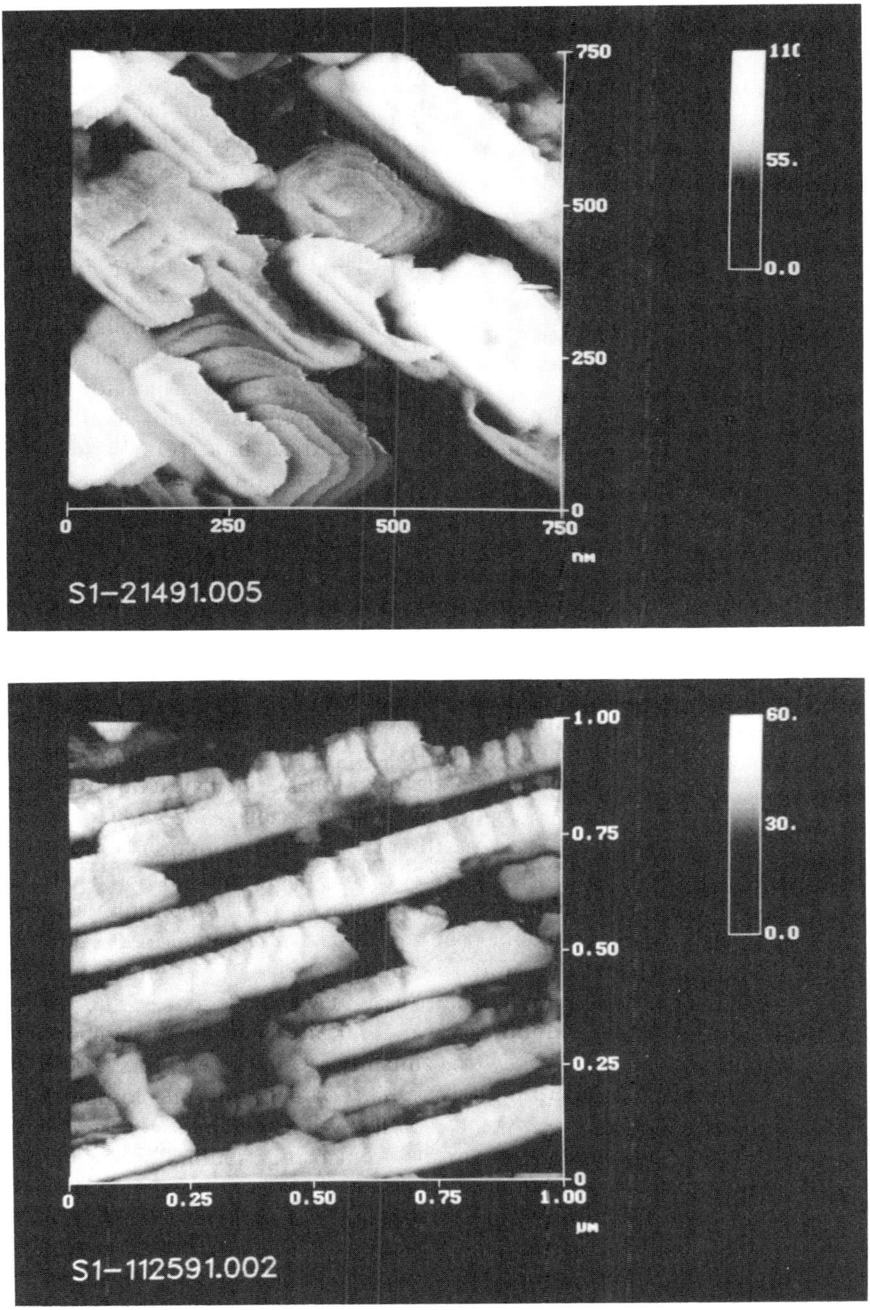

FIGURE 2.23. (a) STM image of a film grown on pseudocubic (100) $NdGaO_3$, showing the nucleation of rectangular grains on top of c-axis growth spirals. It is likely that the rectangular grains are a-axis material, and nucleate on the steps in the c-axis grains. (b) A similar film, grown for twice as long as that in (a), showing that the surface is completely covered by a-axis grains.

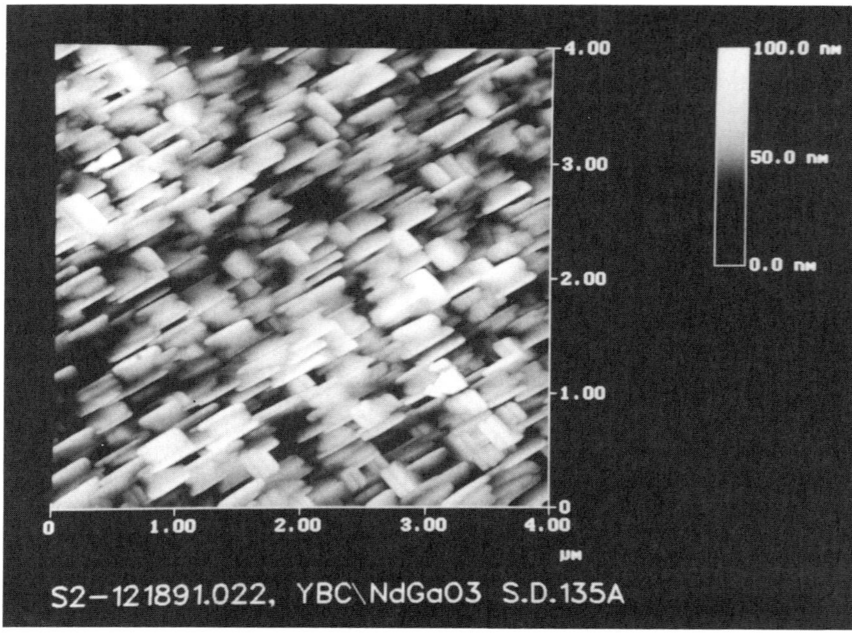
4.00
3.00
2.00
1.00
0
0
1.00
2.00
3.00
4.00
μm
100.0 nm
50.0 nm
0.0 nm
S2–121891.022, YBC\NdGaO3 S.D.135A

(a)

4.00
3.00
2.00
1.00
0
0
1.00
2.00
3.00
4.00
μm
80.0 nm
40.0 nm
0.0 nm
S1–121891.002, YBC\STO S.D.97A

(b)

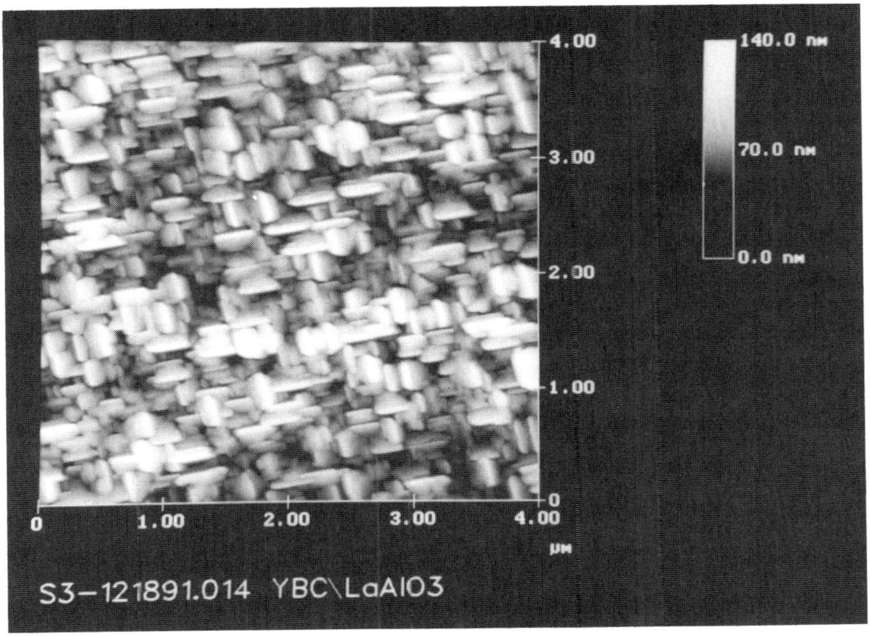

(c)

FIGURE 2.24. Low-resolution STM images of *a*-axis films grown under the same conditions on (a) pseudocubic (100) $NdGaO_3$, (b) (100) $SrTiO_3$, and (c) (112) $LaAlO_3$. Note the clearly preferred growth direction of the film on $NdGaO_3$.

material we do not expect to see spiral (dislocation mediated) growth. The fast growth direction is now perpendicular, rather than parallel, to the substrate, and therefore exposed to the flux of new material. The edge planes of the unit cell are all available for attachment.

Taken together, the STM and x-ray data appear to confirm that the initial stage of deposition must be *c*-axis-normal growth followed by a crossover to an *a*-axis-oriented texture. The epitaxial relationship between *c*-axis spiral grains and the overlying *a*-axis material is suggested in Figure 2.23(a). This film was deposited under conditions identical to those used to obtain well-ordered, *c*-axis spiral grains on MgO. X-ray diffraction indicate that it contains about equal amounts of *a*- and *c*-axis-perpendicular grains. The surface grains (which are presumably *a*- or *b*-axis material) appear to nucleate on the step edges of the *c*-axis spirals, which may be seen at a lower level. This observation immediately explains the shape and size of the *a*-axis grains. The width of the grains in the short direction is limited by neighboring grains growing on an adjacent step edge. Figure 2.23(b) is an image of a film grown for twice as long, but under identical conditions. Clearly the surface coverage with *a* grains is much higher, and their average length has also increased.

There appear to be significant differences among *a*-axis films grown on different perovskite substrates. The most obvious is the nature and degree of in-plane alignment. Young and Sun [35] have found, using x-ray texture analysis, that *a*-axis films grown on the (110) face of $NdGaO_3$ show a strong texturing, with over 90% of the grains aligned along only one direction in the substrate plane. This is in contrast to the situation on $SrTiO_3$, where *a*-axis grains lie in equal amounts in two equivalent directions rotated by 90°. The explanation is presumably that, because of the anisotropy of the (110) $NdGaO_3$ lattice plane, there is a preferred orientation for the superconductor ($h0l$) [*b*-axis perpendicular] or ($0kl$) [*a*-axis perpendicular] planes on the substrate. $SrTiO_3$ has an isotropic (100) face and, therefore, there is no preferred growth direction for the grains.

We have also observed a preferred direction in the orientation of the *a*-axis grains on $NdGaO_3$, which is visible in a comparison of Figure 2.21(a) ($NdGaO_3$) and Figure 2.21(b) ($SrTiO_3$). On $SrTiO_3$ we see almost equal amounts of the two orientations. A further comparison of three films grown simultaneously is given in Figure 2.24. $LaAlO_3$ appears to be much closer to $SrTiO_3$ in grain orientation properties than to $NdGaO_3$. All of the films in this figure contain $>99\%$ *a*-axis growth.

The surface plane anisotropy argument given above would be most convincing if the *a*-axis material grew directly on the substrate. As discussed earlier, however, most of the available evidence suggests that *a*-axis material grows on a layer of *c*-axis material. It would therefore seem to be the case that the anisotropy of the surface plane of $NdGaO_3$ must be reflected in a nonrandom distribution of underlying *c*-axis material on the surface. The surprising implication of this would be that *under the growth conditions* the *a* and *b* lattice parameters differ significantly. On the contrary, we expect the material to be essentially tetragonal at the temperature and pressure of the deposition [36].

2.4 General Considerations of Growth Mechanism and Morphology

It is clear from the images presented in the previous sections, that thin films of *c*-axis-perpendicular $YBa_2Cu_3O_7$, deposited by sputtering, grow by a spiral mechanism over a wide range of conditions. In this section we discuss some general aspects of the growth of crystals, particularly with relation to spiral growth. Our objective is to present a qualitative picture of the material properties and experimental variables that play a role in determining the nature of the growth process. By varying deposition rate, temperature, pressure, substrate surface, etc., it may prove possible to exert improved control over the growth process. Unfortunately, systematic information is not yet available for much of the parameter space. Many of the conclusions presented in this section are the result of theoretical studies, starting with the

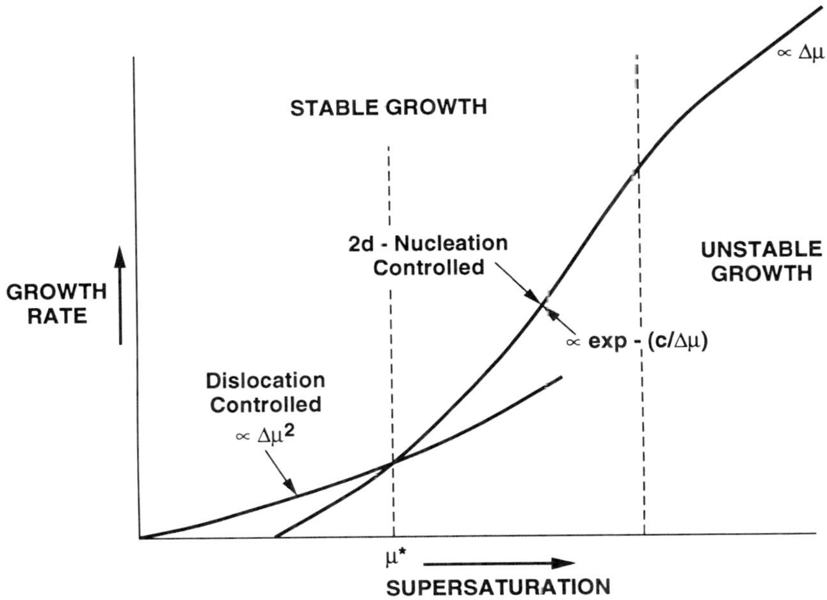

FIGURE 2.25. Schematic dependence of growth regime on supersaturation.

phenomenological continuum approach of Burton, Cabrera, and Frank [27] and Cabrera and Levine [37], and extending through Monte Carlo [38] and recent molecular dynamics simulations [39].

In general, the rate of growth and the morphology of a solid–fluid interface depends on several factors. Most important of these is the effective supersaturation, which is expected to be much lower for a vapor–solid interface, than for a liquid–solid interface. Spiral (or dislocation mediated) growth is characteristic of crystals grown at low effective supersaturations, that is, when the thermodynamic driving force for crystallization from the vapor phase is small. The situation is shown schematically in Figure 2.25. Under conditions of low supersaturation, the probability of nucleation on top of a complete surface layer (*birth and spread* mechanism) is so low that nucleation-mediated growth is very unlikely. If a screw dislocation (or, more precisely, a dislocation with a surface-intersecting screw component) is present, however, nucleation of new growth centers becomes unnecessary, and material can be continuously added to the edge of the step (Figure 2.11). This situation might be especially favorable for the growth of a layered material, such as $YBa_2Cu_3O_7$, oriented such that the fast growth directions (in this case the *a*, *b* directions) are parallel to the substrate.

As the effective supersaturation is increased (Figure 2.25), two-dimensional nucleation on a completed surface becomes more likely and, eventually, a crossover from dislocation-dominated to nucleation-dominated growth is predicted at a critical supersaturation, $\Delta\mu^*$. If supersaturation is increased further, a transition to unstable growth would be expected. In this last regime, uncontrolled nucleation of growth centers on top of each other lead to a dendritic type of growth, which rapidly reaches out into the vapor phase, leading to very rough, and usually undesirable, surfaces. Recently, Järvinen et al. [40] have shown that it is possible to grow whiskers of $YBa_2Cu_3O_7$ on MgO substrates, using sputtering at high gas pressures.

The second important factor in determining the morphology of the surface is the temperature. In this context it is useful to define a normalized bonding energy

$$\sigma = -4\phi_{ss}/2kT,$$

where $-\phi_{ss}$ is the potential energy of a solid-solid nearest-neighbor pair of growth units. Thus, the higher the temperature, the smaller σ. As the temperature is increased, the equilibrium interface becomes rougher, and at a critical temperature, a two-dimensional roughening transition may be defined. This dramatically increases the number of kink sites on the surface, leading to much more rapid growth. In fact, above the roughening transition temperature, there is little or no advantage in having a stepped surface of the type we see here. Below the critical temperature, however, there is still a strong dependence of growth morphology on σ. Principally, the edges of the steps become rougher because of thermal fluctuations. The film shown in Figure 2.15(b) may be an example of this type of roughness.

The temperature also has a strong influence on the polygonization of the grains. Polygonization occurs because the principal crystallographic directions are the most densely packed, and they tend to have the fewest kink sites available for the addition of new material to the step edge. The growth front, therefore, moves most slowly in these directions (for $YBa_2Cu_3O_7$ these are the [100] and [010] directions). The rate of growth is therefore limited by these slow-moving directions, resulting, in this case, in square grains. At high temperatures, however, the presence of roughness in the step edge leads to many more kink sites, and, hence, a more two-dimensionally isotropic growth which results in more circular grains. An example of the effect of temperature on grain shape is shown in Figure 2.26.

From these qualitative observations, it is perhaps not surprising that the microstructures of films of $YBa_2Cu_3O_7$ grown under different conditions or by the different techniques of sputtering and laser ablation are found to be significantly different. Sputtered films are usually grown at rather low rates (i.e., low effective supersaturations) compared with laser deposited material, so they tend to grow by the dislocation-mediated route. In the laser-ablated case, however, nucleation is more likely, and spiral growth is seen less often under typical deposition conditions. At these higher rates, it appears possible to obtain nucleation-dominated layer-by-layer island growth. An example of this is shown in Figure 2.27. Films grown under these conditions show little evidence for spiral growth, but the unit-cell high layers are clearly evident

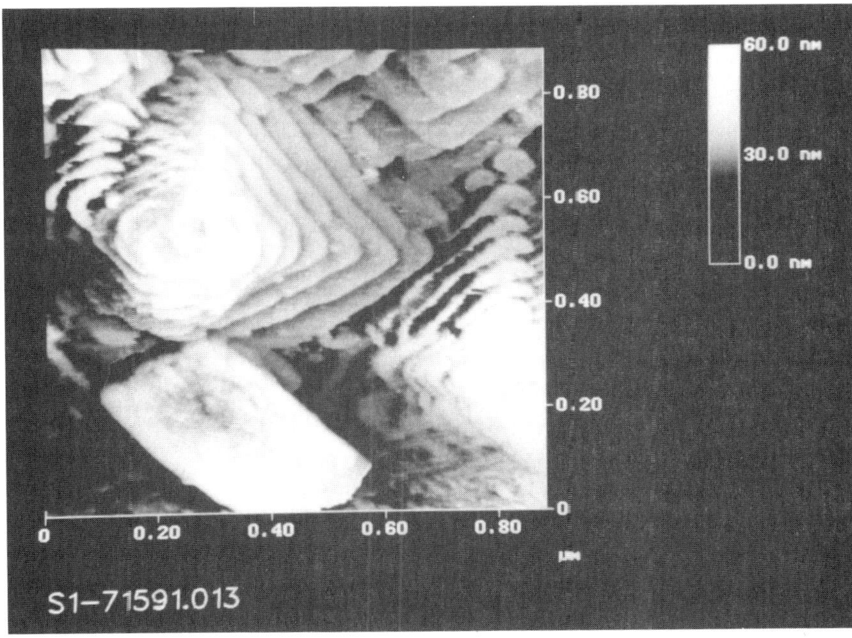

(a)

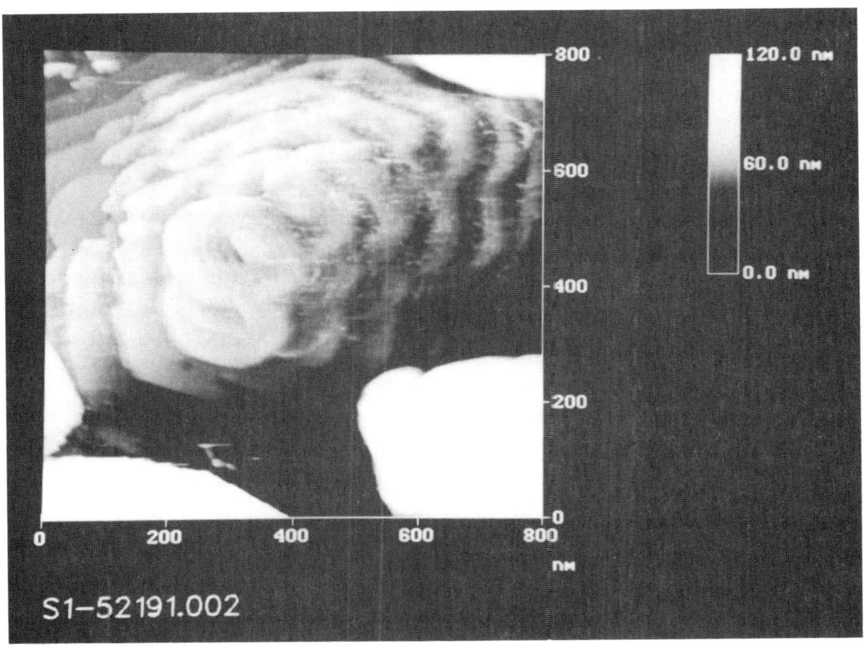

(b)

FIGURE 2.26. Effect of temperature on the growth spiral morphology. (a) At lower temperatures, 728°C, the spirals are square, reflecting the symmetry of the substrate. (b) At higher temperatures, 777°C, the grains are much more circular, as predicted by theoretical models.

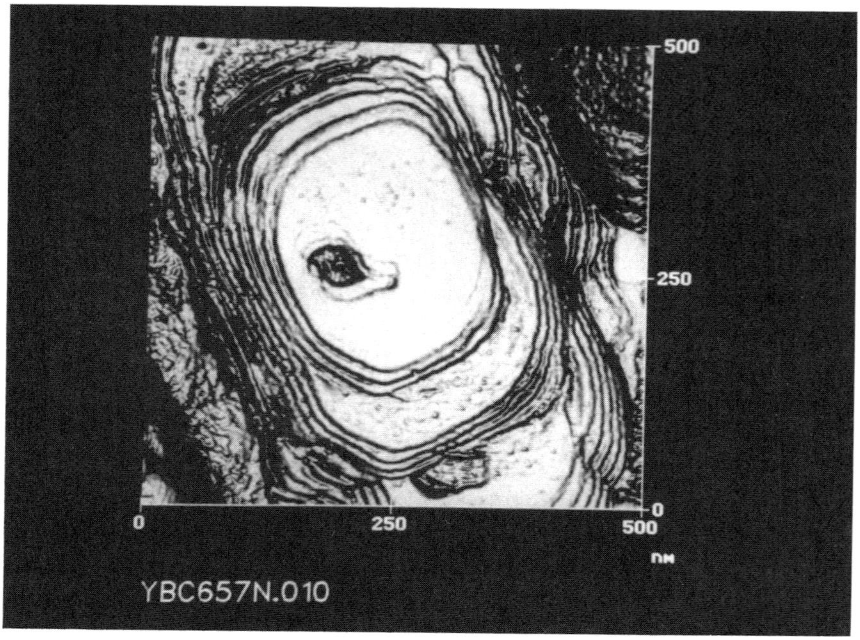

FIGURE 2.27. Typical laser-ablated film, showing well-defined layering but an absence of spiral growth.

both in this image and in the line scan across this grain shown in Figure 2.10(f).

The dependence of rate on vapor–solid chemical potential difference is different in the three regions of Figure 2.25. In the dislocation-dominated region, rate is proportional to $\Delta\mu^2$; in the nucleation controlled region it is proportional to $\exp - c/\Delta\mu$; in the unstable growth regime it is linear in $\Delta\mu$. To date, we are unaware of any verification of these rate laws for films of $YBa_2Cu_3O_7$.

As has been pointed out, the characteristic spiral growth pattern and the resulting helicoid surface result from the intersection of a defect (whose Burgers vector must have a component perpendicular to the lattice planes) with the free surface. The nature and origin of this dislocation, however, have not been determined. The simplest interpretation is in terms of a screw dislocation with $\mathbf{b} = c_0[001]$. Looking at the high-resolution STM images of a 10-nm-thick film [e.g., Figure 2.19(b)], it is clear that many of the grains contain a surface spiral, even at a very early stage of growth. Presumably, such dislocations could propagate all the way through the film. Threading dislocations with the correct **b** have not, however, been unambiguously identified in TEM cross sections. Although dislocations are often seen, they have rarely been well characterized. Eibl and Roas [41], for example, found a high density of dislocations ($10^9/cm^2$, compare with Figure 2.14) in laser-ablated films, but did not determine **b**.

Another possibility is that the growth dislocations are much more local to the surface, that is, they are generated and terminate at specific points in the bulk, and are only active during some part of the growth process.

If this is to be the case, the dislocation must terminate in an appropriate combination of defects in the bulk. A plausible mechanism for generating defects with the correct **b** may be arrived at by considering the types of defects that are commonly seen in electron micrographs. Many of these are defects associated with mistakes in the stacking sequence. For example, Ramesh et al. [42] and others have identified several stacking faults associated with the intercalation of additional cation layers with abnormal cation stoichiometry. The overall impression is that the energetic penalty for mistakes in the *c*-axis stacking sequence is rather small. This is confirmed by the rather small differences between the enthalpies of formation of, say, the 123 and 124 phases, measured by solution calorimetry [43]. Ramesh et al. also showed that one structure is converted to another through the generation of an edge dislocation in the *ab* plane whose extra half-plane corresponds to the cation layer being inserted. Thus, these defects have a Burgers vector in the *c* direction, and the dislocation line along, say, the [100] direction. These authors also found structural edge dislocations in which the extra half-plane corresponds to an entire unit cell of the 123 structure. Such a dislocation has a **b** of $c_0[001]$.

Such fluctuations in the composition and stacking sequence of the growing crystal provide exactly the type of defect necessary for the formation of a surface step and for the generation of (a pair of) screw dislocations. Suppose that a region of the surface of crystal contains a unit-cell high step, and that the step is finite in extent and the ends are pinned. As the step grows (Figure 2.24), by edge-addition, it winds up into two spirals with opposite senses. This growth mechanism will lead to two screw dislocations with opposite Burgers vectors (a Frank–Reed source). Presumably, these active growth sites will continue until they are annihilated by interaction with other defects, including other dislocations.

2.5 Implications of Growth and Microstructure for Superconducting Properties

The film microstructure and associated defects may have significant implications for the superconducting properties of thin films of high-temperature superconductors.

Several types of defects have recently been identified by STM studies, and are present as a consequence of the growth:

1. Growth dislocations, present at least under conditions readily achieved during sputtering, and found predominantly on MgO substrates, although clearly present on lattice matched substrates under some conditions.
2. Granularity evident in the images presented here is a consequence of uncontrolled nucleation. Associated with granularity are low- and high-angle grain boundaries, and antiphase boundaries.
3. Stepped surfaces, resulting in terraces and a variation in film thickness.

The most obvious defect associated with the growth is a screw dislocation, which consists of a cylindrically symmetric stress field and a highly distorted core. A line defect, such as a screw dislocation, with a diameter comparable to the coherence length, and that extends through the entire thickness of the film, probably represents an attractive pinning center for **B**//*c* because it allows the flux line to be pinned along its entire length. The vortex may interact directly with a change in the order parameter in the core of the dislocation, brought about by disruption or distortion of the crystal structure of the superconductor, or with the stress field associated with the dislocation. For a screw, the strain is pure shear, and the effectiveness of the pinning depends on the difference between the shear modulus in the superconducting state and that in the normal state [44, 45]. For a screw dislocation the quadratic coupling may be quite strong, especially when the flux line lies in the same direction as the dislocation vector. The dislocation will pin less effectively when it is perpendicular to the flux line.

The principal objection to the importance of the screw dislocations themselves as effective pinning centers is that there are relatively few of them compared with the vortex density at high fields in the mixed state. From Figure 2.8 we find a dislocation spacing (and, hence, presumably, the vortex spacing) of 100–400 nm. This corresponds to a dislocation density of 10^9–$10^{10}/cm^2$, which would effectively pin a macroscopic field of only a few tenths of a tesla. Experimentally, however, we find that these films have excellent critical currents out to fields of at least 8T (**B**//*c*). Any contribution of the screw dislocations to pinning would, therefore, appear, to be restricted to low fields. In addition, as was suggested in the previous section, it not clear that the dislocation penetrates throughout the entire film. They may start and stop at stacking faults in the interior of the film.

Alternative flux-pinning sites in these films lie at the subgrain boundaries and at the many defects that must be associated with them. Although coupling between the grains is generally excellent, as is shown by the high J_c's and absence of weak-link behavior, the films are clearly not single crystal in nature. Many opportunities exist for mistakes in the orientation and spacing of the unit cells, as is evident from the figures presented here. For example, small amounts of tilt (approximately 1°) are present in some of the *c*-axis grains, and the apparent interlacing of terraces at different levels [Figure 2.7(b)] is indicative of unit cells vertically displaced from each other. Shin et al. [46] and Streiffer [19] have identified certain defects that are a consequence of granularity. Because of the growth mechanism, the grain boundaries form cells with approximately circular or polygonal cross sections, which extend through all or a considerable fraction of the film thickness. Each grain or cell wall could interact with many flux lines. Note that most of these defects do not exist because of spiral growth; rather they are a consequence of the island or granular nature of films.

2.6 Conclusions

We have attempted in this article to summarize recent work on the applications of scanning probe microscopies to understand the growth and microstructure of thin films of $YBa_2Cu_3O_7$ on single-crystal substrates. The principal findings are as follows:

1. Under the deposition conditions appropriate to off-axis magnetron sputtering and the production of c-axis-oriented material, films of $YBa_2Cu_3O_7$ nucleate as two-dimensional islands and grow by the propagation of unit-cell-high ledges. Especially on MgO, but also on perovskites, growth proceeds by a classical spiral mechanism, mediated by a dislocation (probably a screw, with $\mathbf{b} = c_0[001]$).
2. Other growth regimes are accessible. At the higher rates, temperatures and oxygen pressures appropriate to laser ablation, spiral growth is not so evident, and a birth-and-spread layer growth (two-dimensional nucleation) may be more common. It is clear, however, that laser-ablated films can also grow by the dislocation-dominated spiral mechanism under certain conditions.
3. Initial nucleation of the film on the substrate surface is uncontrolled, resulting in a very large density of small nuclei. Some of these nuclei grow faster than others, and as film thickness increases, there is a steady decrease in the number of grains at the surface and a concomitant increase in grain size. A very fruitful area of study may be the effect of deposition conditions on the very early stages of growth.
4. A number of features associated with the growth of c-axis material, such as hollow cores, grain shapes, and rough step edges, are qualitatively understandable in terms of existing theories of spiral growth.
5. Several types of defects are expected to follow from spiral growth originating at many nuclei: (1) the presence at the center of each growth unit (or grain) of a dislocation with a screw component at the surface, and (2) the presence of low-angle boundaries caused by the impingement of neighboring grains. The presence of an appropriate density of threading screw dislocations has not generally been confirmed using transmission electron microscopy, and so the possibility exists that the screw dislocations are only present close to the surface.

6. The origin of the growth dislocations is not established. It has been shown that their density in the films, as measured by STM, is much greater than their density in the substrate (at least in one case), and it seems likely that their facile formation is an intrinsic property of the layered superconductor. It may be associated with the ease with which mistakes can be introduced into the stacking sequence. Some evidence for this comes from the observation of terraces, emanating from neighboring dislocations, but at different levels, indicating the presence of stacking faults beneath the surface.
7. The density of screw dislocations, assuming that they thread through the entire film, is not sufficient to explain the excellent flux pinning exhibited by the sputtered films at high magnetic fields. It is also true that laser-ablated films, which often contain a much lower density of spiral growth centers, have equally desirable critical current carrying properties. Other pinning centers, such as low-angle grain boundary edge dislocation networks, or even the steps in the surface and the voids between grains, might all contribute to flux pinning.
8. *a*-axis-perpendicular material probably grows according to a layer-by-layer process, with material being added to the edges of the layers at the free surface. There appear to be differences in the behavior of different perovskite substrates with respect to the orientation of the *a*-axis grains. On $SrTiO_3$, roughly equal amounts of the two equivalent in-plane orientations are found, whereas on $NdGaO_3$, one of the two directions is strongly preferred.
9. Where *a*-axis material nucleates on top of a layer of *c*-axis material, the *a* grains appear to nucleate at the step edges.

Our understanding of the details of the growth of thin films of $YBa_2Cu_3O_7$ is still relatively primitive, and, consequently, our ability to truly control and optimize properties is very incomplete. Within the last year, scanning probe microscopies have begun to make a significant contribution to our understanding of the growth processes, and, coupled with the more traditional techniques of transmission electron microscopy and x-ray diffraction, we expect a significant improvement in our ability to improve materials over the next year or two.

Acknowledgments. The authors wish to express thanks to the following Los Alamos personnel for experimental assistance and invaluable discussion: Robert Houlton, Fernando Garzon, Jerome Beery, Ross Muenchausen, Xindi Wu, Margarita Piza, Tony Rollett, Terry Mitchell, Marty Maley, Rob Dye, and Steve Foltyn. In addition, we are grateful to Paul Merchant and Renee Jackowitz of Hewlett–Packard Laboratories who allowed us to image one of their *a*-axis films.

References

1. K. Char, N. Newman, S.M. Garrison, R.W. Barton, R.C. Tabor, S.S. Laderman, and R.D. Jacowitz, Appl. Phys. Lett. **57**, 409 (1990); S. S. Laderman, R.C. Taber, R.D. Jacowitz, J.L. Moll, C.B. Eom, T.L. Hylton, A.F. Marshall, T.H. Geballe, and M.R. Beasley, Phys. Rev. B **43**, 2922 (1991).
2. D. Dimos, P. Chaudhari, J. Mannhart, and F.K. Legouse, Phys. Rev. Lett. **61**, 219 (1988); D. Dimos, P. Chaudhari, J. Mannhart, and F.K. Legouse, Phys. Rev. B **41**, 4038 (1990).
3. K. Char, M.S. Colclough, S.M. Garrison, N. Newman, and G. Zaharchuk, Appl. Phys. Lett. **59**, 733 (1991); K. Char, M.S. Colclough, L.P. Lee, and G. Zaharchuk, Appl. Phys. Lett. **59**, 2177 (1991).
4. T.L. Hylton, A. Kapitulnik, M.R. Beasley, J.P. Carini, L. Drabech, and G. Gruner, Appl. Phys. Lett. **53**, 1343 (1988); T.L. Hylton and M.R. Beasley, Phys. Rev. B **39**, 9042 (1989).
5. J. Halbritter, J. Appl. Phys. **68**, 6315 (1990).
6. L. Ji, M.S. Rzchowski, and M. Tinkham, Phys. Rev. B **42**, 4838 (1990).
7. C.B. Eom, A.F. Marshall, S.S. Laderman, R.D. Jacowitz, and T.H. Geballe, Science **249**, 1549 (1990).
8. J.Q. Zheng, M.C. Shih, S. Williams, S.J. Lee, H. Kajiyama, X.K. Wang, Z. Zhao, K. Viani, S. Jacobson, P. Dutta, R.P.H. Chang, J.B. Ketterson, T. Roberts, R.T. Kampwirth, and K.E. Gray, Appl. Phys. Lett. **59**, 231 (1991).
9. T. Terahima, K. Iijima, K. Yamamoto, K. Hirata, Y. Bando, and T. Takada, Jpn. J. Appl. Phys. **28**, L987 (1989).
10. R. Ramesh, A. Inam, D.M. Hwang, T.S. Ravi, T. Sands, X.X. Xi, X.D. Wu, Q. Li, T. Venkatesan, and R. Kilaas, J. Mater. Res. **6**, 2264 (1991).
11. J. Geerk, G. Linker, and O. Meyer, Mater. Sci. Rep. **4**, 193 (1989).
12. Q. Li, O. Meyer, X.X. Xi, J. Geerk, and G. Linker, Appl. Phys. Lett. **55**, 310 (1989).
13. M. Hawley, I.D. Raistrick, J.G. Beery, and R.J. Houlton, Science March 29, 1991, pp. 1537–1589.
14. Ch. Gerber, D. Anselmetti, J.G. Bednorz, J. Mannhart, and D.G. Schlom, Nature **350**, 279 (1991).
15. M.G. Norton and C.B. Carter, *Laser Ablation for Materials Synthesis*, edited by David C. Paine and John C. Bravman, Mat. Res. Soc. Symp. Proc., Vol. 191, p. 165 (1990).
16. M.G. Norton and C.B. Carter, J. Cryst. Growth **110**, 641 (1991).
17. D.M. Hwang, T.S. Ravi, R. Ramesh, S.-W. Chan, C.Y. Chen, and L. Nazar, Appl. Phys. Lett. **57**, 1690 (1990); T.S. Ravi, D.M. Hwang, R. Ramesh, Siu Wai Chan, L. Nazar, C.Y. Chen, A. Inam, and T. Venkatesan, Phys. Rev. B **42**, 10141 (1990).
18. R.W. Balluffi, A. Brokman, and A.H. King, Acta Metall. **30**, 1453 (1982).
19. S.K. Streiffer, B.M. Lairson, C.B. Eom, B.M. Clements, J.C. Bravman, and T.H. Geballe, Phys. Rev. B **43**, 13007 (1991).
20. M.G. Norton, S.R. Summerfelt, and C.B. Carter, Appl. Phys. Lett. **56**, 2246 (1990).
21. I.D. Raistrick, M. Hawley, J. Beery, F. Garzon, and R.J. Houlton, Appl. Phys. Lett. **59**, 3177 (1991).
22. E.B. Eom, J.Z. Sun, B.M. Lairson, S.K. Streiffer, A.F. Marshall, and K. Yamamoto, Physica C **171**, 354 (1990).
23. Enprotech Corp., Pittsburgh, Pennsylvania (1991).

24. G. Leofanti, M. Solari, G.R. Tauszik, F. Garbassi, S. Galvagno, and J.S. Schwank, J. Appl. Catal. **3**, 131 (1982).
25. C.B. Eom, J.Z. Sun, K. Yamamoto, A.F. Marshall, K. E. Luther, T.H. Geballe, and S.S. Laderman, Appl. Phys. Lett. **55**, 595 (1989).
26. F.C. Frank, Discuss. Farad. Soc. **5**, 48 (1949).
27. W.K. Burton, N. Cabrera, and F.C. Frank, Philos. Trans. R. Soc. London A **243**, 299 (1951).
28. H.P. Lang, T. Frey, and H.-J. Güntherodt, Europhys. Lett. **15**, 667 (1991).
29. H.U. Krebs, C. Krauns, X. Yang, and U. Geyer, Appl. Phys. Lett. **59**, 2180 (1991).
30. B.H. Moeckly, S.E. Russek, D.K. Lathrop, B.A. Buhrman, J.Li, and J.W. Mayer, Appl. Phys. Lett. **57**, 1687 (1990).
31. D.P. Norton, D.H. Lowndes, X.-Y. Zheng, S. Zhu, and R.J. Warmack, Phys. Rev. B, **44**, 9760 (1991).
32. J. Moreland, P. Rice, S.E. Russek, B. Jeanneret, A. Rosko, R.H. Ono, and D.A. Rudman, Appl. Phys. Lett. **59**, 3039 (1991).
33. I. Sunagawa and P. Bennema, in *Preparation and Properties of Solid State Materials, Vol. 7. Growth Mechanisms and Silicon Nitride*, edited W.R. Wilcox (Marcel Dekker, New York, 1982), p. 1.
34. A Inam, C.T. Rogers, R. Ramesh, K. Remshnig, L. Farrow, D. Hart, T. Venkatesan, and B. Wilkens, Appl. Phys. Lett. **57**, 2484 (1990).
35. K.H. Young and J.Z. Sun, Appl. Phys. Lett., **59**, 2448 (1991).
36. J.D. Jorgensen, B.W. Veal, A.P. Paulikas, L.J. Nowicki, G.W. Crabtree, J. Claus, and W.K. Kwok, Phys. Rev. B **41**, 1863 (1990).
37. N. Cabrera and M.M. Levine, Philos. Mag. (Ser. 8) **1**, 450 (1956).
38. G.H. Gilmer, J. Cryst. Growth **35**, 15 (1976); R.H. Swendsen, P.J. Kortman, D.P. Landau, and H. Muller-Krumbhaar, J. Cryst. Growth **35**, 73 (1976).
39. F. Falo, A.R. Bishop, P.S. Lomdahl, and B. Horovitz, Phys. Rev. B **43**, 8081 (1991).
40. R.J. Järvinen, E.E. Podkletnov, T.A. Mäntylä, J.T. Laurila, and T.K. Lepistö, Appl. Phys. Lett. **59**, 3027 (1991).
41. O. Eibl and B. Roas, J. Mater. Res. **5**, 2620 (1990).
42. R. Ramesh, D.M. Hwang, J.B. Barner, L. Nazar, T.S. Ravi, A. Inam, B. Duta, X.D. Wu, and T. Venkatesam, J. Mater. Res. **5**, 704 (1990).
43. F.H. Garzon, I.D. Raistrick, D.S. Ginley, and J.W. Halloran, J. Mater. Res. **6**, 885 (1991).
44. H. Ullmaier, *Irreversible Properties of Type II Superconductors*, Springer Tracts in Modern Physics, No. 76, (Springer Verlag, Berlin, 1975), p. 46.
45. A.M. Campbell and J.E. Evetts, Adv. Phys. **21**, 199 (1972).
46. D.H. Shin, J. Silcox, D.K. Lathrop, S.E. Russek, and R.A. Buhrman, *Proc. 47th Ann. Meeting of the Electron Microscopy Soc. of America*, edited by G.W. Bailey (San Francisco Press, San Francisco, 1989), p. 174.

3
Microstructure of Interfaces in $YBa_2Cu_3O_{7-x}$ Thin Films

A.F. MARSHALL and R. RAMESH

3.1 Introduction

Highly oriented, high-quality thin films of the high-temperature superconductor, $YBa_2Cu_3O_{7-x}$ (henceforth referred to as 123 or YBCO), are now successfully and routinely produced by a variety of in situ deposition techniques [1–12]. In particular, the techniques of laser ablation [1, 2] and off-axis sputter deposition [4, 6], both using a single, composite target, have emerged as the most reliable in situ physical processes producing uniform, high-quality films over relatively large substrate areas. The best films are *c*-axis oriented and have T_c's as high as 90 K with transition widths of less than 1 K and J_c's (77 K) > 10^6 A/cm^2. Organometallic chemical vapor deposition has also produced large area films with comparable properties [11, 12]. These thin films, although highly oriented and exhibiting superior superconducting properties, are not single crystals. Rather they exhibit a rich microstructure of grain boundaries, stacking faults, second phases, dislocations, and other crystallographic defects and interfaces, the characteristics of which depend on the orientation and composition of the films, as well as on the processing parameters. Thin-film processes offer the capability of controlling microstructures on a fine scale by varying these synthesis conditions; because of the two-dimensional nature of film growth these processes also allow for synthesis of artificially layered structures. We therefore have the opportunity to design thin films with specific, controlled sets of interfaces. This is important both for fundamental studies probing the physical properties of superconducting materials, and also essential in tailoring these thin-film materials for device applications. As an example, the behavior of individual grain boundaries formed on bicrystal substrates has been characterized in a series of definitive experiments by Dimos et al. [13]. The weak-link nature of many high-angle grain boundaries, clearly demonstrated in these experi-

A.F. Marshall, Center for Materials Research, Stanford University, Stanford, CA 94305; R. Ramesh, Bellcore, Red Bank, NJ 07701.

ments, may actually be utilized to great advantage in forming Josephson junctions for SQUID devices; several groups have already demonstrated ingenious methods for integrating a single grain boundary into a device [14, 15].

It is also of fundamental interest that epitaxial thin films of YBCO have shown the highest transport critical current densities, J_c, of any YBCO material so far. The extent of many defects observed in these materials is commensurate with the superconducting coherence length; hence such defects may play an important role in the flux dynamics and consequently in the attainment of high J_c. It is important to note that point defects, which are difficult to observe by most imaging techniques, may also be significant or even primary pinning defects, as discussed by Hylton and Beasley [16]. The question of which defects in these films are responsible for their superior transport properties remains to be answered by systematic control and characterization of film microstructures and correlated property measurements.

In this chapter we will emphasize the formation and microstructure of various types of interfaces in YBCO thin films; the relationship between microstructure and properties will be discussed where appropriate. This will include typical microstructural defects such as grain boundaries, stacking faults, and second phases, as well as the substrate/film interface and interfaces in artificially layered structures. We will not discuss the (110) tetragonal-to-orthorhombic transformation twins in any detail as these are present in almost all bulk and thin-film material; their structure and properties have been extensively studied and are considered in Chapter 5 of this book. We will begin with a discussion of the factors controlling film orientation, as many of the interface structures discussed here are intimately related to the film orientation.

3.2 Orientation

YBCO thin films have been grown in situ in a variety of orientations normal to the film plane, including (001), or *c* axis [1–12]; (100/010) or *a*, *b* axis [17–20]; (110) [17, 21–24]; (103/013) [17, 22, 25]; and (113) [18]. For simplicity we will not henceforth distinguish between *a*- and *b*-axis notations, referring only to *a*-axis and (103) films, unless appropriate to the discussion. These films also exhibit specific in-plane alignments. The most important factors controlling the orientation relationship are the substrate orientation and lattice match, interfacial chemistry, and the deposition conditions of temperature, rate, and oxygen pressure. Table 3.1 lists conditions resulting in these different orientations for several of the most commonly used substrates. The first consideration is symmetry matching of the pseudocubic sublattice of the YBCO structure to that of the substrate. Thus a (100) orientation of a cubic substrate favors *c*-axis or *a*, *b*-axis growth, usually with the cube axes of the YBCO sublattice and of the substrate aligned in the plane. Similarly

TABLE 3.1. Effect of synthesis conditions on film orientation.

Cubic substrate	Lattice match (cube axes)	Substrate orientation	T_d[a]	Film orientation
$SrTiO_3$, $LaAlO_3$	Good (<3%)	100	H	*c* axis
$SrTiO_3$, $LaAlO_3$	Good	100	L	*a* axis
$SrTiO_3$, $LaAlO_3$	Good	110	H	103
$SrTiO_3$, $LaAlO_3$	Good	110	L	110
$SrTiO_3$	Good	111	H	113
MgO	Poor (>8%)	100	H	*c* axis
MgO	Poor	100	L	*c* axis, Secondary in-plane orientations
YSZ	Poor (>20%)	100	H	*c* axis, 100//100
YSZ	Poor	100	L	*c* axis, 110//100
YSZ	Poor	110, 111	...	No epitaxy

[a] T_d's, or deposition temperatures, reported for a given film orientation vary considerably depending on synthesis and temperature measurement method. We therefore simply list relative T_d for a given range of 123 synthesis: higher (H) or lower (L).

the (110) and (103) orientations form on a (110) cubic substrate, and the (113) on a (111) substrate. A close lattice match and/or compatible interfacial chemistry, for example, perovskite substrates or buffer layers, appears necessary for all but *c*-axis growth.

a-axis films do not form on (100) MgO or Y-stabilized ZrO_2 (YSZ) [18], where the lattice mismatch is large, and neither (110) nor (111) orientations of YSZ produce epitaxial films [26]. The relatively easy growth of *c*-axis films on a variety of substrates is due at least partly to the fact that the (001) surface is the low-energy surface so that (001) texturing is favored. For the same reason factors tending toward equilibrium growth (e.g., favoring increased atomic mobility), such as higher temperatures and lower O_2 pressure, favor *c*-axis over *a*-axis growth [17]. These conditions also favor (103) as opposed to (110) growth [17], apparently because the ideal growth surface for (103) films is the (001) plane (see the following section for further discussion). *a*-axis and (110) growth occur at lower temperatures and higher O_2 partial pressures, when kinetics dominates the growth process. *a*-axis growth by laser deposition has also been optimized at higher temperatures through the use of PBCO buffer layers [19]. Intermediate conditions produce mixtures of *a*/*c* or (110)/(103) films. *a*-axis orientation in mixed films tends to increase with thickness, both due to nucleation of *a*-axis grains on *c*-axis material (possibly defect-induced [27]) and due to overgrowth of *c*-axis by *a*-axis grains. (113) film growth occurs on (111) substrates under conditions similar to *c*-axis growth on (100) substrates [18].

c-axis films on MgO or YSZ, where the lattice mismatch is large, frequently exhibit other in-plane orientations, resulting in polycrystalline, textured films. These secondary orientations are readily identified by planar view transmission electron microscopy [27–29] or by off-axis x-ray phi scans or pole figures [30, 31]. The most frequently observed orientation for both substrates, in addition to $[100]_{YBCO}//[100]_{sub}$, is $[110]_{YBCO}//[100]_{sub}$, giving YBCO grains with a rotation of approximately 45° to grains with the primary orientation. A number of other orientations have been observed on MgO as well, and have been explained in terms of a near coincident site lattice between the substrate and the film [32]. Figure 3.1 shows micro-

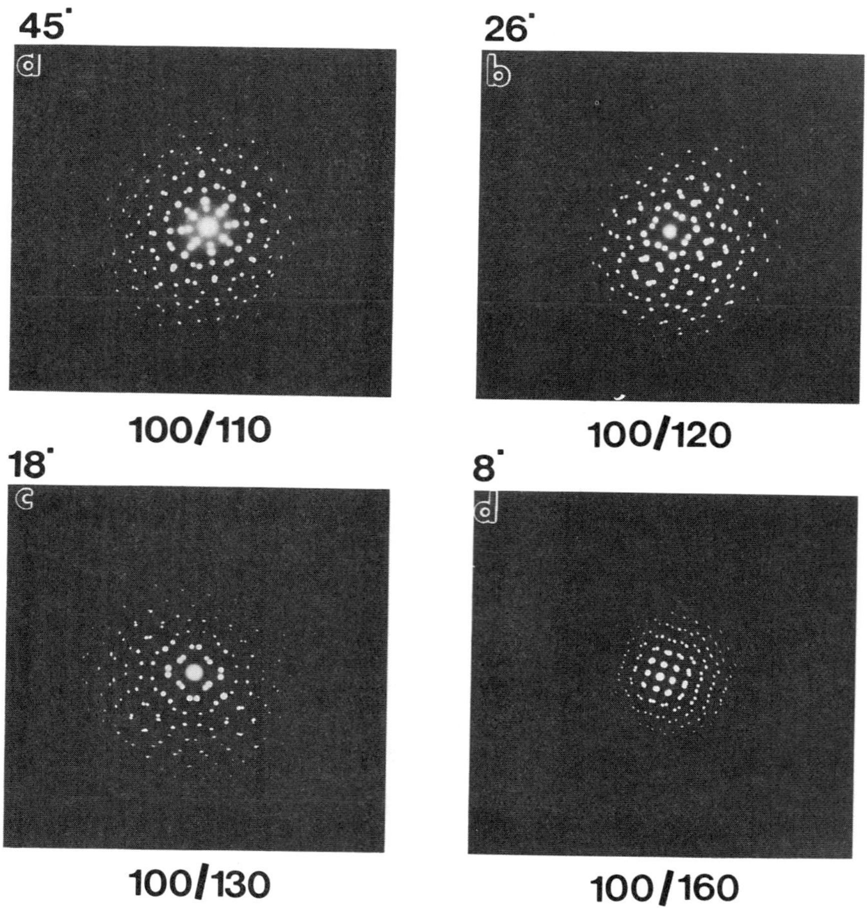

FIGURE 3.1. Selected area electron diffraction patterns showing four of the possible crystallographic orientations between adjacent grains in c-axis oriented YBCO films grown on lattice mismatched substrates such as MgO.

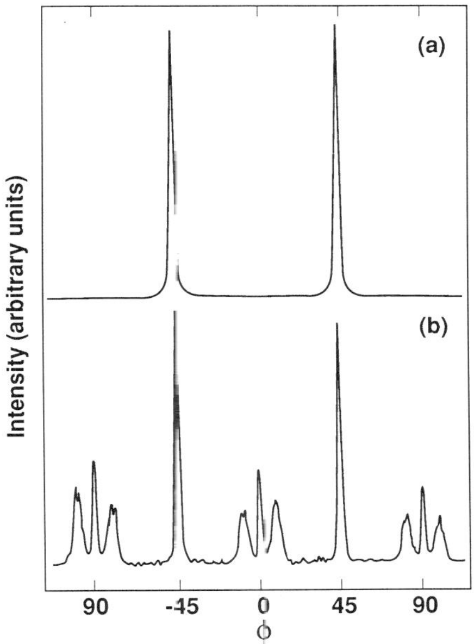

FIGURE 3.2. Off-axis x-ray phi scans of the {103} family of peaks of films grown on (100) YSZ substrates: (a) A film grown on a homoepitaxial buffer layer resulting in a single type of orientation relationship with the cube axes of the film rotated at 45° from those of the substrate. (b) A film with no buffer layer shows the existence of the three principal orientation relationships observed for YBCO on YSZ: in addition to the peaks of (a) there are peaks showing alignment of the cube axes (0° and ±90°) and peaks showing a rotation at ±8° from this orientation (satellite peaks). (Courtesy Dave Fork [30].)

diffraction patterns identifying four of these rotations relative to the primary orientations. These secondary orientations lead to high-angle [001] tilt boundaries between the YBCO grains. Optimization of the deposition temperature [18] or high-temperature annealing of the substrate prior to deposition [31] appear effective in minimizing the secondary orientations and consequent high-angle grain boundaries on MgO substrates. The situation for YSZ substrates is slightly different. Only three orientation relationships are observed (Figure 3.2) [30]; in addition to the two already mentioned there occurs a rotation at 8° from the [100]//[100] orientation. This latter orientation is clearly visible in the phi scan of Figure 3.2(b), showing satellites about phi = 0, ±90°. It has been shown by Garrison et al. [33] and by Fork et al. [26] that the three possible orientations can be controlled both by deposition temperature and by interfacial chemistry through the use of buffer layers and homoepitaxy.

These oriented films, while highly aligned, are not single crystals. High-quality c-axis films, which do not exhibit significant amounts of secondary orientations, nevertheless contain (1$\bar{1}$0) transformation twins, with the a and b axis alternating across the twin boundary, and low-angle grain boundaries. Due to symmetry considerations, films of other orientations are generally characterized by more than one in-plane alignment. These films therefore contain specific sets and types of grain boundaries between orientation domains which depend on the epitaxial relationship with the substrate. The variations in microstructure are evident in SEM images of the film surfaces

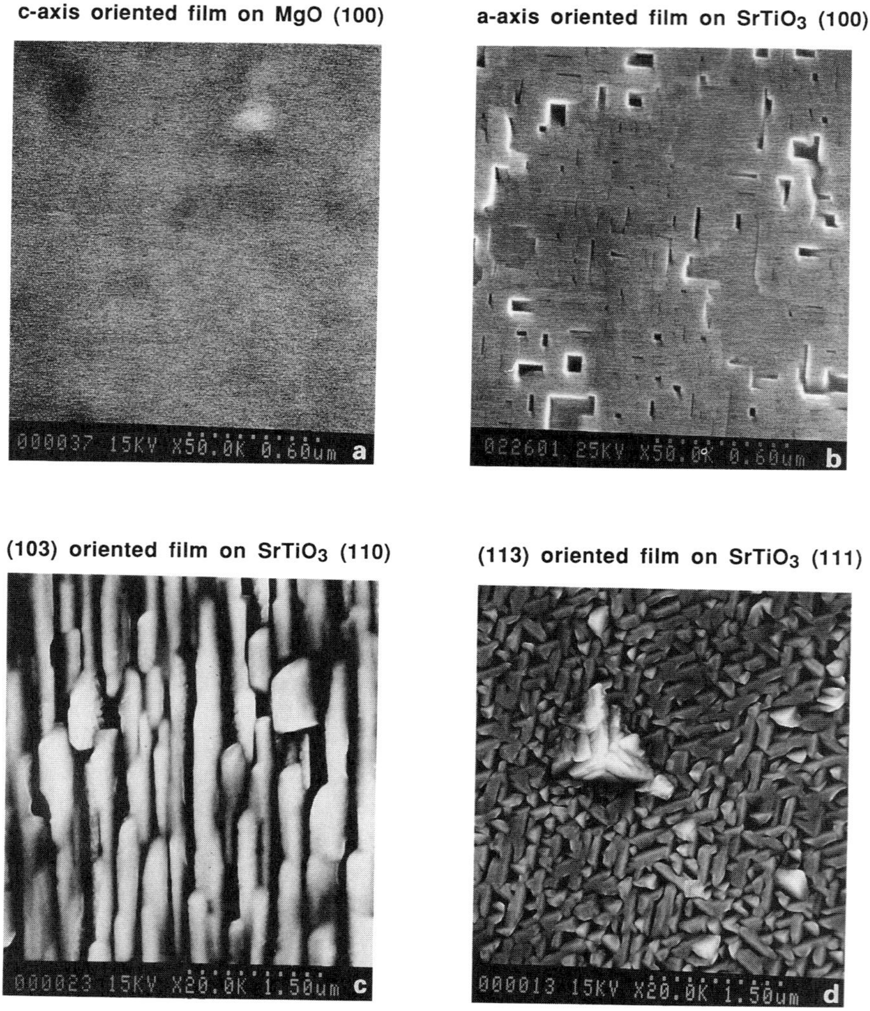

FIGURE 3.3. Scanning electron micrographs of the surface morphology of four types of films: (a) *c*-axis oriented film on (100) MgO showing a smooth surface; (b) *a*-axis oriented film on (100) $SrTiO_3$, showing microscopic pinholes and the blocky growth morphology; (c) (103) oriented films on (110) $SrTiO_3$ showing elongated grains and very rough surface morphology; (d) (113) oriented film, grown on (111) $SrTiO_3$ showing faceted grains and rough surface morphology.

shown in Figure 3.3 for *c*-axis, *a*-axis, 103, and 113 films. Transmission electron microscopy gives detailed information on the nature and structure of the interfaces in these films, including the film–substrate interface, the domain or grain boundaries, stacking faults and related dislocation structures, and second phase interfaces. Since the properties of the film are controlled by the interfaces, we begin with a discussion of the primary interface, namely, the film–substrate interface.

3.3 Film/Substrate Interface

The microstructure of the film/substrate interface can be described by the epitaxial relationship between the two, and by the actual identity of the layers comprising the interface. As in the case of semiconductor heterostructures, the formation of the first few layers of the film is very critical for subsequent growth. In the case of layered structures such as YBCO, grown on ionic substrates such as $LaAlO_3$, in addition to structural matching, chemistry matching (which includes an ionic component) is also important. Understanding the atomic structure of such interfaces requires imaging or spectroscopy with high resolution and confirmation of the interpretation through detailed simulations [34]. High-resolution electron microscopy in conjunction with image processing and simulations has proven to be extremely useful in characterizing the defect structures in these cuprates [35], as will be illustrated in a later section. In the case of heterophase interfaces, simulation programs have been available only recently and hence not much has been reported in the literature. Results of preliminary experiments will be presented here.

The chemistry of the [100] surface of $SrTiO_3$ and $LaAlO_3$ is illustrated schematically in Figure 3.4(a) to 3.4(d). It is to be noted that in the case of $SrTiO_3$, both types of surfaces are charge balanced. However, both surfaces in the case of [100] $LaAlO_3$ have charged layers (the LaO layer has a net $+1$ charge and the AlO_2 layer has a net -1 charge, Figure 3.4). Figure 3.4(e) shows schematically the interface between one unit cell of *c*-axis oriented YBCO and the substrate, which is either [100] $SrTiO_3$ or [100] $LaAlO_3$. The two types of surfaces on the substrate are termed type A and type B, while there are four possible interface terminations in the YBCO layer, indicated as 1–4 in Figure 3.4(e).

Figures 3.5(a) and 3.5(b) show representative high-resolution images of the interface region for the two types of substrates. Image simulations were then carried out for the eight different combinations of interface terminations, for each substrate. For simplicity, the simulations were carried out for a constant thickness of 30 Å and an objective lens defocus of -500 Å (close to the experimental imaging conditions). The results of the image simulations are shown in Figures 3.6(a)–3.6(d). The image contrast at the interface is quite different for the four types of YBCO interface terminations, for a given termi-

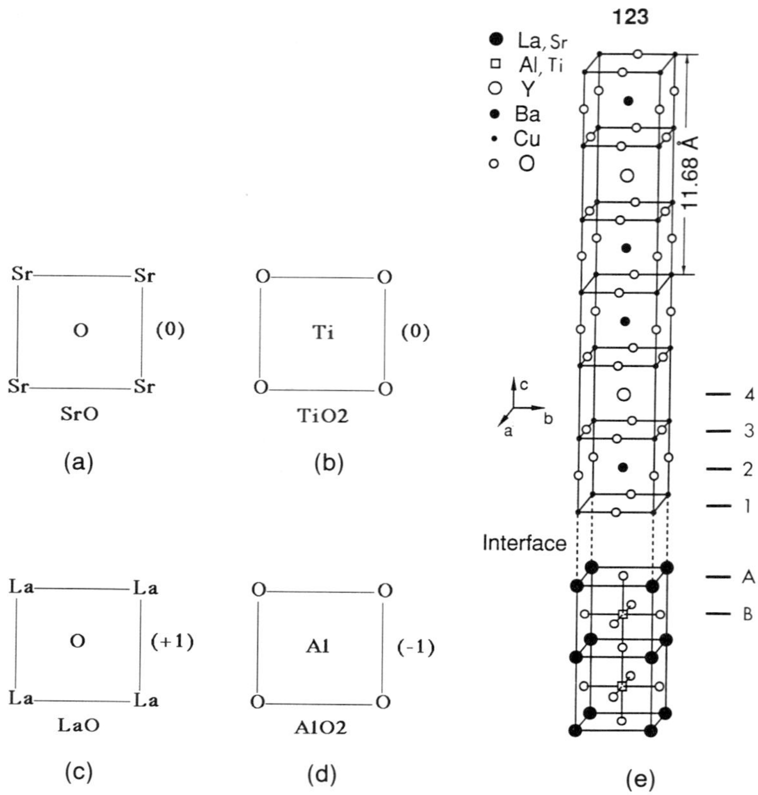

FIGURE 3.4. (a)–(d) Schematic illustrations of the possible surface structures in the case of (100) $SrTiO_3$ and $LaAlO_3$. Note that in the case of $SrTiO_3$, both types of surfaces are charge balanced, while in the case of $LaAlO_3$ both the surfaces are charged. (e) A schematic illustration of the interface between the $SrTiO_3$ (or $LaAlO_3$) substrate and the *c*-axis oriented YBCO film grown on top. The two types of surface terminations on the substrate are termed as A and B while the four possible surface terminations in YBCO are termed 1–4.

nation on the substrate side. For example, in the case of the $LaAlO_3$ substrate, comparison of the experimental image contrast in Figure 3.5(b) with that in Figures 3.6(c) and 3.6(d) suggest that Figure 3.6(d-1) is the most possible interface structure, i.e., a Cu–O chain termination. Although these results are preliminary, such experiments clearly have immense potential in not only revealing the atomic nature of the interfaces, but also in designing and controlling the crystallographic orientation and interface chemistry. This can be achieved through the use of buffer layers in a manner akin to what has been demonstrated very successfully in the case of compound semiconductors.

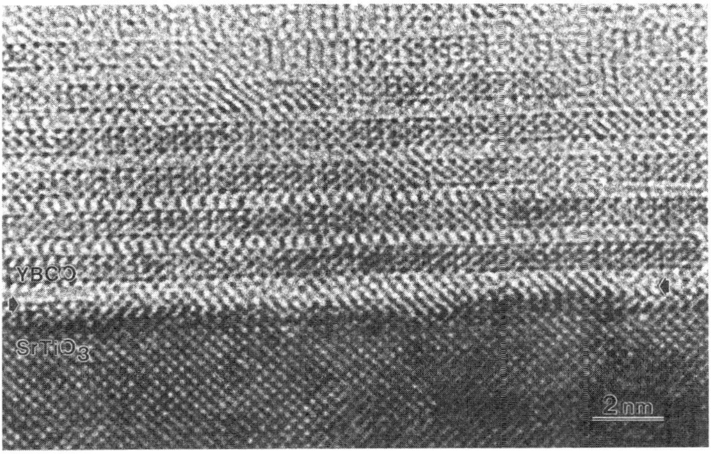

(a)

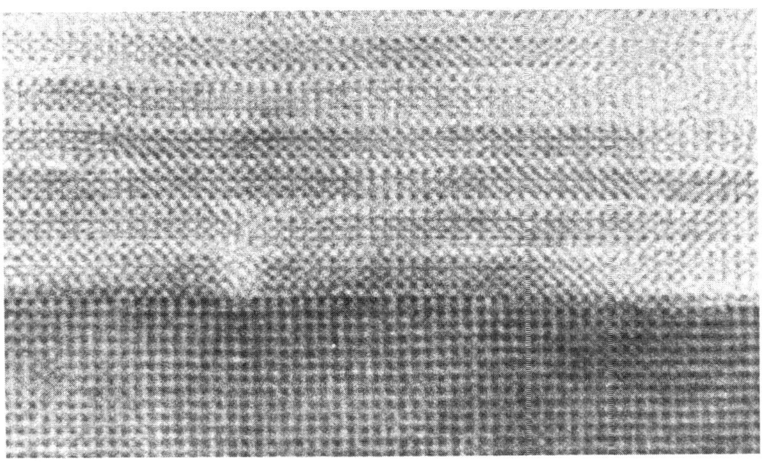

(b)

FIGURE 3.5. High-resolution electron micrographs of the interface between (a) $SrTiO_3$ and YBCO and (b) $LaAlO_3$ and YBCO. From image symmetry considerations, the interface layer in the case of $SrTiO_3$ was deduced to be a SrO layer while in the case of $LaAlO_3$ it was deduced to be a CuO chain layer.

The finite lattice mismatch between the substrate and the a–b plane of the YBCO film is generally relaxed through the formation of interface dislocations. Such dislocations can be identified through direct lattice imaging or through strain contrast. As an example, Figure 3.7 shows a lattice image of a film grown on $LaAlO_3$, which has an a-axis lattice parameter which is about 1% smaller than that of YBCO. Thus, one expects an extra half-plane in the

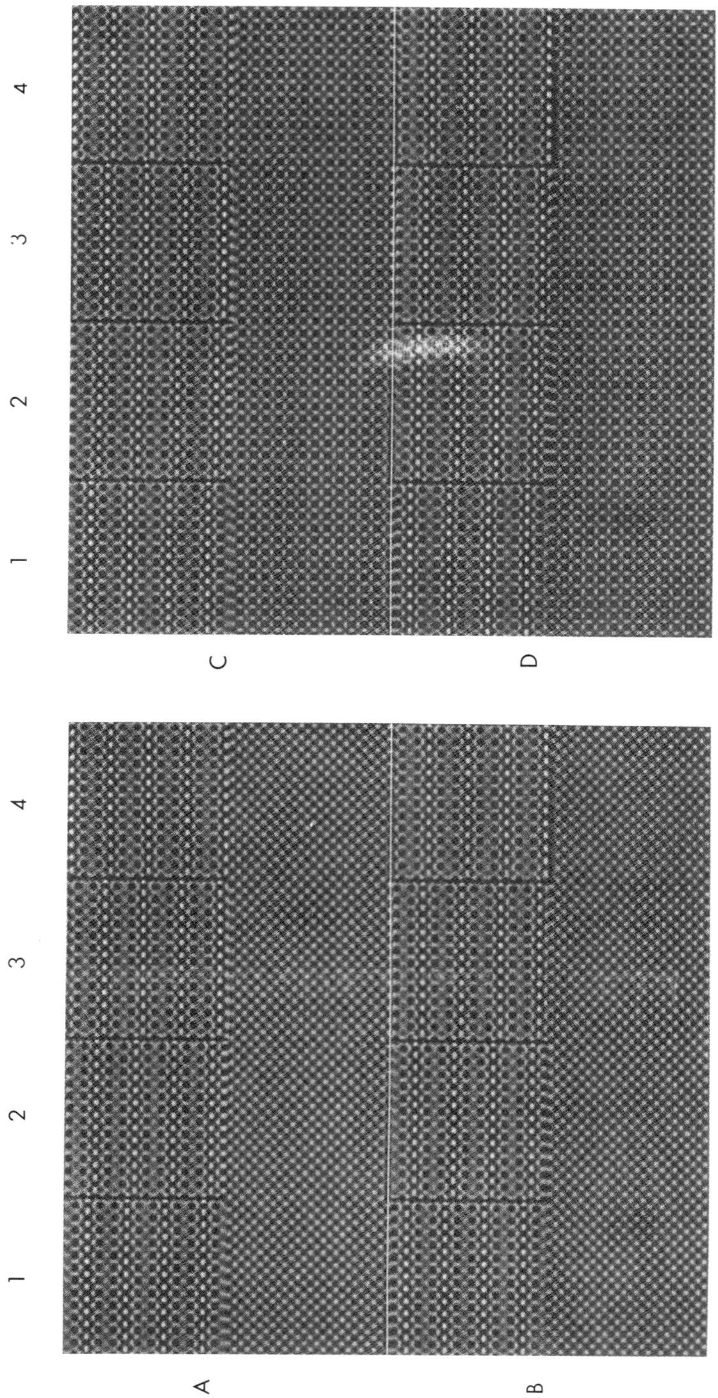

FIGURE 3.6. (a) and (b) Calculated images in the case of $SrTiO_3$ substrate for the eight possible surface terminations; (c) and (d) the same set of images, in the case of a $LaAlO_3$ substrate. In all the cases, the simulations were carried out for a foil thickness of 30 Å and an objective lens defocus of −500 Å.

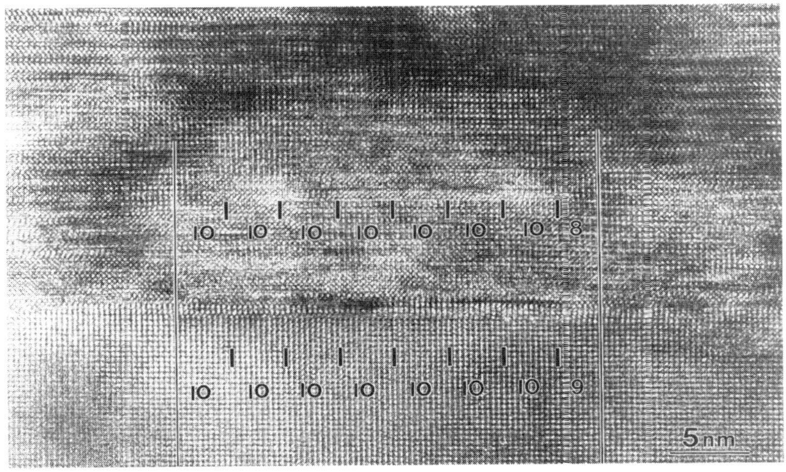

FIGURE 3.7. A high-resolution lattice fringe image that depicts the formation of interface dislocations to accommodate the lattice parameter mismatch between the substrate and the *a–b* plane of the YBCO film.

$LaAlO_3$ lattice for approximately every 100 lattice planes, as is observed in this lattice image. Also outlined in this image is the effect of the mismatch, which leads to strain contrast (in this case, the semicircular region which appears darker). Very thin films (i.e., those below the critical thickness) are known to grow pseudomorphically, although films thicker than a few nanometers start relaxing. For the typical thicknesses of technological importance, i.e., a few thousand Å, the films are completely relaxed, as illustrated in Figure 3.7. Relaxation by the formation of misfit dislocations is possible only for small misfit values, such as in the case of growth on $LaAlO_3$ or $SrTiO_3$. Even in these cases, one frequently observes small-angle grain boundaries. However, when the misfit is much larger, as in the case of MgO and ZrO_2, the films are no longer single crystalline. Such films, although still *c*-axis oriented, have grain boundaries in them. This is the focus of the next section, which will also discuss the various types of domain boundaries.

Defects on the substrate surface propagate as defects in the *c*-axis oriented films. Several types of defects can form on polished substrate surfaces. Depending upon the quality of the polish, macroscopic steps (of height 0.2–1.0 μm) can be observed. Such substrates generally yield poor quality films. On a microscopic scale, unit-cell-high steps always occur and are difficult to avoid, even by prolonged annealing at high temperatures (which only produce a stable array of surface steps). Figure 3.8 shows one example where unit-cell-high ($\sim$3.8 Å) steps has nucleated a translational boundary in the *c*-axis

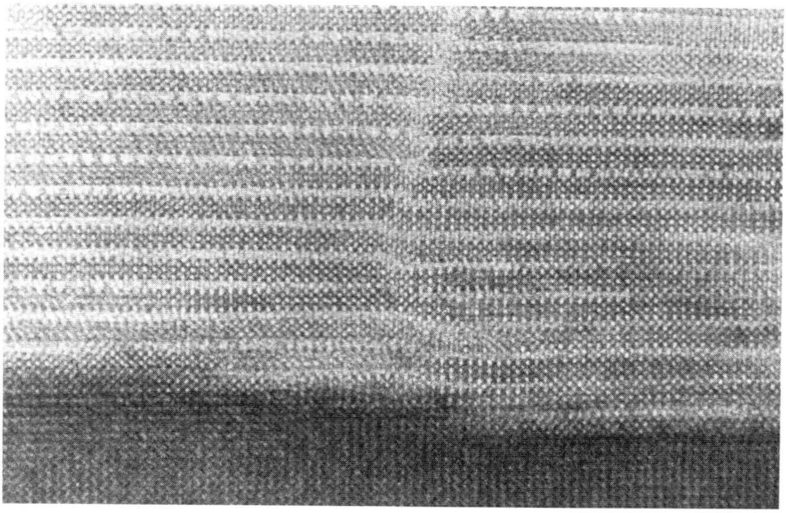

FIGURE 3.8. A HREM cross-sectional image showing the nucleation of a translational boundary at a unit-cell-high step on the substrate surface. Note that the misfit in the boundary is progressively reduced as the film grows.

oriented film. Since these films grow layer-by-layer, they can accommodate the misfit on either side of the translational boundary by compositional fluctuations, that is, by polytypoidic variations. Hence, such translational boundaries anneal themselves out.

3.4 Grain Boundaries

It is now clear that the presence and structure of grain boundaries has a major influence on superconducting properties, as first indicated by the very low critical currents measured in bulk polycrystalline material as compared with those of highly aligned thin films. Dimos et al. [13] carried out a careful analysis of the relationship between grain-boundary structure and critical current by synthesizing individual grain boundaries on bicrystal substrates and characterizing their transport behavior as a function of boundary misorientation. For grain boundary angles greater than approximately 5°–10°, the critical currents dropped significantly as compared with intragranular currents. Although their work concludes that intrinsic grain-boundary structural disorder coupled with the very short coherence lengths of YBCO is responsible for the degradation observed in their thin-film bicrystals, work on bulk materials has shown that extrinsic characteristics such as compositional disorder, which would further degrade properties, may also exist at grain boundaries [36]. A combination of these factors is likely responsible for the very low J_c's of conventionally processed bulk material.

The grain-boundary structure of the best *c*-axis films, as mentioned, is comprised mainly of low-angle boundaries. (110) twin boundaries, which form during the post-deposition cooldown due to the tetragonal-to-orthorhombic phase transformation, are also present. A study by Lairson et al. indicated that these boundaries do not contribute significantly to pinning in thin films [37]. Figure 3.9(a) is a typical planar view of a *c*-axis film showing the (110) twins; Figure 3.9(b) is a cross section of a 150-nm film. Low-angle grains are not clearly visible in planar view, because there is no consistently strong contrast between such grains. Vertical boundaries are visible in the cross-section view with a spacing of roughly 30–80 nm; while some of these may be low-angle grain boundaries, e.g., the 2° tilt boundary marked by arrows, most cannot be unambiguously identified [38]. Again the imaging conditions are not ideal, in this case because neither the low-angle boundaries nor the ubiquitous (110) twin boundaries are generally normal to the imaging direction. Grain sizes have been reported in ultrathin films, where the grains are more visible in planar view due to moiré fringes which magnify the angular variations [39, 40], and in scanning tunneling microscopy (STM) images of film surfaces [41, 42]. Grain sizes as small as several tens of nanometers have been reported in ultrathin films and up to 1 μ in STM images of films on $SrTiO_3$. A more typical grain size is on the order of a few hundred nanometers.

The STM images indicate a spiral growth mechanism for the grains with a screw dislocation at the center of each grain. The height of the step of the screw dislocation is generally one *c*-axis lattice parameter. These steps also act as ledges [see also Figure 3.29(b)] that propagate laterally during the film thickening process. Other growth mechanisms may also occur depending on the synthesis conditions [43]. It seems likely that grain sizes are varying with deposition temperature, substrate, growth mechanism, and perhaps also with thickness of the film. The average mosaic spread of the grains can be determined by x-ray rocking-curve measurements, which measure the angle of tilt, ω, out of the plane, and by off-axis peak width measurements about the rotation angle, ϕ, giving rotation angles in the plane. $\Delta\omega$ scans on a variety of substrates, including MgO, $LaAlO_3$, $SrTiO_3$, etc., give angular spreads of less than a degree, although still typically an order of magnitude larger than that of the substrate [38, 44–46]. Streiffer et al. [38], for example, report YBCO rocking-curve widths of 0.2° on MgO substrates having 0.02 $\Delta\omega$. However using TEM they observe out-of-plane tilt boundaries with misorientation angles as high as 2°. These higher angle tilts apparently are not of sufficient density to contribute significantly to the rocking-curve width. $\Delta\phi$ scans which measure rotation angles in the plane are reported less frequently than $\Delta\omega$ scans; however, it appears that these values tend to be somewhat

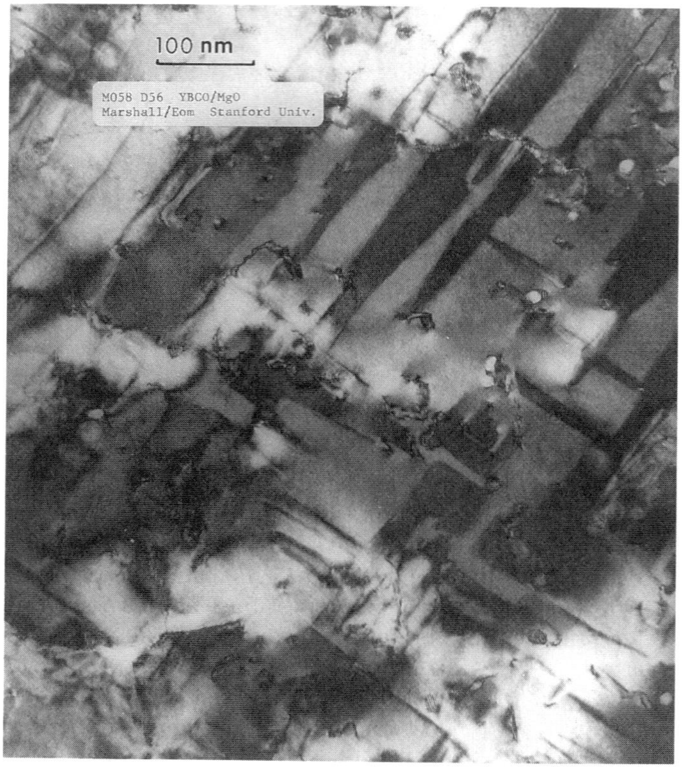

(a)

FIGURE 3.9. (a) A typical low magnification, diffraction contrast micrograph of a planar section of a c-axis film showing the transformation twins and other defects such as 132 precipitates.

higher, on the order of 1° to 4° depending on the substrate and buffer layers [26, 40, 46, 47].

These highly oriented c-axis films, with the superconducting CuO_2 planes aligned in the plane of the film, exhibit the highest critical currents observed for YBCO superconducting materials, greater than 10^7 A/cm^2 at 4.2 K and greater than 10^6 A/cm^2 at 77 K. This is consistent with the results of Dimos et al. [13] since the angular misorientation of the grains is well below the threshold for weak-link behavior of the grain boundaries. It has been suggested that dislocation structures of low-angle boundaries [47], and/or the screw dislocations relating to spiral growth [41, 42], may be responsible for the good pinning and high J_c's of these films. Mosaic spread alone, as determined by $\Delta\phi$ and $\Delta\omega$ measurements does not generally correlate with J_c [48]. Of course grain size must be also considered along with mosaic spread to

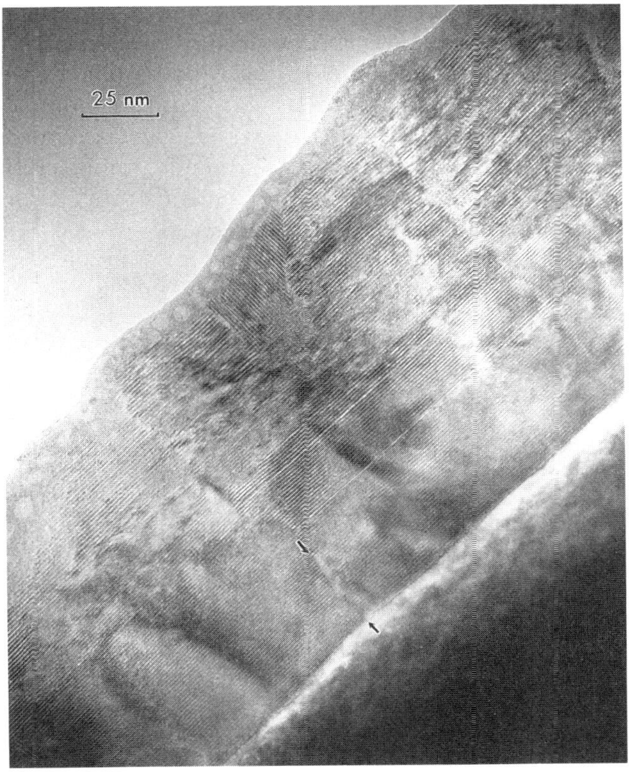

(b)

FIGURE 3.9. (b) a cross-section image of the same c-axis oriented film, showing stacking defects and low-angle grain boundaries. A 2° tilt boundary is marked by arrows. (Courtesy S.K. Streiffer.)

determine grain boundary dislocation density, but such detailed characterization of c-axis films in conjunction with J_c measurements has not, to our knowledge, been carried out. In addition, not all films with good J_c's appear to grow by the screw dislocation mechanism [43]. These observations do not rule out the possibility that these dislocation structures contribute to pinning, but suggest that the high critical currents of thin films are related to more than one pinning defect.

When secondary orientations occur, resulting in textured films with high-angle grain boundaries, the transport properties are degraded. Garrison et al. [33] showed that the fraction of $[100]_{YBCO}//[100]_{sub}$ vs $[110]_{YBCO}//[100]_{sub}$ alignment could be controlled by varying temperature and that J_c's could drop by as much as four orders of magnitude for films comprised of roughly equal mixtures of both orientations, whereas J_c was maximized for films

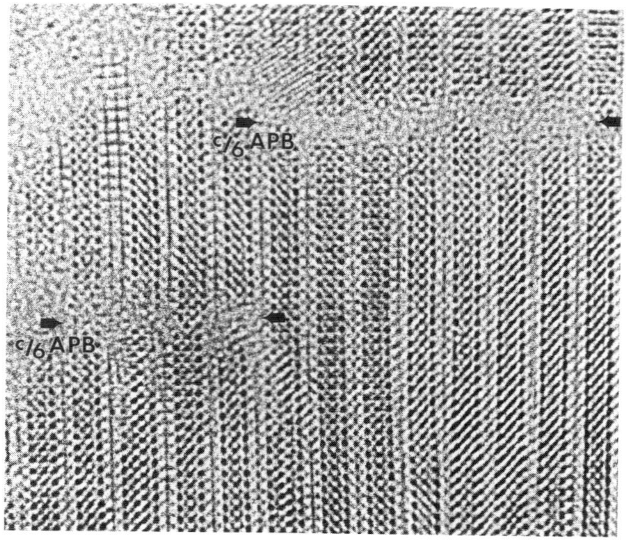

FIGURE 3.10. A high-resolution electron micrograph showing the formation of a **c**/6 antiphase (or translational) boundary in a *c*-axis oriented film.

containing greater than 90% volume fraction of a single orientation. Small fractions of misaligned material have been correlated with higher surface resistive losses by Laderman et al., suggesting that microstructural alterations can affect both the dc and the ac properties of the material [44].

Antiphase boundaries (APB's) are also observed in *c*-axis films; these are boundaries normal to the interface with a shift across the boundary of **c**/3 or **c**/6 between the (001) planes (Figure 3.10). The **c**/3 boundaries, in particular, have been observed to nucleate at steps along the substrate surface [34, 38]. Note that across a **c**/3 boundary only one out of every two CuO_2 superconducting planes connects with another such plane. Although these boundaries may extend for some distance into the film as it grows, they tend to eventually heal themselves through the intersection of a stacking fault which compensates the APB shift (Figure 3.10). In more faulted films, such as those made by laser ablation, they are observed to also nucleate at stacking fault dislocations, although again they have a tendency to heal themselves after some distance rather than propagating through the film thickness [49]. The transport behavior of such boundaries is not clear: if they are deleterious, they appear to be sufficiently isolated so as not to degrade transport properties.

In contrast to *c*-axis films, *a*-axis films contain many 90° domain boundaries [20, 34]. These are a result of two possible in-plane orientations of the *c* axis, along either of the cube axes of the substrate, resulting in a domain structure with a 90° misorientation of the *c* axis between domains. *a*-axis

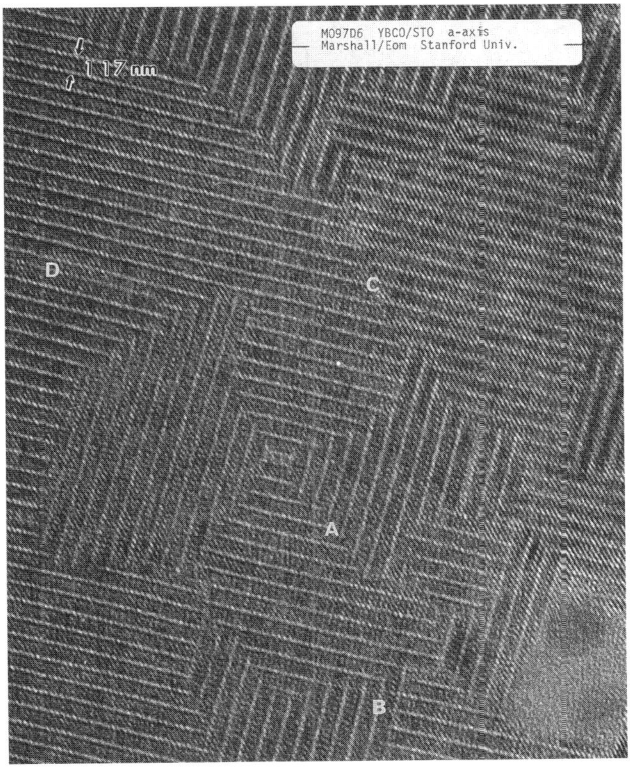

FIGURE 3.11. Planar view of an *a*-axis film showing the domain structure resulting from the two in-plane orientatations of the *c* axis. 90 degree (100) tilt boundaries occur between domains: these may be either approximately symmetrical (A) or basal-plane-faced boundaries (B). **c**/3 antiphase boundaries are also observed (C and D).

films also contain antiphase boundaries due to out-of-registry nucleation of adjacent domains. X-ray diffraction of sputter-deposited films on $SrTiO_3$ and $LaAlO_3$ indicate that the *b* axis lies entirely in the plane, that is, the *a*–*b* twins, which typically have a (110) boundary plane and are present in almost all other YBCO materials, are absent in these *a*-axis films [20]. This is not necessarily true, however, of all films for which the *c*-axis lies in the plane.

A TEM planar view of a typical *a*-axis film is shown in Figure 3.11. The 90° grain boundaries can be characterized as one of two types: either basal-plane faced boundaries, with the interface parallel to the (001) planes of one of the grains, or boundaries at approximately 45° between the two sets of (001) planes. We will refer to these latter boundaries as (ideally) symmetrical (100) tilt boundaries, following standard nomenclature (e.g., see Ref. [50]). The TEM images show that these boundaries are often not perfectly symmetrical,

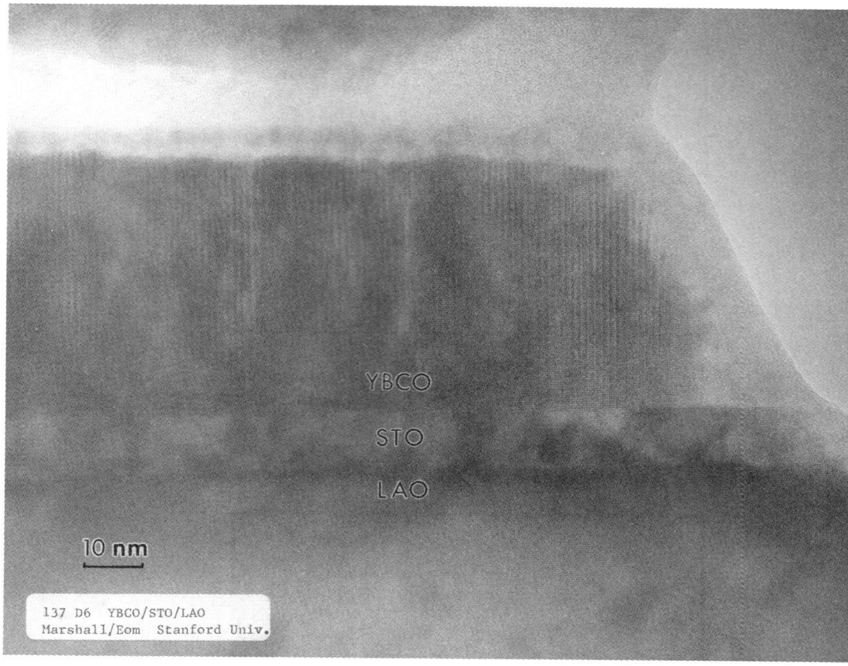

FIGURE 3.12. A lattice image from a cross-sectioned *a*-axis oriented film illustrating the columnar growth and exceptionally smooth surface in this orientation.

or even straight, and may actually include some basal-plane faceting on a fine scale. The columnar growth of the grains and the exceptionally smooth surface of these *a*-axis films is evident in the cross-section TEM image of Figure 3.12.

Antiphase boundaries in the *a*-axis films can also be separated into two categories: those parallel to the (001) planes and those at an oblique angle to (001) planes; these boundaries are seldom exactly normal to the (001) planes. The former type of APB creates a stacking fault along the (001) planes as indicated in Figure 3.10. This type of boundary is parallel to the superconducting current direction and therefore is not likely to interfere significantly with transport behavior. The latter type of boundary is also observed in *c*-axis films as previously discussed.

The boundary structure of these fine-grained *a*-axis films clearly has an effect on transport behavior. This is seen in comparing the resistivity-temperature curves of *a*-axis with *c*-axis films, both made by off-axis sputtering (Figure 3.13) [20]. The normal state resistivity, slope, and zero intercept are much higher for *a*-axis than for *c*-axis films; also the critical

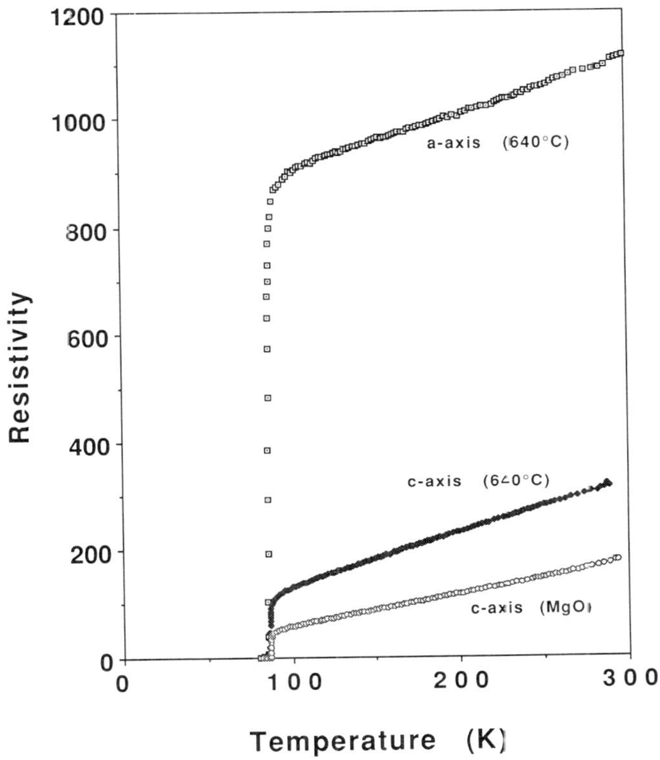

FIGURE 3.13. Resistivity-temperature plots for *a*- and *c*-axis oriented films. Note the large difference in the normal state resistivity although the superconducting transition temperature is about the same. Also shown is the resistive transition for a *c*-axis oriented film grown on MgO at the same temperature at which the *a*-axis film is grown. Note that the normal state resistivity in this case is also slightly higher than the optimum *c*-axis film, probably due to a higher density of defects and in-plane grain boundaries.

currents of the best *a*-axis films are typically an order of magnitude below that of *c*-axis films, in the range of 10^6 A/cm^2 (4.2 K). Since there are essentially two types of grain boundaries as well as the APB's, however, we cannot say anything about the properties of a single type of boundary. Also, 90° boundaries could not be synthesized in the experiments of Dimos et al. [13], because of the cubic symmetry of the substrate. There is recent evidence that *a*-axis films can achieve high J_c's and low resistive losses comparable to *c*-axis films [51]. However, no microstructural characterization of the grain-boundary structure is yet available.

Although pure *a*-axis films do not contain (110) twins, it is still not possible to measure transport properties in a unique direction, because of their domain structure. Young et al. [52] have reported the synthesis of an *a*-axis film with approximately 80% of the grains aligned in one direction in the plane. This was achieved by depositing the film on a substrate with twofold symmetry in the plane. A high degree of in-plane alignment offers the possibility of measuring anisotropic properties unambiguously; however, to our knowledge no transport measurements of these aligned films have yet been reported.

Films deposited in a {103} orientation also have a domain structure with a 90° orientation relationship between the grains. These films are deposited on (110) cubic substrates; since (110) is equivalent to (101) we will use the latter in analogy with {103} of the YBCO. The two major in-plane directions of the substrate are the [010] and the $[\bar{1}01]$. The grains are elongated in the [010] direction which is the faster growth direction; the grain structure is shown in planar view in Figure 3.14 [25, 53]. There are two possible orientations of the grains during film synthesis, (103) and $(\bar{1}03)$, with the CuO_2 planes at approximately 45° to the substrate surface and at 90° to each other. These grains contain (110) twins which form during the cooldown process, giving additional (013) and $(0\bar{1}3)$ orientations. In the $[010]_{sub}$ direction, the boundaries are predominantly of a single type, shown in Figure 3.15. These are [010] 90° twist boundaries (again not distinguishing the *a* and *b* axes) with the CuO_2 planes intersecting in a grid across the boundary. In the $[\bar{1}01]_{sub}$ direction there are two types of 90° boundaries, the same as those observed in *a*-axis films. This can be seen in Figure 3.16, which shows a typical cross section of a nominally 800-Å-thick film. The grains are very small at the interface, but quickly become larger as the film thickness increases. The film surface is very rough and three-dimensional; clearly growth is along the [100] and [001] directions rather than along [301] resulting in triangular-shaped grains. There is indication that the grain size stabilizes and the grain boundaries are all of the symmetrical [100] tilt type as the films thicken beyond the first few thousand Angstroms. This corresponds to termination of (100) surfaces by intersection of grains, and continued growth on (001) surfaces only, with grain boundaries normal to the interface. Such a microstructure offers the possibility of characterizing the symmetrical tilt boundary in measurements of thicker {103} films.

Transport measurements of the {103} films in the $[010]_{sub}$ and $[\bar{1}01]_{sub}$ directions are highly anisotropic as shown in Figure 3.17. ρ vs T in the [010] direction, for a 4000-Å-thick film, is the same as that of a good *c*-axis film synthesized in the same deposition run! This suggests that the [010] 90° twist boundary, the predominant grain-boundary type in this direction, is not a weak-link boundary [25]. Other types of planar interfaces in this direction are the previously discussed (110) twin boundaries and **c**/3 antiphase boundaries. In the $[\bar{1}01]_{sub}$ direction the normal state resistivity, its slope and zero intercept, are much higher, indicating boundary resistance. The resistivity

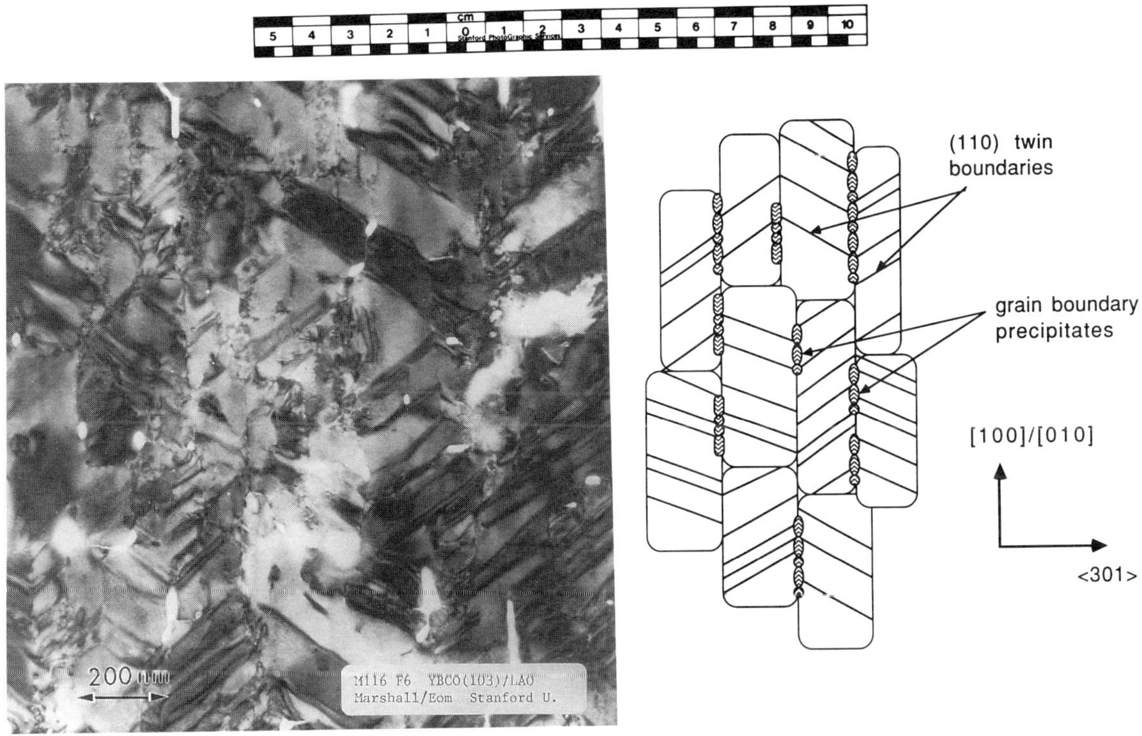

FIGURE 3.14. A planar section electron micrograph of a (103) oriented film grown on [110] $LaAlO_3$. The schematic diagram to the right illustrates the elongated morphology of the grains. The long axis of the grains is along the [010] direction and two growth variants are observed. The schematic diagram also shows the formation of second phase particles at the boundaries along the [301] direction.

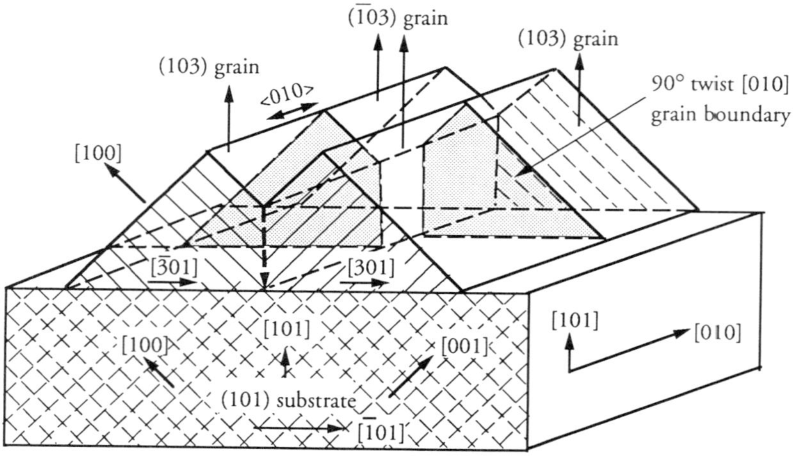

FIGURE 3.15. A schematic illustration of the different crystallographic variants that form during growth and during subsequent cooldown in the case of a {103} oriented film grown on a (110)-type substrate.

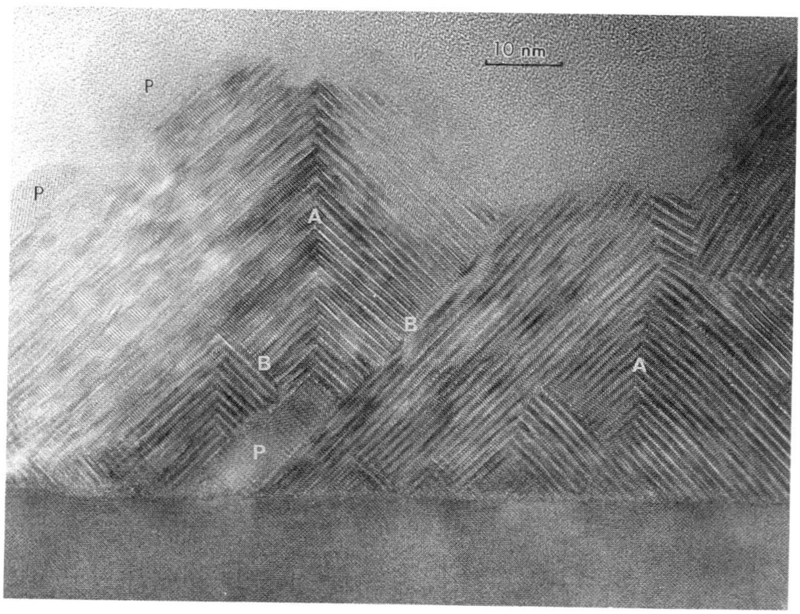

FIGURE 3.16. A cross-section TEM micrograph of a (103) oriented film, illustrating the formation of both symmetrical (A) and basal-plane-faced (B) 90° (100) tilt boundaries. Second phase precipitates (P) are also visible at the interface and along the (001) surfaces of the film.

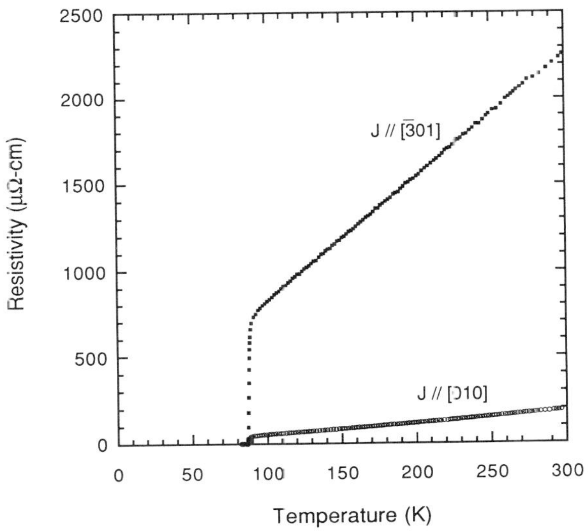

FIGURE 3.17. Resistivity-temperature plots for a (103) oriented film showing the anisotropy in the normal state properties along the [301] and [010] in-plane directions. Note that the T_c and ΔT_c are the same for the two directions.

measurements in both directions are a strong function of film thickness. The implications of these measurements have been discussed in detail by Eom et al. [53].

The observation of weak-link-free behavior in high-angle boundaries, specifically the [010] 90° twist boundary, is in contrast to the results of Dimos et al. for high-angle boundaries up to 40° misorientation [13] and the measurements on individual 45° boundaries synthesized for device applications [15]. Weak-link-free behavior across grain boundaries separating grains with a 90° misorientation of the *c* axis has also been reported by Chan et al. in post-annealed films [54], and by Babcock et al. in bulk material [55]. The boundaries observed in these materials were wavy or faceted, involving more than one crystallograpic plane in the boundary structure. We conclude, therefore, that at least one type of 90° *c*-axis boundary, and possibly more, exhibit relatively weak-link-free behavior. Furthermore *a*-axis films have recently been synthesized which exhibit J_c's as good as those of *c*-axis films in contrast to the *a*-axis films measured by Eom et al. which exhibit lower J_c's [20]. These differences may relate to differences in interface structures, which include various types of 90° grain boundaries and antiphase boundaries.

The characterization of grain boundary structures in YBCO films of different orientation indicates that we can use these films to study the properties of grain boundaries and other interfaces. Systematic control of the interface structures of *a*-axis and (103) films (e.g., by controlling grain size or in-plane orientation) may shed further light on the properties of 90° c-axis boundaries and APB's. The boundary structure of (110) and (113) films have not yet been extensively analyzed. The former will ideally have no high-angle grain boundaries; however they have exhibited cracking along the [001] direction due to thermal expansion mismatch which has caused problems in measuring more intrinsic properties [24]. Eventually the latter films may also allow for specific grain-boundary studies as well as measurements of intrinsic anisotropies. We suspect that in the future, research needs to be focused on further understanding the details of the structure and transport properties of controlled grain boundaries, both natural and artificial (e.g., bi-crystals). The fact that bi-crystal grain boundaries have already shown promise for commercial applications as weak links in squids, means that there is a considerable commercial payoff for such a systematic study. It is also clear from the results obtained so far on the structure, chemistry and transport properties of the grain boundaries in thin films, that the poorer transport properties obtained in sintered polycrystalline samples is primarily due to the crystallographic misorientations between grains and the chemical disorder/impurities associated with them. The next section deals with the phase stability in the thin films and the types of second phases that form during thin-film deposition.

3.5 Second Phases

Second phases can occur in YBCO thin films due to departures from stoichiometry as well as to non-equilibrium deposition conditions. The presence of second phases can both positively and negatively affect the superconducting properties. Very small second phase precipitates may act as pinning sites for flux lines, thereby enhancing critical current properties if evenly and densely distributed. Even larger precipitates might be beneficial in influencing the formation of associated defects, such as dislocations, which may serve as pinning sites. Second phase regions could serve as nucleation sites for other phases or for other crystallographic orientations of the 123 phase. Second phase precipitates might be a source of rf loss in the films. Precipitates clustered at grain boundaries could interrupt current paths. It is thus essential to understand the phase stability and the dynamics of formation of such second phase regions.

The phase diagram for the bulk synthesis conditions of 950°C and ambient pressure has been well established, at least for the region surrounding the 123 phase (e.g., see Beyers et al. [56], and references therein). The phases in equilibrium with 123 are Y_2BaCuO_5 (211), CuO, and $BaCuO_2$; these are observed in post-annealed films as well as in bulk material. In situ deposition of the 123 phase is carried out at lower temperatures and lower O_2 pressure, in the range of 600 to 800°C and 1 to 200 mTorr, sometimes using activated oxygen sources. Equilibrium phase diagrams for lower temperatures and O_2 pressures have been determined by careful and sensitive measurements of oxygen

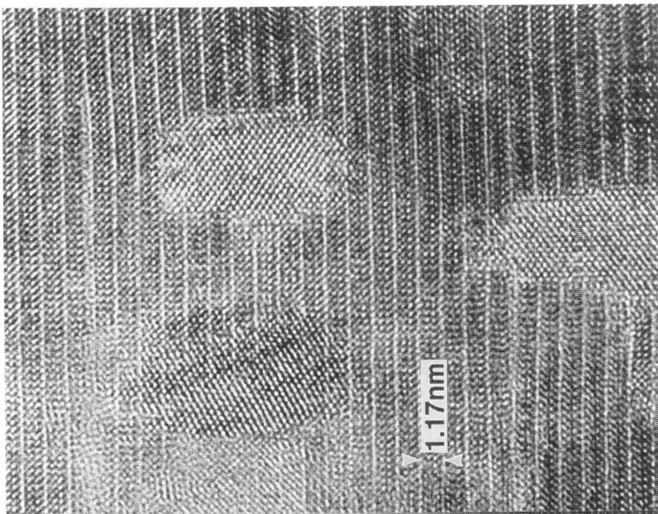

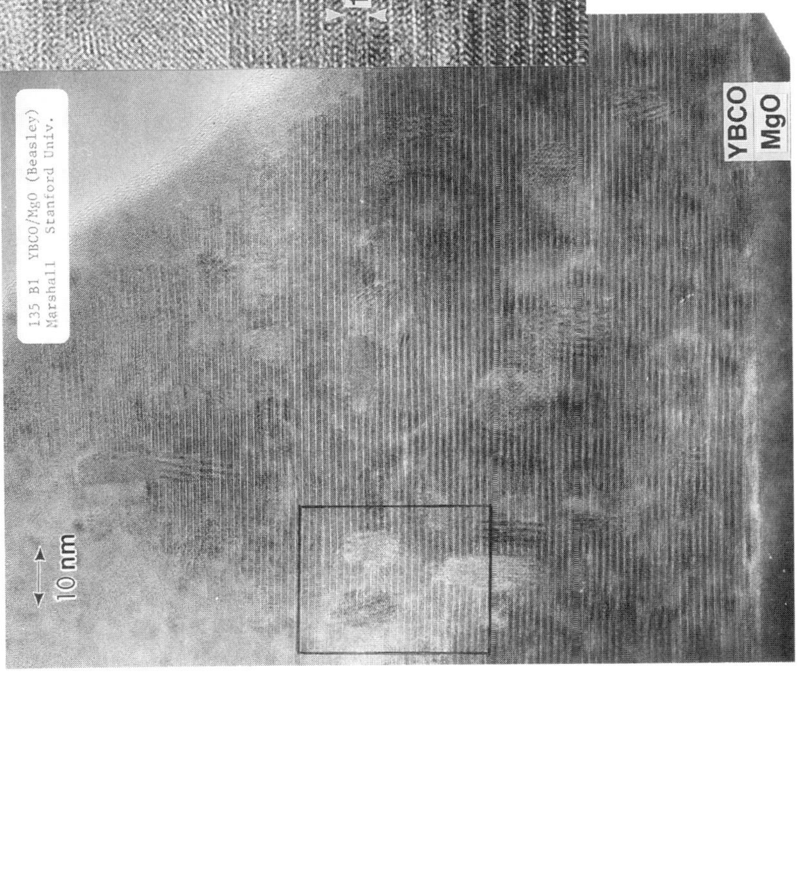

FIGURE 3.18. A cross-section lattice image showing the existence of $YCuO_2$ precipitates in the 123 matrix of a *c*-axis oriented film.

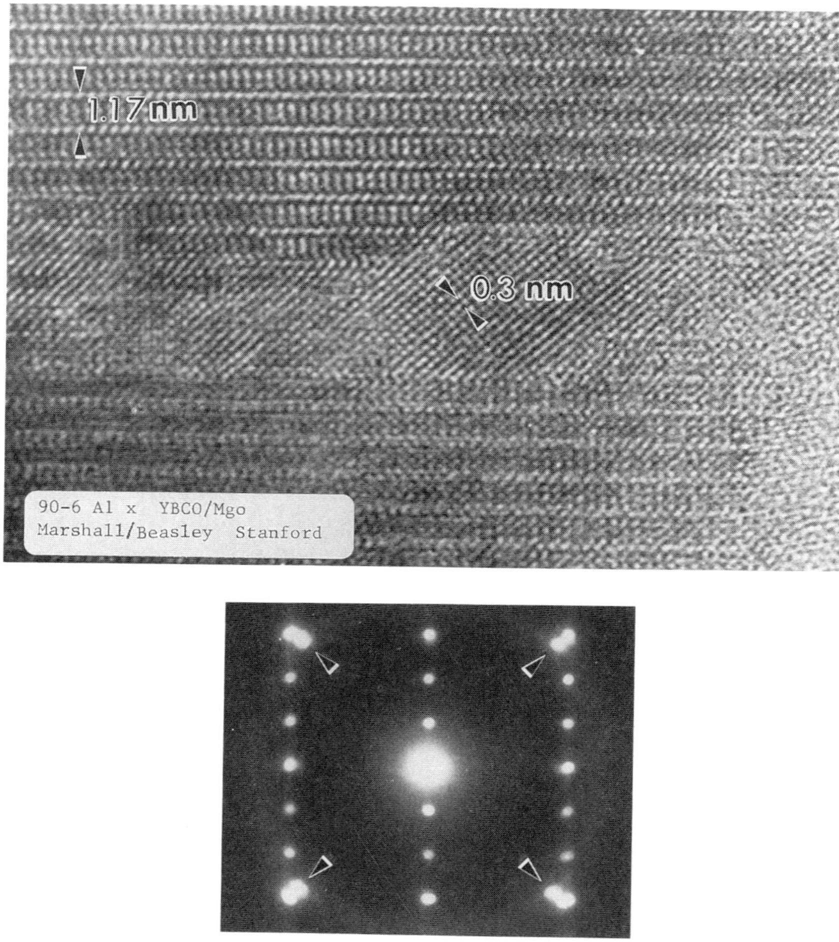

(a)

titration through solid-state electrolytic cells by Ahn et al. [57]. As the pressure is lowered the phases in equilibrium with 123 go through a variety of changes, in number and structure, with Cu tending from the +2 oxidation state to the +1 state. The conditions explored by Ahn et al. go as low as 750°C in temperature and approximately 30 mTorr in O_2 pressure; they completely encompass the range of conditions for laser deposition and partially encompass those used for sputter deposition and electron-beam evaporation. It is therefore useful to refer to these equilibrium phase diagrams as a first step in characterizing second phases in in situ thin films. At the point where 123 reaches its stability limit, the phase diagram shows it is in equilibrium with 211, $BaCu_2O_2$, and "$YBa_3Cu_2O_{6+y}$" (nominal composition of the

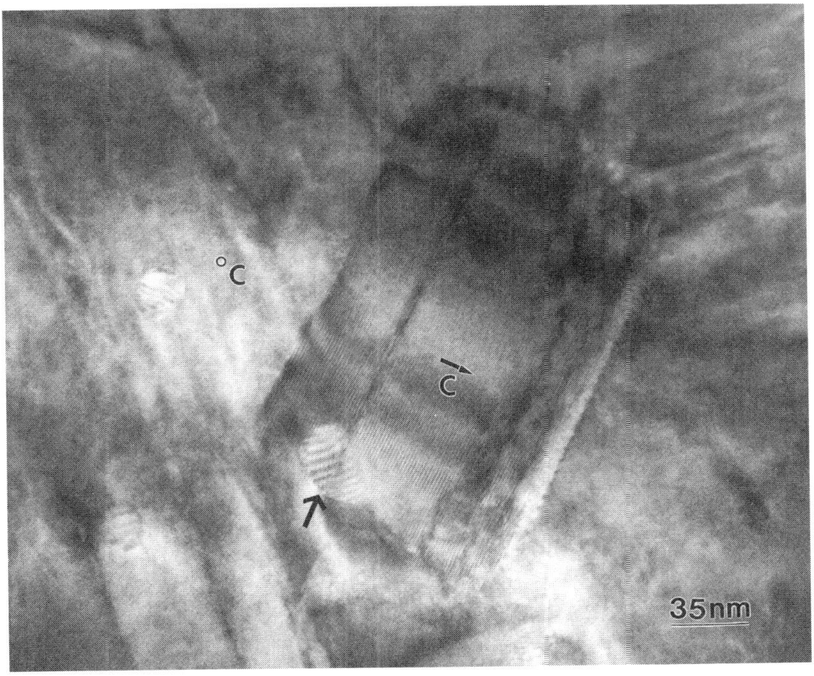

(b)

FIGURE 3.19. (a) A high-resolution micrograph that shows the formation of a 132 precipitate in a Ba-rich film. The lower part shows the microdiffraction pattern obtained from this region, and indicates the extra reflections from the 132 phase; tilting studies confirm that the phase is cubic, with a lattice parameter of approximately 4.1 Å, and aligned with the YBCO psuedocubic subcell. (b) A lattice image obtained from a planar section of a *c*-axis oriented film showing the formation of a 132 precipitate that has heterogeneously nucleated *a*-axis oriented 123 material.

"other" perovskite in this system, also referred to as "132"). However, since deposition processes are by their nature non-equilibrium processes, it is entirely possible that second phases will occur that are not predicted by the phase diagram.

In contrast to second phases observed in bulk material, which tend to be large, incoherent particles, second phases in in situ films are often quite small, highly oriented, coherent, or semicoherent particles. Several such precipitates have been characterized; the phases formed are not necessarily those predicted by the low-pressure phase diagram. Marshall et al. [58] identified the $CuYO_2$ phase in films made by electron-beam evaporation at pressures of 1–10 mtorr and about 700°C—these conditions are below the range available to be measured by Ahn et al. [57], so the question of phase stability remains open. These precipitates were quite small and densely distributed (less than 100 Å in size with a mean separation of 100–200 Å), as shown in Figure 3.18, suggesting a promising microstructure for pinning. However, comparison with stoichiometric films did not immediately indicate improved transport properties—the question of other possible pinning mechanisms in these films overlapping or dominating the precipitate effect is discussed in more detail by Matijasevic et al. [59].

Several researchers have identified 100-Å-sized precipitates of Y_2O_3 in Y-rich YBCO films made by MOCVD [60]. As these films are deposited at temperatures and pressures well within the range measured by Ahn et al. [57], where Y_2O_3 is not in equilibrium with 123, these precipitates occur by a non-equilibium process. These precipitates are small and finely distributed suggesting a possible pinning microstructure. However other microstructural defects occur in these films and the role of the precipitates is again ambiguous. Eibl and Roas [61] identified a number of phases in laser deposited films synthesized under different O_2 pressure conditions. Most of these were relatively large incoherent phases (a few hundred to few thousand angstroms in size); CuO, Y_2O_3, $Y_2Cu_2O_5$, and Y_2BaCuO_5 were identified for oxidizing conditions and Y_2O_3, Cu_2O, $BaCu_2O_2$, and Y_2BaCuO_5 for reducing conditions. In addition, very small, unidentified precipitates were identified for reducing conditions, with a size of less than 100 Å and a mean spacing of 200 Å.

In Ba-rich films prepared by electron-beam evaporation or sputtering, and in laser-deposited films, precipitates which appear to be the "132" phase are observed by transmission electron microscopy. A high-resolution image and diffraction pattern are shown in Figure 3.19(a). This phase typically appears as a cubic or pseudocubic structure with a lattice parameter of about 4.1 Å and is therefore a disordered version of "132", for which several structures and stoichiometries have been reported (e.g., see discussion and references in Ref. [56]). It is also visible in the *c*-axis film of Figure 3.9(a) and the (103) films of Figures 3.14 and 3.16. In the latter films it occurs mainly at grain boundaries and at the surface of the film. In evaporated films made farther off-stoichiometry it appears as a more ordered version with additional reflections in electron diffraction; this latter structure was also observed in non-stoichiometric films of Dy–Ba–Cu–O made by MBE [62].

One reported structure for the pseudocubic "132" phase is unique in the sense that it has Cu–O chains along all the three crystallographic directions. In a *c*-axis oriented film, this could provide a possible nucleation site for *a*-axis oriented material. Indeed, this has been observed in laser-deposited YBCO films, as illustrated in the planar section micrograph in Figure 3.19(b). The *a*-axis oriented 123 region is always attached to the "132" inclusion, supporting the above hypothesis. Using observations from cross-section micrographs, it has been suggested that *a*-axis regions can also nucleate at antiphase boundaries. While such a nucleation process is definitely possible, making such inferences from only cross-section micrographs can be erroneous, since they do not provide information over large areas and are but one section of the thin film. It would thus be preferable to use both cross-section information and planar section information to deduce the type of nucleation site.

Such *a*-axis oriented outgrowths are detrimental in the case of *c*-axis oriented YBCO–PrBCO–YBCO heterostructure junctions, since they provide easy conduction paths. The effect of *a*-axis oriented grains on other properties such as microwave surface resistance or flux noise is not known. From the film growth point of view, a better understanding of all possible conditions under which *a*-axis oriented material nucleates in a *c*-axis oriented film is essential.

3.6 Stacking Defects

The HTSC cuprates have layered, two-dimensional, anisotropic crystal structures; they exhibit a high degree of structural complexity with large unit cells and a variety of poltypoidic stacking sequences. The latter can result both in distinct phases with their own set of properties, such as the 248 and 247 superconducting phases [63–67], or in a defect structure of stacking faults, and associated dislocations and antiphase boundaries within the 123 structure [35, 49, 68–76]. High-resolution electron microscopy has been extensively utilized by many groups to understand the atomic structure of stacking defects. Many such defects observed in YBCO films occur by intercalation of extra layers along the Cu–O chain layer, leaving the central **perovskite block** containing the CuO_2 plane layers unaltered [35, 75]. The most common of these is the 248 defect shown in Figure 3.20(a) where a double rather than single Cu–O chain layer occurs between the perovskite blocks. Such an intercalation may terminate as an edge dislocation separating a single and double chain layer, with consequent bending of the lattice planes in the area of the dislocation. In addition to HREM imaging, 248 defect structures may be imaged at lower magnification by diffraction contrast of the associated dislocation structures [Figure 3.20(b)] [73, 74]. Alternatively an antiphase boundary, where there is a rigid shift of the lattice planes across the boundary, may nucleate at the termination of a 248 layer (Figure 3.21) [49].

Other intercalated structures have been observed, particularly in *c*-axis films made by laser ablation, which are typically highly faulted. These include the 224 defect [35, 49], the 125 defect [75], and a cation vacancy ordered defect, related to the 248 structure but with the 123 stoichiometry [76]. For example, as illustrated in Figure 3.22(a), the structure of the 224 defect differs from that of 123 by intercalation of a second Cu–O chain layer and another layer which was deduced as a Y–O layer. This conclusion was the result of extensive image simulations, image processing, and comparison of the intensity distribution in the actual HREM image and the computed image. Under kinematical imaging conditions, that is, when the foil is very thin and the

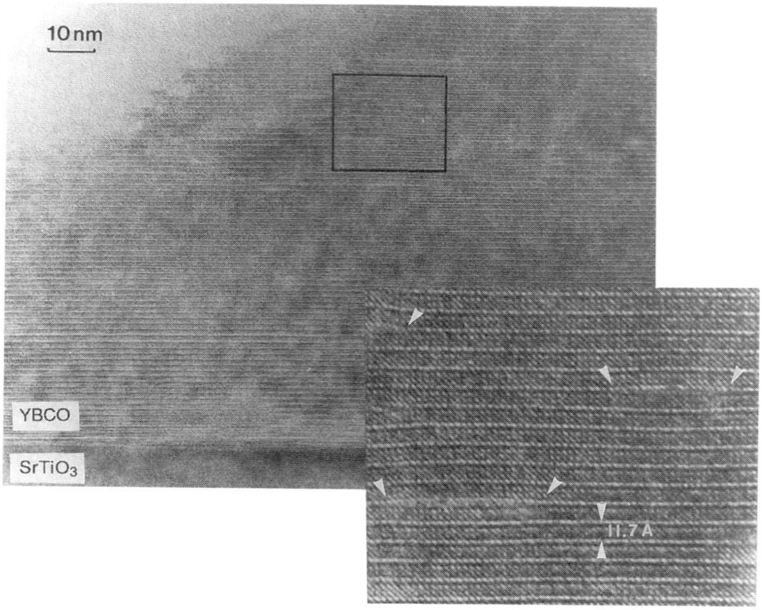

(a)

(b)

FIGURE 3.20. (a) A HREM image showing the intercalation of extra CuO chain layers forming 248 stacking faults in the 123 structure. This is the most commonly observed defect in c-axis oriented 123 films. The stacking faults may be very short in extent, with a dislocation where the stacking faults terminate (arrows); (b) diffraction contrast image in planar view showing dislocations (arrows) at stacking fault terminations, as well as twin boundaries (T) and second phase precipitates (P).

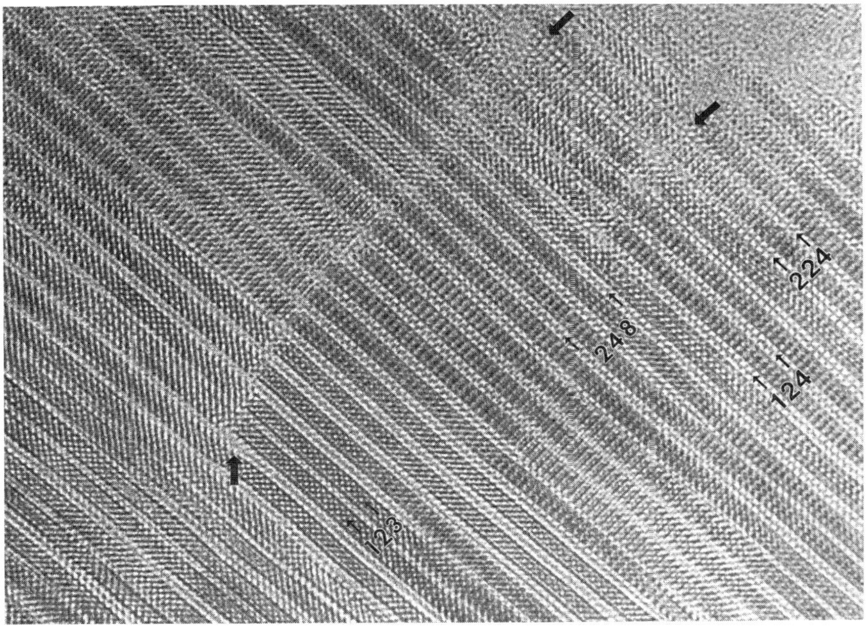

FIGURE 3.21. A HREM image illustrating the accommodation of the stacking mismatch between the 248 region and the 123 region through the formation of a **c**/6-type translational boundary (antiphase boundary). **c**/3 translational boundaries are observed more frequently, since they involve an isochemical shift from one Ba site to another.

weak-phase approximation holds, the image intensity distribution is a direct projection of the atomic potential along that crystallographic direction. In Figure 3.22(a), the perovskite blocks can be clearly identified. In order to identify the chemical species in the intercalated layers, detailed image simulations were carried out for all possible combinations of cations, that is, Y, Ba, and Cu [35]. The best matching of the experimental image and the computed image was obtained when the intercalated layers were comprised of two Cu–O layers separated by a Y–O layer, as illustrated in the structural model in Figure 3.22(b).

Under suitable conditions, the image intensity distribution and symmetry can be sensitive to the presence of weakly scattering species such as oxygen. For example, in the case of the Y–O layer in the 224 structure, the matching of the computed image to the experimental image is better if there is an oxygen in the Y layer. Figure 3.23 shows simulated images for two structural models, viz., with and without oxygen in the Y layer between two Cu–O chains. Figure 3.23(a) is for the model in which there is no oxygen in the Y layer while the image in Figure 3.23(b) has oxygen in the Y layer. When

(a)

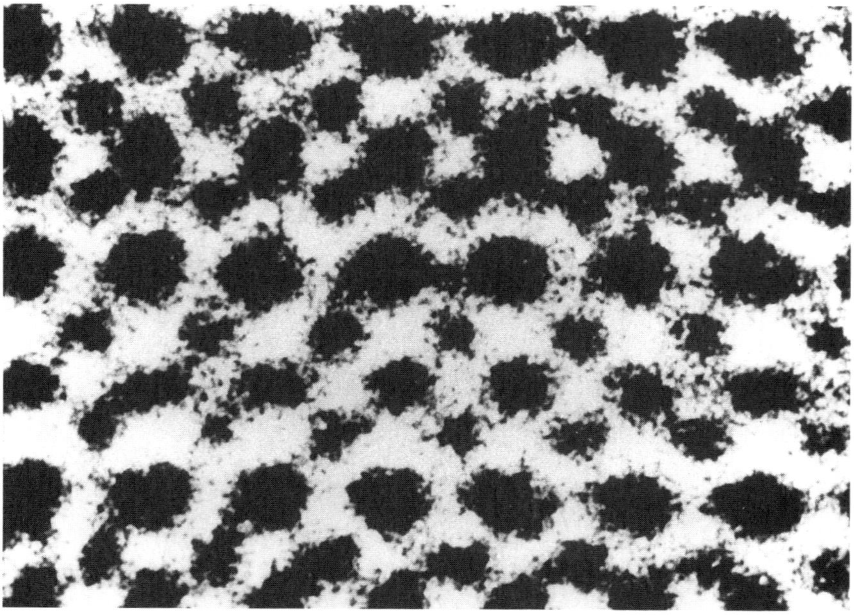

(b)

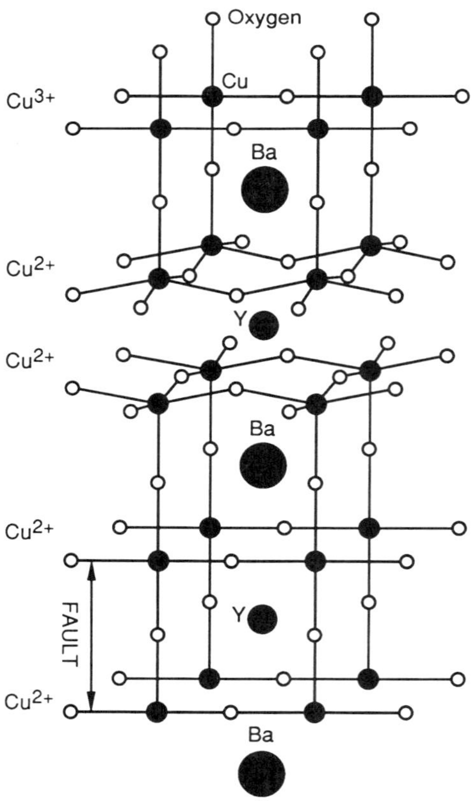

FIGURE 3.22. (a) An atomic resolution image showing the structure of the 224 defect. It differs from the 123 structure by the intercalation of a Y–O layer and a Cu–O chain layer, as is schematically illustrated in the structural model in (b).

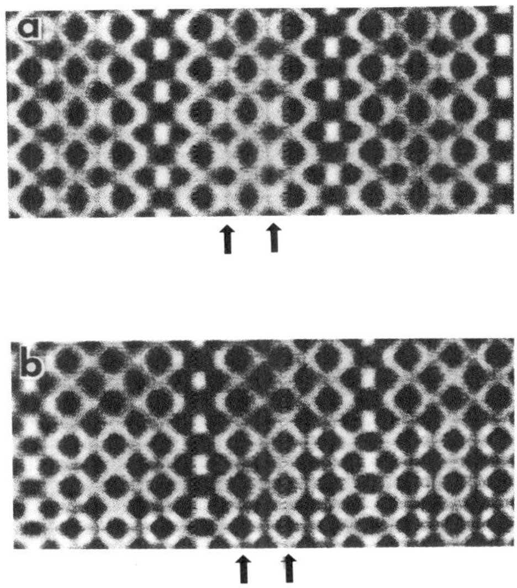

FIGURE 3.23. Simulated images for two possible structural variants of the 224 phase. (a) Without any oxygen in the intercalated Y layer; (b) with oxygen in the intercalated Y layer. The intensity distribution and symmetry in the experimental image shown in Figure 3.22(a) matches better the simulated image in (b) than that in (a).

oxygen is included in this layer, the intensity distribution for the Y-atom position is oblong with the long axis along the [100] direction, which resembles the experimental image. When there is no oxygen in this layer, the intensity distribution for the Y layer has circular symmetry, which is not observed in the experimental image. The 125 and cation-vacancy-ordered 248 defects were similarly identified using a combination of HREM and image simulations [75, 76]. Occasionally layering defects within the perovskite block have also been observed, for example, an extra cation layer in the center of the block, adjacent to the Y layer and translated by a glide plane [49, 68]. However, the 248 stacking fault is by far the most commonly observed, with the 224 defect also observed frequently in laser-deposited films.

The 248 stacking variation is also the basis for two new superconducting phases, the 248 phase with the double chain layer occurring every unit cell and the 247 phase with the double chain layer alternating with single chain layers. Both of these phases have been synthesized in bulk using overpressures of O_2 and their stability relative to the 123 phase has been investigated extensively [77, 78]. Although there is some controversy as to the exact position of the stability lines, it is clear that these phases are not stable under the low pressures of in situ film deposition processes. Their formation in thin-film form has only been achieved by post-annealing processes, which is how the 248 phase was first identified and its superconducting properties determined [63, 64]. The properties of post-annealed films and bulk material containing large densities of 248 stacking faults, achieved by processing under conditions near the phase stability line, give further indication that these defects may have beneficial effects on properties such as J_c and T_c [64, 79, 80].

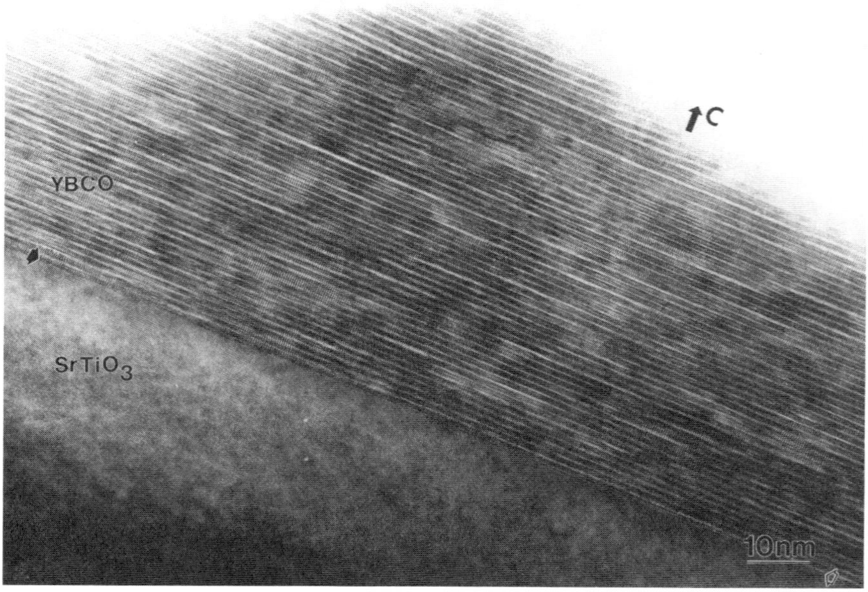

FIGURE 3.24. A lower magnification image of the same film as Figure 3.21, in which a wavy contrast is observed due to the frequent intercalation of stacking defects.

In situ films made by laser ablation contain high densities of 248 and other stacking faults. They also contain regions over which the defects are ordered, giving small areas (<200 Å) which are a distinct phase. The typical microstructure consists of layers of the 123 structure interspersed with a high density of polytypoidic stacking defects, as illustrated in Figure 3.21; a lower magnification image of the same region is shown in Figure 3.24 in which the wavy structure is due to the intercalated stacking defects. Films made at lower pressures by sputtering or evaporation tend to show a less wavy structure, suggesting fewer stacking defects. For in situ c-axis films made by various processes and with various microstructures, a clear correlation of J_c and T_c with 248 stacking faults and other faults has not been reported. However, we note that these defects generally occur in conjunction with other crystallographic defects. Furthermore, the growth process itself may influence the characteristics of these defects; for example, 248 faults are observed to terminate both as dislocations (causing curvature and possible compression of surrounding planes), and as nucleation sites for antiphase boundaries. These two structures might have quite different effects on properties.

It is interesting to note that a-axis films made by laser ablation have relatively few stacking faults compared with c-axis films [19]. This phenomenon

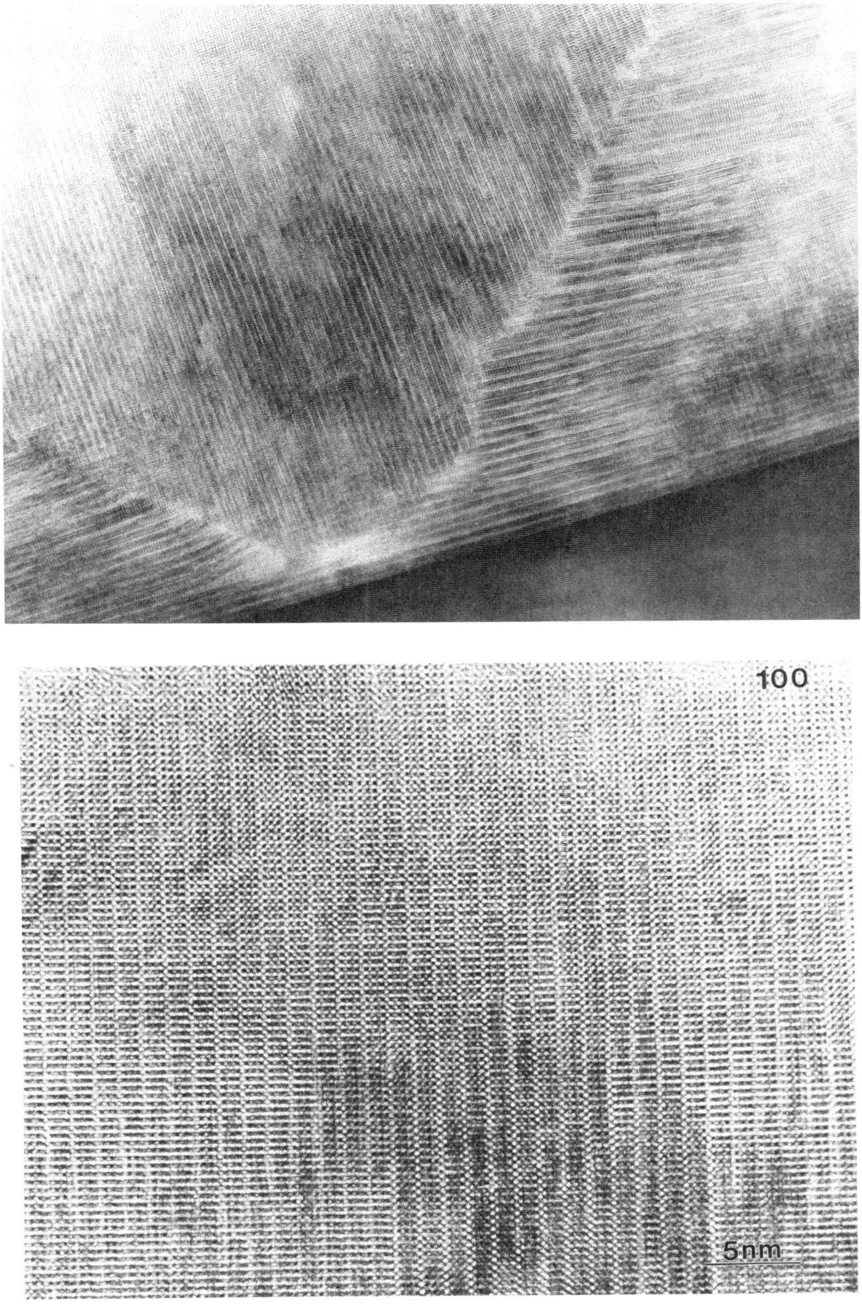

(a)

(b)

FIGURE 3.25. (a) A lattice image from a cross section of a film which has both c- and a-axis oriented regions showing that while the c-axis region is highly defective, the a-axis region is relatively defect free, as is better illustrated in (b), which is a magnified view of the a-axis region.

can be seen in adjacent *a*- and *c*-axis regions of the same film [Figure 3.25(a)], and is therefore clearly related to differences in the growth behavior of these two orientations under the same processing conditions. Figure 3.25(b) is a magnified view of the *a*-axis part and is evidently less defective, in comparison to the image shown in Figure 3.21. *a*-axis growth does not require long-range ordering of incoming species and can therefore adjust the local composition to overall stoichiometry fairly quickly. Additionally, *a*-axis growth is constrained by the symmetry of the substrate (which is typically a [100] oriented perovskite structure such as $LaAlO_3$) and hence the typical 248 defect cannot form. *c*-axis growth, on the other hand, locks the stoichiometric variations into the layered structure, without any penalty due to structural differences (all the polytypoidic structures have approximately the same *a*–*b* plane structure and lattice parameter).

3.7 Heterostructures and Multilayers

The early experiments were primarily aimed at growing single *c*-axis oriented layers. Such films are important by themselves for applications such as microwave filters and interconnects, where single layer thin films of the highest quality are required. However, the realization of a viable superconducting Josephson electronics technology relies heavily on our ability to engineer heterostructures using the superconducting compounds. This involves the fabrication of superconductor–normal metal–superconductor (SNS) or S–insulator–S (SIS) junctions. In the case of the cuprate superconductors, the small coherence length (4–30 Å depending on the crystallographic direction), puts severe constraints on the quality of the heterointerfaces. The normal metal or insulating barrier layers have to be continous (i.e., no pinholes). This means that the wetting properties of the barrier layer on YBCO needs to be studied. Additionally, the metal contact to the top layer also has to be of low resistance and ohmic in nature. Finally, the whole heterostructure should have smooth surfaces. Consequently, considerable attention is being given to optimizing the growth conditions and the structural and chemical quality of the heterointerfaces.

In order to create such heterostructures, many routes have been adopted. For example, the insulating properties and the excellent lattice match of $SrTiO_3$ to YBCO are being used to create S–I–S-type junctions, while the intrinsically similar chemistry and lattice match of PrBCO with YBCO is being utilized to grow S–N–S junctions [81–84].

Successful heterostructures and multilayers of the YBCO superconductor were first made in *c*-axis orientation by sputtering or laser deposition, using PrBCO or $SrTiO_3$ as the sandwiching layers [81–84]. Intact continuous layers of the PBCO/YBCO structures can be grown down to single-unit-cell thicknesses (i.e., 24-Å bilayers) [81, 84]; in contrast MgO layers are not continuous below a thickness of 50 Å [85]. These heterostructures have been characterized by a variety of techniques, including x-ray diffraction [81–84], cross-section TEM [81, 83, 86], ion channeling spectroscopy [82, 83], secondary ion mass spectroscopy [83], and Auger profiling [83]. In particular the position and width of the superlattice x-ray peaks, as compared with those of the fundamental peaks, give accurate information on the superlattice spacing, the coherence of the modulation, and the possibility of intermixing of layers. Cross-section TEM complements these results giving similar information on localized areas, and shows the morphology of the layers. The results, and the corresponding transport measurements and comparison with alloys of similar overall composition, all confirm the continuity and flatness of the layers. They further indicate that the interfaces are compositionally sharp to within a unit cell, with no observable interdiffusion between compounds. However, in addition to the intrinsic problem of a small coherence length along the *c* direction, the *c*-axis oriented heterostructure junctions suffered from the formation of *a*-axis oriented outgrowths, as described earlier. These two issues stimulated a stronger emphasis on growing high-quality *a*-axis oriented films and heterostructures. This is the current trend in film growth as well as in device fabrication.

Heterostructures and multilayers made with *a*-axis orientation allow for utilization of the relatively long coherence length in the *a*–*b* plane (15–30 Å) for better tunneling and proximity effects. YBCO *a*-axis growth, as previously discussed, is optimized at lower temperatures or by deposition on PBCO buffer layers. Early studies of *a*-axis oriented films were hampered by poor superconducting properties (low T_c), but these problems were overcome through the use of a novel deposition scheme involving PBCO as a buffer layer and multistage growth [87]. Recent studies show that *a*-axis oriented films of the same electrical quality as *c*-axis films, that is, $T_c = 91$ K, $J_c >$ 1 MA/cm^2, microwave surface resistance of the order of 50 $\mu\Omega$ and very low flux noise, can be obtained by this technique. A typical YBCO–PBCO–YBCO *a*-axis heterostructure along with the silver top contact is illustrated in Figure 3.26(a), while Figure 3.26(b) shows a closeup of the silver electrode–YBCO interface. As in the case of the *c*-axis oriented heterostructures, the contact electrode–YBCO interface is free of second phases and/or amorphous material and is structurally abrupt.

In addition to bilayer and trilayer heterostructures, *a*-axis multilayers of YBCO/PBCO have been fabricated by Eom et al. [88]. Multilayers have been fabricated with bilayer periods as small as 20 Å; ideally the period, for equal layers of YBCO and PBCO, will be in multiples of twice the *a*-axis lattice parameter. Otherwise one expects some intermixing at the interfaces and a drift of the interface composition, as a function of its position throughout the structure, as discussed in more detail by Triscone et al. [89]. Cross-section TEM of the multilayers (Figure 3.27) shows that the layers are continuous and the interfaces sharp; however the layers are often curved upward

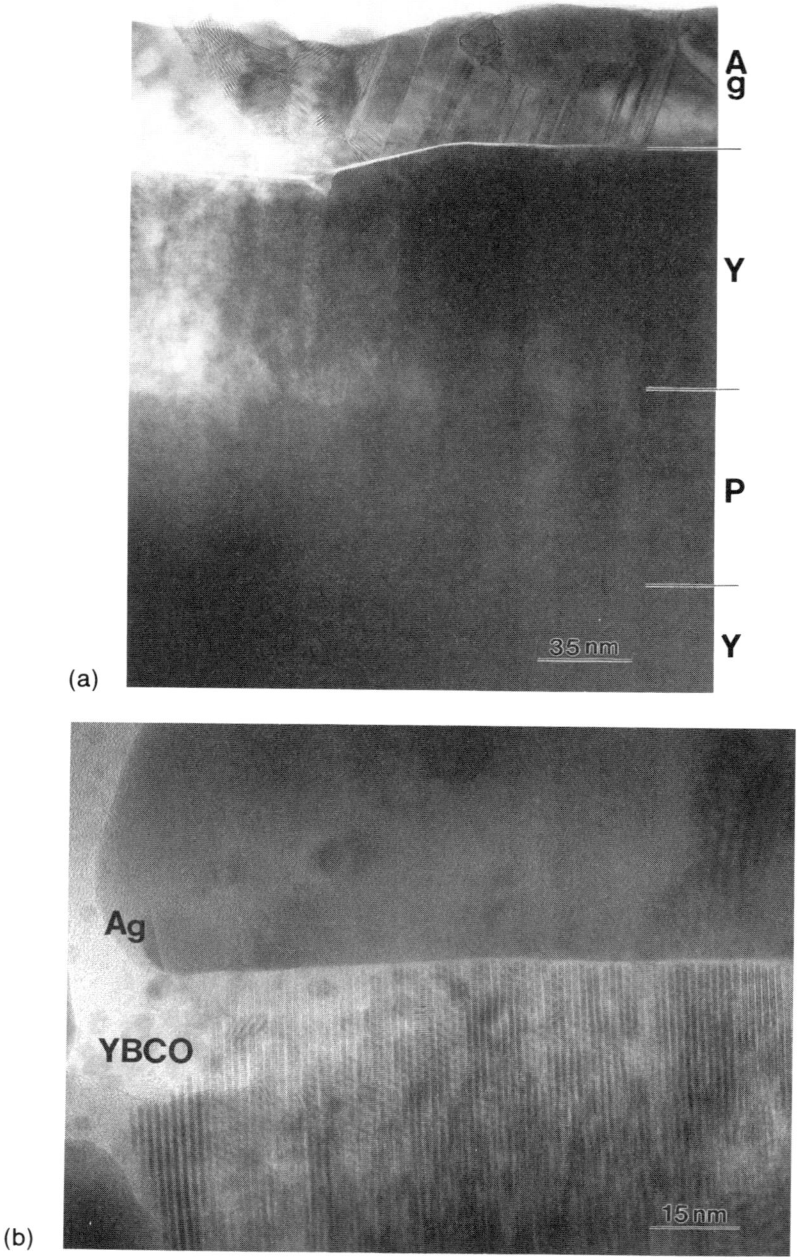

FIGURE 3.26. (a) A cross-section TEM photograph of a YBCO–PrBCO–YBCO heterostructure along with the silver top ohmic contact which reveals that the *a*–*b* planes are continuous through the interfaces; (b) a closeup of the top YBCO–silver interface showing that it is free of chemical segregation or second phases. In the case when the ohmic contact is grown ex situ, generally some amorphous material is observed at this interface.

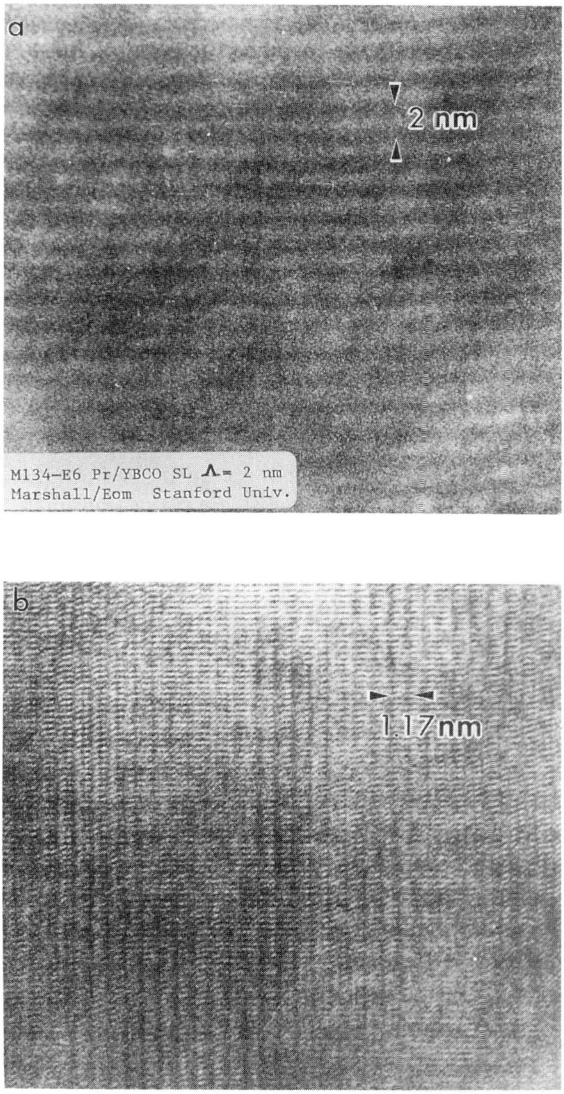

FIGURE 3.27. *a*-axis oriented YBCO–PrBCO superlattice imaged under (a) mass contrast conditions showing that the alternating layers are continuous and well-defined; (b) a lattice image showing the crystalline perfection of the lattice through the layers which are now only faintly visible.

toward the center of a grain although continuous across the grain boundary. The lattice image, on the other hand, shows that the crystal structure is highly aligned both in the plane and normal to it. These latter results are confirmed by the small rocking-curve width observed by XRD (0.6°). As with the pure YBCO films, both the *a*-axis heterostructures and multilayers discussed here are characterized by extremely smooth surfaces, with fluctuations of <40 Å as measured by STM.

Several growth and microstructure related problems still remain to be solved in *a*-axis oriented heterostructures. The main problem is cracking due to thermal expansion mismatch during cooldown. The microcracks can be observed in SEM images (Figure 3.3) as well as in planar section transmission electron micrographs, as illustrated in Figure 3.28. Microcracking is also observed in [110] oriented films and YBCO/PrBCO heterostructures, for the same reasons [90]. The experimental observations agree very well with theoretical predictions. One solution for this problem may be growing at a lower temperature, where the critical thickness is larger. However, this would yield inferior transport properties. Another possible solution would be to grow the *a*-axis oriented heterostructure on top of a *c*-axis oriented film, with the expectation that the *c*-axis base layer would absorb the strain due to the thermal expansion mismatch. A related problem is pinholes, such as that illustrated in Figure 3.28. They form in *a*-axis oriented films due to the fact that lateral thickening is much slower compared to thickening along the film

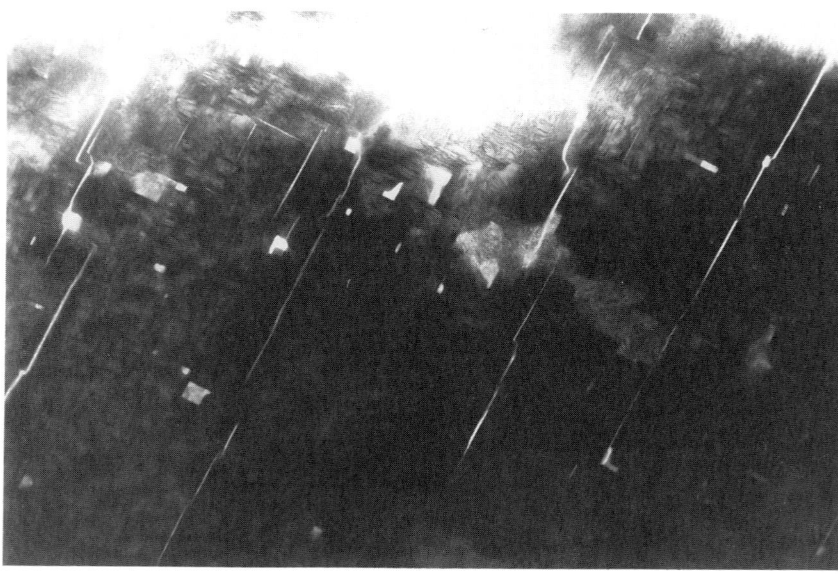

FIGURE 3.28. A planar section TEM micrograph of an *a*-axis oriented heterostructure (approximately 2000 Å thick) showing the existence of microcracks, spaced 0.5–1.0 μm apart. Also seen in this image are rectangular pinholes.

normal (easy growth direction). This also explains the rectangular morphology of the pinholes. (Pinholes are also observed in *c*-axis oriented films and heterostructures, but have no specific morphology. They are also found to be deleterious to the microwave surface resistance of single layer films [91].)

3.8 Properties of Multilayers

In both *c*- and *a*-axis oriented heterostructures, such as the YBCO–PrBCO–YBCO heterostructure, the main focus has been to study the vertical transport properties that are of importance in fabricating Josephson junctions. Typically, current–voltage measurements are made with and without microwave irradiation or magnetic fields. The earlier *c*-axis oriented heterostructures showed weak microwave modulation; however, the more recent *a*-axis oriented heterostructures showed very clear modulation of the of the zero-field critical current. Details of these measurements are published elsewhere [92]. The response of the junctions to dc magnetic fields is not consistent with the conventionally expected Fraunhoffer-type pattern, as is observed, for example, in low T_c Nb junctions [93]. This indicates the existence of inhomogeneities which may be related to the microstructure, such as the presence of growth domains, pinholes, or even microcracks, as discussed earlier. This aspect needs further detailed investigation, both at the microstructural level as well as detailed measurements of the magneto-transport properties.

In addition to device applications heterostructures are used to study effects of dimensionality in the behavior of the HTSC's and therefore give insight into the basic physics of these materials. PBCO/YBCO multilayers have been used to investigate coupling behavior within and between superconducting unit cells as the thickness of both layers is varied. Several independent studies yield similar results, showing that a single-unit-cell layer of YBCO can be superconducting, although with the T_c suppressed relative to that of the bulk material. It is not clear if this is an intrinsic effect, or due to suppression of T_c by the adjacent layers which might occur by a number of mechanisms. SC layers with greater than eight unit cells are approaching the bulk value (>85 K); it is possible than the increasing T_c is due to coupling effects between unit cells which enhance the T_c. When the PBCO layer is decreased below about 30 Å, significant coupling also occurs across this layer increasing the T_c for thin SC layers.

3.9 Ohmic Contacts to *c*- and *a*-Axis Oriented Heterostructures

In many cases, especially in the case of junction-type heterostructures, the final layer on top is a metal, which is the ohmic contact to the superconductor. There have been a few studies of the structure and interface chemistry of

(a)

(b)

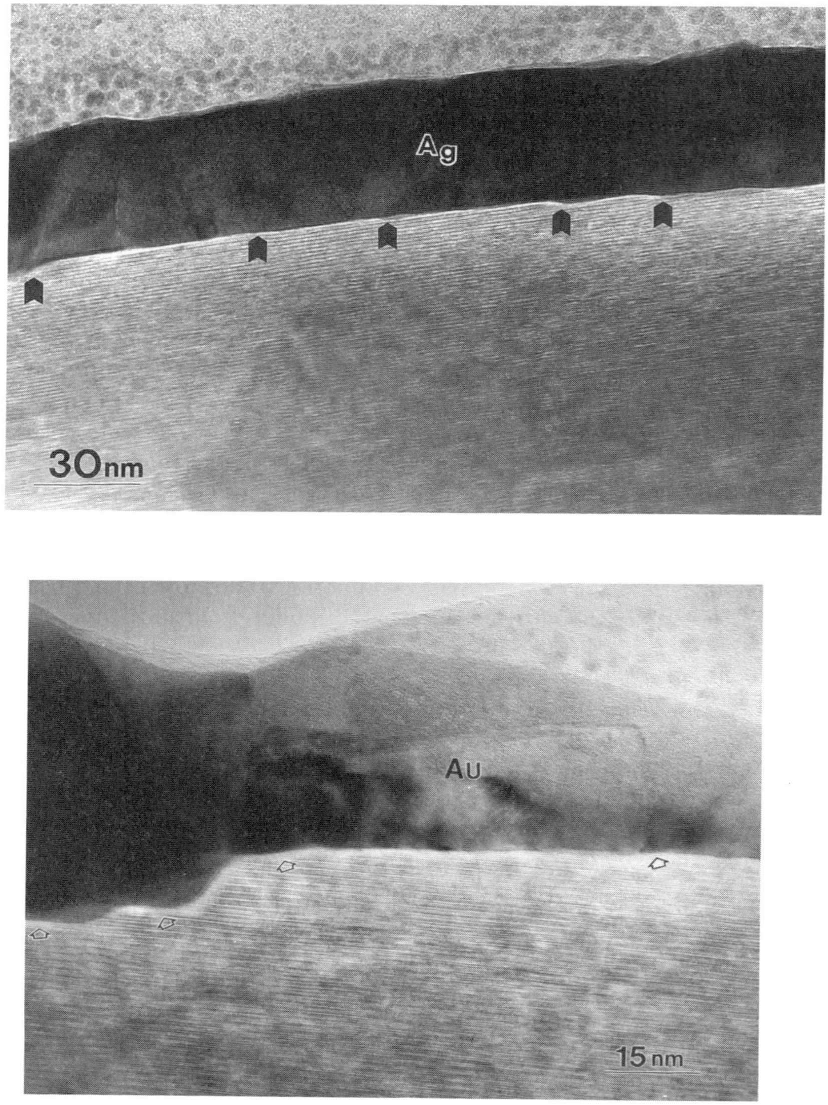

FIGURE 3.29. (a) A low magnification cross-section image of the YBCO–Au–metal interface which reveals two interesting aspects: (i) the interface is free of second phases and amorphous material; (ii) there are steps on the YBCO surface, indicating that it is growing by a ledge mechanism, with a ledge height equal to the *c*-axis lattice parameter; (b) another similar micrograph for a case where the surface step height is an integral multiple of the *c*-axis parameter.

metal–superconductor contacts, but considerable further studies will be required in order to optimize the properties of the contact–superconductor interface. In the case of both *c*- and *a*-axis oriented YBCO–PrBCO heterostructures, it was found that ex situ deposition of the contact material, which is generally silver, led to higher contact resistance (approximately 200 ohms) compared to in situ deposition, which now routinely yields contact resistance of the order of a few ohms. Examination of the metal-superconductor interfaces, in the case of the *a*-axis oriented heterostructures [Figures 3.26(a) and 3.26(b)] and in the case of *c*-axis oriented heterostructures, Figure 3.29(a), shows that these interfaces are free of second phases (both crystalline and amorphous). This image also reveals interesting details of the growth mechanism of *c*-axis oriented films. The almost periodic array of arrows point out unit-cell-high steps on the surface of the YBCO film, indicating that *c*-axis oriented films grow by a ledge mechanism along with the associated screw dislocations. The ledge height, in general, is equal to the *c*-axis lattice parameter, although ledge heights that are multiples of the *c*-axis parameter have also been observed, as illustrated in Figure 3.29(b). The spacing and height of the ledges are likely to be dependent upon the deposition conditions. A systematic study using both cross-section TEM and STM studies would be of great value in understanding the dynamics of growth on both lattice matched and mismatched substrates.

3.10 Summary

A complex array of microstructural interfaces and defects occurs in YBCO thin films as a function of the deposition conditions. A variety of artificial interfaces have also been synthesized by thin-film processes with the aim of developing device applications. Electron microscopy studies coupled with techniques such as x-ray diffraction and scanning tunneling microscopy lead to detailed characterizations of the interface structures and give insight into the structure/property relations. As we become more knowlegable of the complex variation in structure down to the atomic level and of ways to control this structure during synthesis, we can carry out even more systematic studies of the effect of individual defects on properties and of ways to control defect densities and artificial structures to our advantage. Of equal importance is the insight into the fundamental aspects of high-T_c superconductivity which may be gained from knowledge of the influence of specific microstructural variations on superconducting properties.

References

1. A. Inam et al., Appl. Phys. Lett. **53**, 908 (1988); T. Venkatesan et al., Appl. Phys. Lett. **54**, 581 (1989).
2. S. Witanachchi et al., Appl. Phys. Lett. **53**, 234 (1988).

3. H.C. Li et al., Appl. Phys. Lett. **52**, 1098 (1988).
4. R.L. Sandstrom et al., Appl. Phys. Lett. **53**, 444 (1988).
5. X.X. Xi, Z. Phys. B **74**, 13 (1989).
6. C.B. Eom et al., Appl. Phys. Lett. **55**, 595 (1989).
7. D.K. Lathrop et al., Appl. Phys. Lett. **51**, 1554 (1987).
8. T. Terashima et al., Jpn. J. Appl. Phys. **27**, L91 (1988).
9. N. Missert et al., IEEE Trans. Mag. **25**, 2418 (1989).
10. J. Kwo et al., Appl. Phys. Lett. **53**, 234 (1988).
11. S. Matsuno et al., Jpn. J. Appl. Phys. **29**, L947 (1990).
12. Y.Q. Li et al., Appl. Phys. Lett. **58**, 648 (1991), and references therein.
13. D. Dimos et al., Phys. Rev. B **41**, 4038 (1990), and references therein.
14. R. Gross et al., Appl. Phys. Lett. **57**, 727 (1990).
15. K. Char et al., Appl. Phys. Lett. **59**, 733 (1991).
16. T.H. Hylton and M. R. Beasley, Phys. Rev. B **41**, 11669 (1990).
17. G. Linker et al., Solid State Commun. **69**, 249 (1989); J. Fujita et al., J. Appl. Phys. **64**, 1292 (1988).
18. C.B. Eom et al., Physica C **171**, 354 (1990).
19. R. Ramesh et al., Appl. Phys. Lett. **57**, 1064 (1990).
20. C.B. Eom et al., Science **249**, 1549 (1990).
21. Y. Enomoto et al., Jpn. J. Appl. Phys. **26**, L1248 (1987).
22. T. Terashima et al., Appl. Phys. Lett. **53**, 2232 (1988).
23. A. Gupta et al., Physica C **162–164**, 127 (1989).
24. E. Olsson et al., Appl. Phys. Lett. **58**, 1682 (1991).
25. C.B. Eom et al., Nature **353**, 544 (1991).
26. D Fork et al., J. Mater. Res. **7** (1992).
27. R. Ramesh et al., Appl. Phys. Lett. **56**, 2243 (1990).
28. L.A. Tietz et al., J. Mater. Res. **4**, 1072 (1989); M.G. Norton et al., Mater. Res. Soc. Symp. Proc. **169**, 513 (1990).
29. D.H. Shin et al., Appl. Phys. Lett. **57**, 508 (1990).
30. D.K. Fork et al., Appl. Phys. Lett. **57**, 2504 (1990).
31. B.H. Moeckly et al., Appl. Phys. Lett. **57**, 1687 (1990).
32. D.M. Hwang et al., Appl. Phys. Lett. **57**, 1690 (1990).
33. S.M. Garrison et al., Appl. Phys. Lett. **58**, 2168 (1991).
34. R. Ramesh et al., J. Mater. Res. **6**, 2264 (1991).
35. R. Ramesh, et al., J. Mater. Res. **5**, 704 (1990).
36. S.E. Babcock et al., Appl. Phys. Lett. **55**, 393 (1989).
37. B.M. Lairson, S.K. Streiffer, and J.C. Bravman, Phys. Rev. B **42**, 10067 (1990).
38. S.K. Streiffer et al., Mater. Res. Soc. Symp. Proc. **183**, 363 (1990).
39. M.G. Norton et al., Appl. Phys. Lett. **55**, 2348 (1989).
40. S.K. Streiffer et al., Phys. Rev. B **43**, 13007 (1991).
41. M. Hawley et al., Science **251**, 1588 (1991).
42. Ch. Gerber et al., Nature **350**, 279 (1991); D.G. Schlom et al., Z. Phys. B **86**, 163 (1992).
43. T. Terashima, et al, Phys. Rev. Lett. **65**, 2684 (1990).
44. S.S. Laderman et al., Phys. Rev. B **43**, 2922 (1991).
45. N. Newman et al., Appl. Phys. Lett. **57**, 520 (1990).
46. D. Fork et al., IEEE Trans. on Appl. Superconduct. **1**, 67 (1991).
47. V.M. Pan et al., AIP Conf. Proc. No. 251, pp. 603–614 (1992).
48. D. Fork (private communication).

49. R. Ramesh et al., Science **247**, 57 (1989).
50. J. Hirth and J. Lothe, *Theory of Dislocations* (McGraw Hill, New York, 1967).
51. A. Inam (private communication).
52. K.H. Young et al., Appl. Phys. Lett. **59**, 2448 (1991).
53. C.B. Eom et al., submitted to Phys. Rev. B **46**, 11902 (1992).
54. S.W. Chan et al., AIP Conf. Proc. No. 200, pp. 172–185 (1990).
55. S.E. Babcock et al., Nature **347**, 167 (1990).
56. R. Beyers and B.T. Ahn, Annu. Rev. Mater Sci. **21**, 335 (1991), accepted, and references therein.
57. B.T. Ahn et al., Physica C **167**, 529 (1990).
58. A.F. Marshall et al., Appl. Phys. Lett. **57**, 1158 (1990).
59. V. Matijasevic et al., J. Mater. Res. **6**, 682 (1991).
60. A. Catana et al., Appl. Phys. Lett. **60**, 1016 (1992); P. Lu et al., Appl. Phys. Lett. **60**, 1265 (1992).
61. O. Eibl and B. Roas, J. Mater. Res. **5**, 2620 (1990).
62. E. Hellman et al., J. Mater. Res. **4**, 476 (1989).
63. A.F. Marshall et al., Phys. Rev. B **37**, 9353 (1988).
64. K. Char et al., Phys. Rev. B **38**, 834 (1988).
65. P. Marsh et al., Nature **334**, 141 (1988).
67. P. Bordet, et al., Nature **334**, 596 (1988).
68. Y. Matsui, et al., Jpn. J. Appl. Phys. **26**, L777 (1987)
69. A. Ourmazd et al., Nature **327**, 308 (1987).
70. J. Narayan et al., Appl. Phys. Lett. **51**, 940 (1987).
71. H.W. Zandbergen et al., Phys. Status Solid A **105**, 207 (1988).
72. H.W. Zandbergen et al, Nature **331**, 596 (1988).
73. A.F. Marshall et al., Mater. Res. Soc. Symp. Proc. **169**, 785 (1990).
74. A.F. Marshall et al., J. Mater. Res. **5**, 2049 (1990).
75. R. Ramesh et al., Nature **346**, 420 (1990).
76. R. Ramesh et al., Mater. Lett. **10**, 23 (1990).
77. J. Karpinski et al., Physica C **160**, 449 (1989).
78. D.E. Morris et al., Physica C **159**, 287 (1989).
79. S. Jin et al., Appl. Phys. Lett. **56**, 1287 (1990).
80. D.E. Morris et al., Mater. Res. Soc. Symp. Proc. **169**, 245 (1990).
81. J.M. Triscone et al., Phys. Rev. Lett. **63**, 1016 (1989); **64**, 804 (1990).
82. U. Poppe et al., Solid State Commun. **71**, 569 (1989)
83. C.T. Rogers et al., Appl. Phys. Lett. **55**, 2032 (1989); T. Venkatesan et al., Appl. Phys. Lett. **56**, 391 (1990).
84. D.H. Lowndes et al., Phys. Rev. Lett. **65**, 1160 (1990).
85. S. Tanaka et al., Adv. Superconduct. **3**, 1183 (1990).
86. O. Eibl et al., Physica C **172**, 365 (1990); **172**, 373 (1990); Phys. Status Solidi A **122**, 589 (1990).
87. A. Inam et al., Appl. Phys. Lett. **57**, 2484 (1990).
88. C.B. Eom et al., Science **251**, 780 (1991).
89. J.M. Triscone et al., submitted to J. Less Common Metals.
90. E. Olsson et al., Appl. Phys. Lett. **58**, 1682 (1991).
91. R. Ramesh et al., Mater. Sci. Eng. B **14**, 188 (1992).
92. J.B. Barner et al., Appl. Phys. Lett. **59**, 742 (1991).
93. T. VanDuzer and C.W. Turner, *Principles of Superconductive Devices and Circuits* (Elsevier North Holland, New York, 1981).

4 Interfacial Interactions Between High-T_c $YBa_2Cu_3O_{7-x}$ Thin Films and Substrates

EVA OLSSON and SUBHASH L. SHINDÉ

4.1 Introduction

Thin-film applications were immediately anticipated as the high-T_c superconductors were developed. Prior to the application of these films, however, a number of materials issues had to be investigated since the superconducting properties of these materials are strongly influenced by the microstructure. Individual microstructural features may be either beneficial or detrimental depending on their fine scale microstructure and distribution as well as on the specific application. A well-known example is the grain boundaries which can constitute weak links that limit the critical current density [1–3] but they can also be used in Josephson junction based devices [4–12]. Other defects may act as pinning centers that improve the critical current densities j_c, but depending on their distribution and morphology they could also decrease the j_c.

The effect of grain boundaries on the behavior of the films in combination with the pronounced anisotropy of the CuO-based superconductors most often makes it necessary to grow epitaxial films. A good lattice match between substrate and film is therefore essential. In addition, a good match between the thermal expansion coefficients is also important since the high-T_c supercondcutors are produced at high temperatures in oxidizing atmospheres. A badly matched thermal expansion coefficient may result in severe cracking of the film [13, 14]. The high deposition temperature also puts high demands on the inertness of the substrate relative to the film. Interdiffusion should be avoided in order to prevent degradation of the superconducting properties.

The knowledge of what controls the epitaxial relationships between film and substrate can be used not only to optimize j_c but also to introduce grain boundaries at specific locations in integrated circuits [11, 12]. The effect of

E. Olsson, Department of Physics, Chalmers University of Technology, S-412 96 Göteborg, Sweden; S.L. Shindé, IBM, T.J. Watson Research Center, Yorktown Heights, NY 10598.

[1] S.L. Shindé's present address is the IBM Corporation, B 300-40E, Hopewell Junction, NY 12533.

the other interfacial interactions between the substrate and the film mentioned above are also instrumental in determining the performance of the devices. Here we address the effect of lattice match, growth characteristics, thermal expansion coefficient match, and chemical interaction between the film and substrate on the microstructure and the superconducting characteristics. The discussion is limited to the growth of $YBa_2Cu_3O_{7-x}$ but many parallels can be made to other CuO-based high-T_c superconductors based on the similarities between them. This is especially true for c-axis oriented films. The text is restricted to the mechanisms during in situ growth while most recent work concentrates on in situ deposition techniques.

4.2 Initial Interactions at the Film/Substrate Interface

Several parameters of the deposition procedure have a critical influence on the microstructure of the film and thus also on the superconducting characteristics. The choice of substrate will put a natural limit on the obtainable microstructure due to interfacial interactions between the film and substrate. Initial interactions in the early stages of in situ growth will determine the crystallographic orientation of the film. Subsequently, chemical reactions, interdiffusion, and mechanical interactions may ensue and cause substantial changes to the microstructure. The use of polycrystalline materials as substrates is an illustration of the limiting nature of the substrate.

4.2.1 *Polycrystalline Substrates*

The work on polycrystalline substrates was initiated due to the interest in developing films for use in large scale power applications, for example, as wires, cables, magnets, and electromagnetic shields. These applications require a flexible substrate and attention was thus paid to metals. The pronounced dependence of j_c on the crystallographic direction in YBCO with the highest j_c in directions perpendicular to the c axis [e.g., Ref. 15] made it necessary to align the grains in order to increase the j_c. The c-axis oriented films were preferred since the current transport should be in the plane of the film. A further increase in j_c could be achieved if the grains could also be aligned in the plane of the films. This would eliminate the grain boundaries which act as weak links thus reducing the j_c [1–3]. The exact behavior of each individual boundary depends on its detailed microstructure and some boundaries would not be detrimental to the j_c. High j_c could therefore be achieved if the grain-boundary structures could be controlled. A rational route to increase j_c, however, is rather to obtain films with a single c-axis orientation.

4.2.1.1 Metals

Metal alloy substrates, such as stainless steels and Ni–Cr–Mo alloys, were found to give randomly oriented YBCO thin films and c-axis orientation could not be achieved [16, 17]. In addition, interdiffusion between the metal substrates and the films deteriorated the T_c and j_c. The solution was to deposit a buffer layer of yttria stabilized ZrO_2 (YSZ) on the metals prior to deposition of the YBCO. The buffer layer was polycrystalline but tuning of the deposition conditions could result in alignment of the YSZ grains [17]. The temperature was instrumental for the grain alignment on stainless steel where the (001) orientation was achieved at 70°C with a gradual change in alignment with increasing temperature. A lack of oriented growth was observed at intermediate temperatures (300°C) while (111) oriented YSZ was obtained at 700°C. The structure of the YBCO deposited on top of these buffer layers was influenced by the YSZ orientation with the (001) layers being the best promoter of *c*-axis YBCO films. Recently, ion beam assisted deposition of the YSZ buffer layer has been reported to result in in-plane orientation of (001) YSZ films [18]. During the deposition an assisting O + Ar ion beam concurrently bombarded the films. The subsequent YBCO film was grown epitaxially and the obtained j_c was about 10 times higher than for films on (001) YSZ buffer layers that were randomly oriented in the plane of the film.

The interdiffusion between the substrate and the films was also reduced by the presence of the YSZ buffer layer. The thickness of the buffer layer was vital for the effectiveness of the layer [16]. In general, the minimum thickness will depend on the species that diffuse and the temperature at which the diffusion takes place. It is not only the deposition temperature of the YBCO film that needs to be taken into account but also possible subsequent annealing or deposition in multilayer structures. As will be discussed in a following section concerning buffer layers, there may also be a maximum buffer layer thickness above which the film quality will degrade [19].

4.2.1.2 Polycrystalline YSZ

Polycrystalline YSZ has also been proposed as a candidate substrate for the large scale applications due to its flexibility [20, 21]. Experiments have shown that it is possible to achieve *c*-axis oriented films even on randomly oriented substrates resulting in j_c of the order of 10^4 A/cm^2 at 77 K with no magnetic field applied. However, the grain boundaries in these polycrystalline substrates may contain contaminants, for example Si, which diffuse into the YBCO film and degrade the superconducting characteristics. Again a thin layer of YSZ deposited prior to the growth of the YBCO inhibits the degradation of the film by acting as a buffer layer.

4.2.2 *c-Axis Orientation*

Most work concerning high-T_c superconducting thin films up to date has been carried out on *c*-axis oriented films. The reason for this is that the large anisotropy of growth kinetics promotes this orientation of the YBCO [22] and it has been difficult to isolate other orientations of the c axis with respect to the substrate and simultaneously obtain high-quality films. Epitaxial films with one unique orientation relationship between the film and the substrate is desired for applications where a high critical current density is required due to the anisotropy of YBCO and the weak-link behavior of most grain boundaries. Most films are tetragonal during deposition which is usually carried out at temperatures between 600°C and 800°C. The films transform to the orthorhombic structure during cooling provided that the films thickness exceeds a critical thickness [23, 24] and (110) twins form as a result of this transformation. The twin planes are perpendicular to the plane of the film and separate regions where the a and b axis exchange positions when passing a twin boundary. The adjacent regions thus have different epitaxial orientation relationships. Since the epitaxy is established at higher temperatures when YBCO is tetragonal this twinning effect will not be considered as an additional epitaxial relationship in the following text.

4.2.2.1 Lattice Match Between Film and Substrate

A good lattice match at the deposition temperature and similar crystal structures favor epitaxial growth. The YBCO consists of two primitive unit cells of perovskite-like $BaCuO_{3-x}$ with a unit cell of $YCuO_{3-y}$ in between. The perovskite structure $SrTiO_3$ has therefore been frequently used as substrate for YBCO thin films. The lattice mismatch between the $[100]_{YBCO}$ and $[010]_{YBCO}$ and the $\langle 100 \rangle_{SrTiO_3}$ is about 1% at 700°C [25, 26] , see Figure 4.1. This is a relatively good match and consequently high-quality epitaxial c-axis oriented thin films with a j_c well in excess of 10^6 A/cm^2 at 77 K in zero magnetic field can be grown on (001) $SrTiO_3$ substrates using different deposition techniques. The in-plane orientation relationship is $[100]_{YBCO}$ and $[010]_{YBCO}$ parallel to $\langle 100 \rangle_{SrTiO_3}$. The search for alternative substrates has been intense despite the favorable lattice match between $SrTiO_3$ and YBCO. The severe disadvantage of $SrTiO_3$ is that the dielectric constant and loss tangent, $\tan\delta$ (277 and 0.0125, respectively, at 1 MHz [27], see Figure 4.2) are higher than what can be accepted for microwave applications. Other considerations are the cost and the availability of substrates with large diameters. Future needs of incorporating the high-T_c films in integrated circuits emphasize the aspect of compatibility of the substrate to other circuit components and have initiated the research concerning Si, Al_2O_3, and GaAs substrates. Selected aluminates and gallates have also attracted attention because of their distorted perovskite structure, good lattice match, low dielectric loss, and $\tan\delta$. It should be noted that $LaAlO_3$ has a gradual phase transition from rhombohedral to cubic structure at 400°C to 500°C during heating [28, 29]. The transition can result in heavy twinning which influences the YBCO film quality. It is possible, however, to grow high-quality films on $LaAlO_3$ [e.g., Ref. 30]. $LaGaO_3$ exhibits an abrupt first-order transition at 145° [29] which also could be disadvantageous for the YBCO films [31]. A

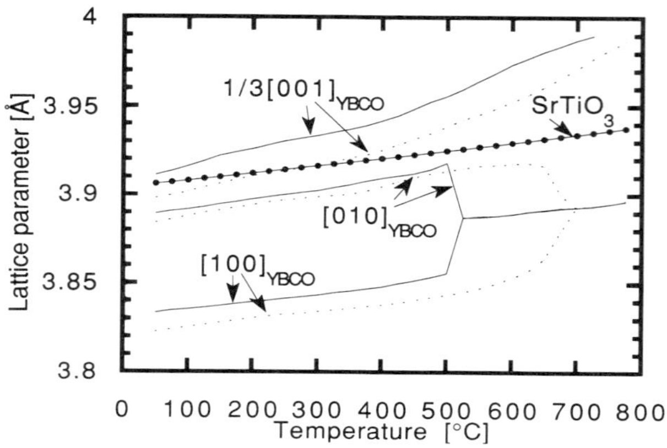

(a)

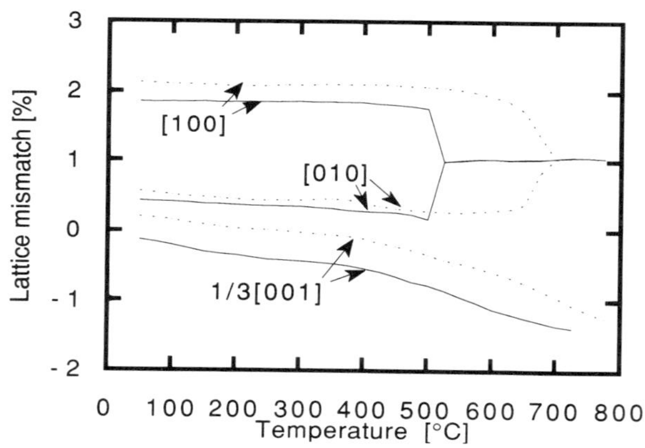

(b)

FIGURE 4.1. (a) A comparison of the lattice parameters of $SrTiO_3$ [25] and YBCO [26] as a function of temperature. The transition temperature where YBCO transforms from tetragonal to orthorhombic structure depends on the the oxygen partial pressure (solid line—0.005 atm, broken line—1 atm). (b). The lattice mismatch between YBCO and $[100]_{SrTiO_3}$ as a function of temperature for two oxygen partial pressures (solid line—0.005 atm, broken line—1 atm). Negative values correspond to larger YBCO parameters in comparison to $SrTiO_3$. The absolute value of the lattice mismatch decreases with decreasing temperature for *a*-axis oriented films which have the *c* axis in the plane of the film while the lattice mismatch for *c*-axis oriented films is nearly constant at frequently used deposition temperatures.

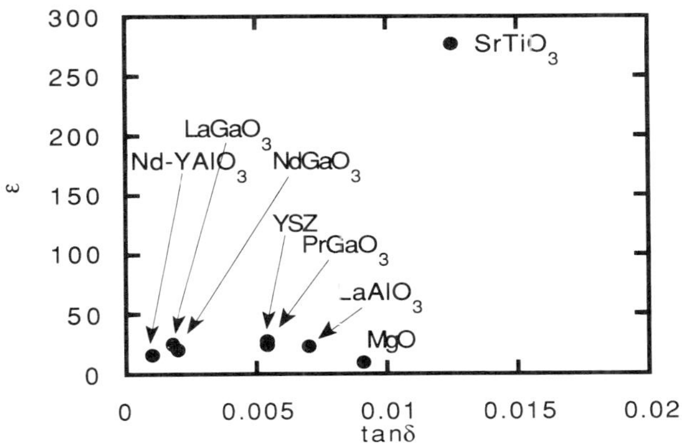

FIGURE 4.2. Dielectric constant ε and loss tangent tan δ at 1 MHz for selected candidate substrate materials.

comparison between $NdGaO_3$ and $LaGaO_3$ showed that indeed the $LaGaO_3$ renders lower j_c [32]. The lattice parameters of some candidate substrate materials for YBCO thin films as a function of temperature are shown in Figure 4.3. Epitaxy with one single orientation with respect to the substrate has been observed in all these systems.

4.2.2.2 Epitaxial Orientation Relationships

The materials can be divided into two groups where in one group the $[100]_{YBCO}$ and $[010]_{YBCO}$ align with $[100]_{substrate}$ and $[010]_{substrate}$. In the other group the *a* and *b* axis of the YBCO align with $\langle 110 \rangle_{substrate}$ instead. Theses two groups are clearly distinguished in Figure 4.3 where it also can be seen that the YSZ and MgO substrates differ from the other substrates in that the lattice mismatch is significantly larger for these two substrates. Detailed investigations of the growth of YBCO films on these substrates have shown several possible in-plane orientation relationships between the film and the substrate. Typical rotation angles of the YBCO with respect to the MgO were 45°, 27°, 16°, 9°, and smaller angles [34, 35]. The orientation relationships between the lattices of YBCO and MgO were investigated in terms of configurations with a high density of near-coincidence sites, near-coincidence-site lattices (NCSL), as opposed to the coincidence-site lattice theory used for high-angle grain boundaries [35]. The cube-on-cube orientation relationship had the largest misfit (8.89% at room temperature) of the observed orientations. It was therefore surprising that this orientation was the most frequent.

4.2.2.3 Graphoepitaxy

The structure of YBCO films deposited on vicinally polished surfaces of MgO substrates was another surprising observation [36]. The *c* axis of the

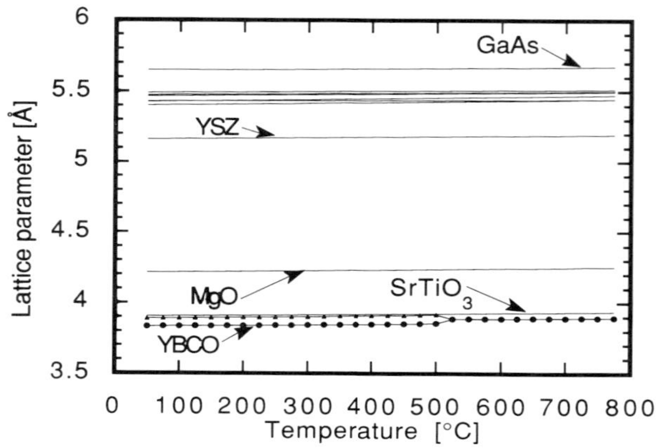

(a)

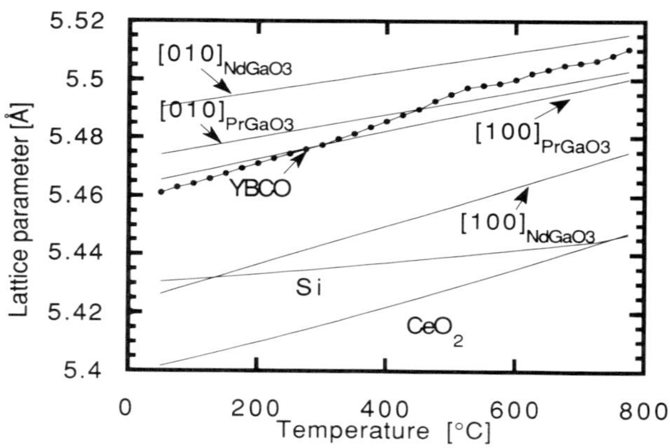

(b)

FIGURE 4.3. (a) Lattice parameters ([100] and [010]) as a function of temperature for selected candidate substrate materials [25–33]. Two groups of materials can be distinguished in the diagram. The group at about 4 Å have a cube-on-cube orientation relationship. $LaAlO_3$ is not shown in the figure but has a lattice parameter of 3.832 Å at room temperature. Several epitaxial orientation relationships have been observed in *c*-axis oriented YBCO films on (001)MgO and (001)YSZ due to the relatively large lattice mismatch for these systems. Si, $NdGaO_3$, $PrGaO_3$, and CeO_2 have lattice parameters of about 5.5 Å which are shown in detail in Figure 4.3(b). (b) A comparison between $[110]_{YBCO}$ and [100] and [010] of Si, $NdGaO_3$, $PrGaO_3$, and CeO_2. The observed in-plane orientation relationships in *c*-axis oriented YBCO films on (001) substrates have been $[110]_{YBCO}//[100]_{substrate}$ and $[110]_{YBCO}//[010]_{substrate}$. All materials, except Si, have a good match between the thermal expansion coefficients.

film aligned with the substrate normal and not with the $[001]_{MgO}$ crystallographic direction. There was, however, a distribution of the *c*-axis orientation with a tail towards the $[001]_{MgO}$ crystallographic orientation. The intensity of the tail increased as the vicinal angle decreased. The in-plane orientation was still cube-on-cube for vicinal angles up to 10°. Similar experiments were carried out on $SrTiO_3$ where indeed the *c* axis of the film aligns with the $[001]_{SrTiO_3}$ crystallographic direction [37]. The observations implied that there is a significant difference in the mechanism determining the orientation relationships between film and substrate on MgO and $SrTiO_3$.

The effect of the surface preparation was also investigated by comparing mechanically polished, chemically etched (hot phosphoric acid), and annealed (1100 to 1200°C in oxygen) substrates [34, 38]. The chemically etched substrates were considered to provide the most atomically flat substrate while the others were characterized by stepped surfaces. The annealing at higher temperatures resulted in $[001]_{MgO}$ terraces with edges along the $\langle 100\rangle_{MgO}$ crystallographic directions. The YBCO was shown to nucleate on the terrace steps and graphoepitaxy gave rise to the cube-on-cube orientation that was favored by the anneal of the MgO substrate at elevated temperatures prior to deposition of the YBCO.

4.2.2.4 Homoepitaxial Layers and Initial Monolayers

The growth of homoepitaxial layers on substrates prior to deposition of the subsequent films is a well-established procedure for fabrication of semiconductor heterostructures based on, for example, GaAs and Si. The homoepitaxial buffer layers improve the substrate surface quality and result in reproducible structures. The same technique has been evaluated for YBCO films on YSZ substrates with better quality films as a result [39]. Three main orientation relationships (0°, 9°, and 45°) have been reported for YBCO on YSZ substrates [39–42]. The use of homoepitaxial YSZ layers before depositing YBCO improved the epitaxy and favored the $\langle 110\rangle_{YBCO}//\langle 100\rangle_{YSZ}$ (45°) orientation [39]. A similar effect was achieved by using approximate monolayers of CuO, Y_2O_3, and $BaZrO_3$ instead of YSZ. The less common orientation observed on YSZ was the 9° which could be detected by the presence of corresponding broad peaks in x-ray diffraction phi scans. The width of the peaks reflected that the rotation of the YBCO relative to the YSZ substrate was not well-defined but small deviations from the exact 9° rotation broadened the peaks. Interestingly a monolayer of BaO at the YSZ substrate interface could induce a full 9° orientation of the YBCO films.

The temperature has also been shown to be crucial for the established epitaxy on YSZ [39–42]. The 45° orientation was favored at lower temperatures, while the cube-on-cube orientation relationship could prevail at higher temperatures [39, 41]. The reason for the cube-on-cube orientation is not immediately obvious from the lattice match shown in Figure 4.3 and has been suggested to be explained by the NCSL theory used for predicting the orientation relationships in MgO. It has also been suggested that the higher temperatures promote the separation of Ba to the interface during nucleation and thus promotes the 0° and 9° orientations [39].

4.2.3 *a-Axis Orientation*

Even though the *c*-axis oriented films are suitable for applications where a high j_c is required in the plane of the film there are additional applications where it would be advantageous to have the *c* axis in the plane of the film. Planar superconductor–normal-superconductor (SNS) junctions are examples where such films are desired since the YBCO exhibits the highest j_c along the (001) planes and the longest coherence length in the same planes [43–46]. The *a*-axis films are an attractive alternative and efforts have been made to find reproducible deposition techniques. The success has been limited and at present edge junctions constitute an alternative. However, the edge junctions require more processing steps still making the planar junctions interesting alternatives. *a*-Axis films also offer the possibility to characterize the anisotropy of transport properties and optical properties in YBCO thin films.

4.2.3.1 *a*-Axis Orientation by Lower Deposition Temperatures

The *a*-axis growth on (001) $SrTiO_3$ and (001) $LaAlO_3$ substrates is promoted by lowering the deposition temperature [45, 47]. The resulting films are epi-

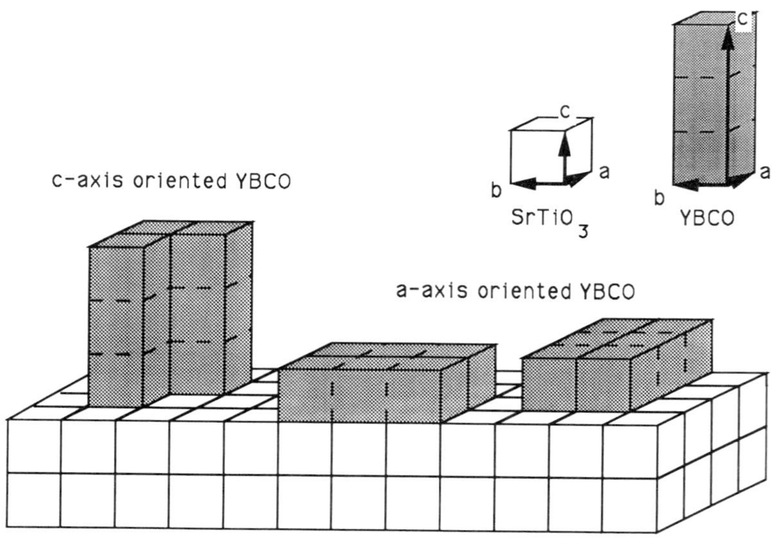

FIGURE 4.4. A schematic illustration of three observed epitaxial orientation relationships for YBCO films on (001)$SrTiO_3$ substrates. The *a*-axis oriented films are most often obtained at lower deposition temperatures. The symmetry of the (001)$SrTiO_3$ substrate interface induces a domain structure of the *a*-axis YBCO films.

taxial but the c axis in the plane of the film experiences the two orthogonal $[100]_{substrate}$ and $[010]_{substrate}$ directions to be identical. There are thus two possible epitaxial orientations and the films adopt a domain structure where the c axis rotates 90° when passing a domain boundary [47], see Figure 4.4. The domain size increases with the deposition temperature and sizes between 0.01 and 0.1 μm have been obtained.

It would be beneficial to avoid the domains in order to have one unique orientation of the c axis in the film. $NdGaO_3$ has been reported to be an alternative that would give a-axis films without domains [48]. The $NdGaO_3$ has a perovskite unit cell with $a = 5.431$ Å, $b = 5.499$ Å, and $c = 7.710$ Å (please note that the lattice parameters differ slightly between different sources and these are the numbers from Ref. [48]), see also Figure 4.3. The a-axis films were grown by lowering the temperature from that used for c-axis films. The substrate interface showed a twofold symmetry instead of fourfold due to the difference between a and b and this induced the alignment of the c axis in the plane of the film.

4.2.3.2 a-Axis Orientation by Use of $PrBa_2Cu_3O_{7-x}$ Template Layers

Lower deposition temperatures generally render lower T_c and j_c than for films deposited at temperatures where c-axis films are grown [49]. A different approach designed to overcome this disadvantage and also to improve the a-axis growth was to deposit a template layer of $PrBa_2Cu_3O_{7-x}$ (PBCO) on (001) $SrTiO_3$ at the lower temperature [50]. The PBCO has similar crystal structure to that of YBCO with slightly larger lattice parameters [51]. Previous experiments had shown that the PBCO had a more pronounced tendency of forming a-axis oriented films at lower temperatures than the YBCO. The PBCO layer established the a-axis orientation and provided a template layer for growth of YBCO. The a-axis orientation was also retained in the YBCO films grown at higher temperatures where the superconducting characteristics were comparable to those of high-quality c-axis oriented films. The a-axis films had, however, the same domain structure as the single a-axis YBCO films deposited at lower temperatures, see Figure 4.4.

4.2.4 *(110) $SrTiO_3$ Substrates*

The (110) $SrTiO_3$ substrates offers an alternative route to obtain the c axis in the plane of the film. The fourfold symmetry of the (001) $SrTiO_3$ substrate interface that gives rise to the domain structure of the a-axis oriented films is eliminated in (110) substrates, see Figure 4.5. There is still one [100] direction which can be used for obtaining the c axis in the plane of the film and the $[110]_{YBCO}$ aligns with the $[110]_{SrTiO_3}$.

Several attempts have been made to grow (110) films by lowering the deposition temperature in direct analogy with the a-axis films on $SrTiO_3$ [14, 52–54]. It was shown to be difficult to obtain pure (110) films reproducibly and instead a mixture of (110) and (103) usually grew on these substrates. The

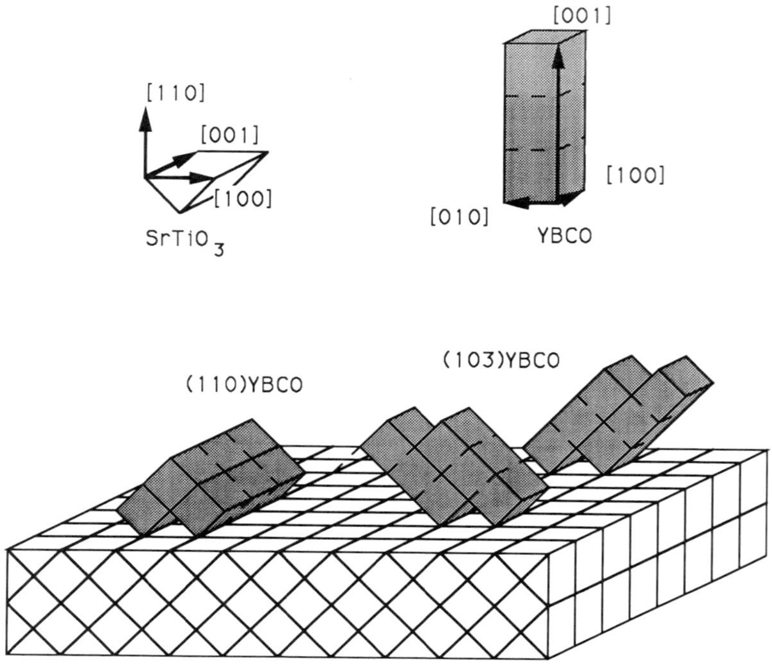

FIGURE 4.5. A schematic illustration of observed orientation relationships for YBCO on (110)$SrTiO_3$. The (103)YBCO films will obtain domain structures.

lower temperature would also induce degradation of the T_c and j_c. Reproducible results were instead obtained by using a PBCO template layer deposited at lower temperatures and subsequently depositing the YBCO at higher temperatures [14] in analogy with the *a*-axis oriented films [50]. The experiments illustrated that the orientation of the YBCO films can be manipulated using knowledge about the initial interactions between the substrate and the film. In particular, the use of template layers can improve the reproducibility of the film growth.

It is clear from Figure 4.5 that there are two possible (103) epitaxial orientations of the YBCO. This results in domain structures in the (103) films and has been observed using TEM. The use of vicinal (110) $SrTiO_3$ substrates has been illustrated to eliminate one of the (103) orientations in $La_{2-x}Sr_xCuO_4$ films [56]. The same principle can most probably be applied to other high-T_c superconductors although the specific mechanism introduced by vicinal surfaces may be different depending on the individual substrate. This is illustrated by the comparison of growth of *c*-axis oriented YBCO films on vicinal (001)MgO and (001)$SrTiO_3$ substrates. The $[001]_{YBCO}$ aligned with the surface normal and not with the $[001]_{MgO}$ crystallographic direction on (001)MgO [36]. For the corresponding (001) $SrTiO_3$ substrates the crystallographic directions of film and susbtrate aligned irrespective of the surface normal [37].

4.3 Mechanical Interactions

The films will experience mechanical interactions with the substrate that will introduce defects as the film thickness increases. The epitaxial film will be strained in the initial stages of the growth. Depending on the magnitude of the lattice mismatch and on the film thickness, part of the strain can be relaxed by the introduction of misfit dislocations [57–59], see Figure 4.6. The

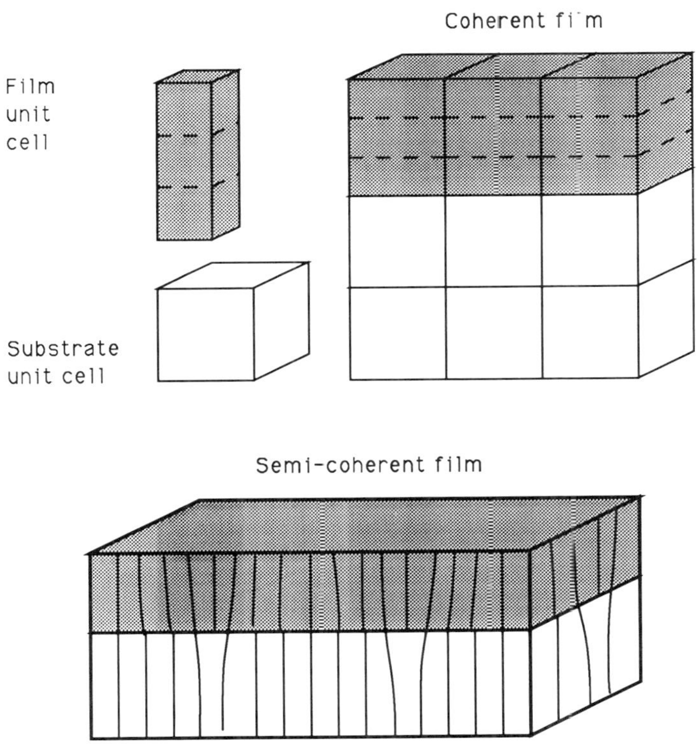

FIGURE 4.6. A schematic illustration of coherent and semicoherent film–substrate interfaces where it is assumed that the film thickness is much smaller than the thickness of the substrate. The lattice mismatch is accommodated by strain in the film at the coherent interface and the film unit cell attains the in-plane lattice parameters of the substrate. The strain energy increases with increasing film thickness and dislocations are introduced in films with thicknesses that exceed the critical film thickness resulting in a semicoherent interface.

dislocations can provide pinning centers for the magnetic flux lines and can thus increase the j_c [60, 61].

During cooling a difference in thermal expansion coefficient will introduce further strain in the film. This strain will be referred to as thermal strain in the following text. The strain can either be an accommodated lattice strain or more or less be relaxed by plastic flow through the introduction of dislocations. The introduction of twins is another mechanism that relieves strain [24]. If the difference in the thermal expansion coefficient is large, inducing a lattice mismatch that cannot be accommodated by either of these mechanisms, crack formation and crack propagation may occur [13, 14].

4.3.1 *Epitaxial Strain*

It will be assumed that the epitaxial film is very thin in comparison to the substrate. The lattice mismatch between the film and the substrate will then be accommodated by a uniform strain in the film and a slight bending of the substrate. The elastic strain is

$$\varepsilon = \Delta a/a \approx (a_s - a_f)/a_s, \tag{1}$$

where a_f and a_s are the lattice parameters of the film. The strain energy E increases with increasing film thickness according to

$$E = Mh\varepsilon^2, \tag{2}$$

where M is the biaxial elastic modulus of the film and h is the film thickness. The strain can be relaxed by the introduction of plastic flow through dislocation nucleation and motion. The relaxation cannot, however, occur below a critical thickness, h_c, since the energy associated with the created dislocations is greater than that relieved by the lattice relaxation. According to the equilibrium theory for formation of misfit dislocation the dislocation energy per unit area E_{dis} is approximately

$$E_{\text{dis}} = \mu b^2 \ln(\beta h/b)/(4\pi[1 - v]S), \tag{3}$$

where μ is the shear modulus of the film and substrate and is considered to be the same for both. v is Poisson's ratio and β is a numerical constant. The factor $2/S$ represents the misfit dislocation length per unit film area. The total energy E of thin films containing misfit dislocations is given by adding the two terms in Eqs. (2) and (3):

$$E = Mh(\varepsilon - b/S)^2 + \mu b^2 \ln(\beta h/b)/(4\pi[1 - v]S). \tag{4}$$

The energy has a minimum in the equilibrium state and the critical thickness can be determined by finding the minimum energy for $b/S = 0$. The corresponding expression is

$$h_c/\ln(\beta h_c/b) = \mu b/(4\pi[1 - v]M\varepsilon), \tag{5}$$

and below h_c the coherently strained film is thermodynamically stable. Com-

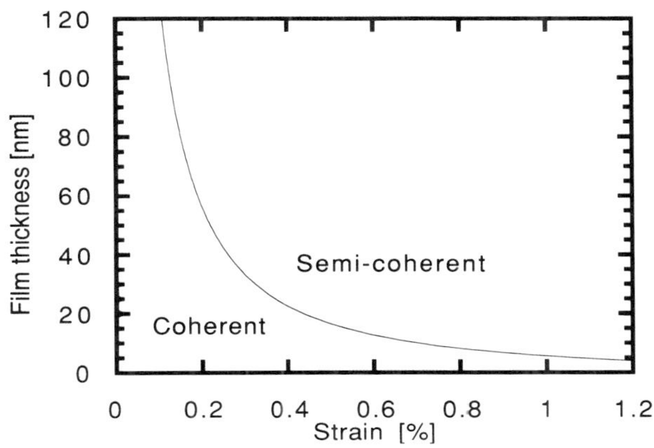

FIGURE 4.7. Critical film thickness for (001)YBCO films as a function of strain (lattice mismatch) according to Eq. (6). The Burgers vector is 3.89 Å at 700°C, M is 248 GPa, v is 0.281, and μ is 70 GPa [64].

parison of the predictions of Eq. (5) with experimental observations show that the predicted critical thicknesses are always smaller. The discrepancy is due to the kinetics of nucleation, motion, and multiplication of the dislocations. The model was subsequently modified in order to better account for the formation of misfit dislocation by considering the energy associated with an incremental extension of the misfit dislocation instead of the overall energy change associated with a periodic array of misfit dislocation [62]. The resulting expression for pure edge dislocations would then be

$$h_c/\ln(\beta h_c/b) = 2b\mu/(4\pi[1 - v]\sigma) = 2b\mu/(4\pi[1 - v]M\varepsilon), \tag{6}$$

where b is 0.701 and assuming that μ is the same for both film and substrate. The critical thickness as a function of misfit e can thus be obtained from Eq. (6). The misfit in epitaxial c-axis oriented YBCO films will be accommodated by dislocations with the Burgers vectors $[100]_{YBCO}$ and $[010]_{YBCO}$ [63]. The relevant Burgers vector to be used in Eq. (6) is then 3.89 Å for YBCO at 700°. The shear modulus μ is 70 GPa, Possion's ratio v is 0.281, and the biaxial elastic modulus M is 248 GPa for a (001) YBCO film [64]. The critical thickness as a function of lattice misfit ε is shown in Figure 4.7. It can be noted that the lattice mismatch between c-axis oriented YBCO and (001) $SrTiO_3$ is about 0.010 which gives a critical thickness of about 60 Å. This is an indication that the strain is relaxing for very thin films.

4.3.2 *Thermal Strain*

As the YBCO contracts during cooling, any difference in thermal expansion coefficients between the film and the substrate will introduce strain in the film. The magnitude of the resulting strain after cooling depends on the residual strain at the deposition temperature and the changes in lattice parameters of the film and substrate during cooling. The strain can be relaxed by misfit dislocations and also by twinning which most often is observed in the films. If these mechanisms cannot accommodate the strain, cracking may ensue and this has indeed been observed in (001) films on (001) Si substrates [13] and in (110) films on (110) $SrTiO_3$ substrates [14]. The cracks propagate along specific crystallographic planes of the YBCO where the planes are the (100) and the (010) for the (001) Si and the (001) planes for the (110) $SrTiO_3$.

Dislocations and stacking faults completely relieved the strain in the (110) films on the (110) $SrTiO_3$ substrates at the deposition temperature [14]. The dislocations accommodated the strain along the [110] direction and since the lattice mismatch along this direction did not vary with temperature the dislocation density thus remained constant during cooling. The cracks formed during cooling due to the thermal expansion mismatch along the c axis in the plane of the film. According to an analytical solution of the crack spacing λ as a function of film thickness h, the relation is

$$\lambda \approx 5.63\sqrt{hK_f/M\varepsilon}, \tag{7}$$

where K_f is the mode-I fracture toughness along the cleavage plane of the film [65]. The analysis provided a critical film thickness, H_c, which is given by

$$H_c = 0.50(K_f/M\varepsilon)^2. \tag{8}$$

No cracks could propagate thermodynamically below H_c. The critical thickness for crack propagation as a function of strain in a (110) film is shown in Figure 4.8. The fracture toughness is 0.8 $MPa^{1/2}$ [65] and M is 102 GPa for the (110) YBCO films.

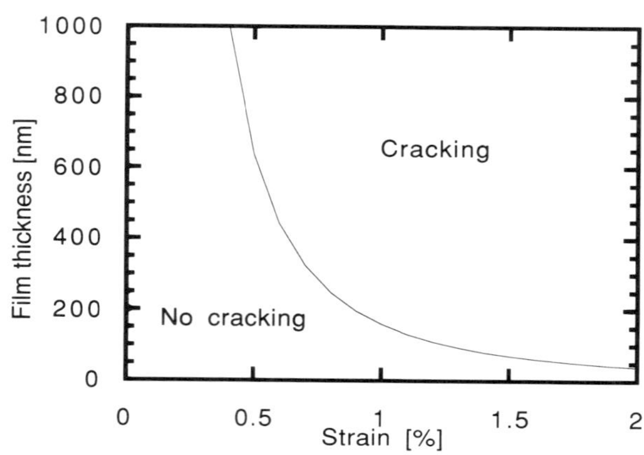

FIGURE 4.8. Critical film thickness for crack propagation as a function of strain for a (110)YBCO film according to Eq. (8). The fracture toughness K_f is 0.8 $MPa^{1/2}$ [65] and M is 102 GPa for this orientation of the film.

4.4 Chemical Reactions and Interdiffusion

The film depositions are made at elevated temperatures which promote possible chemical reaction and interdiffusion between the film and the substrate. It is not unusual that the first layer of multilayer structures that is in contact with the substrate is exposed to temperatures in the range of 600°C to 750°C for as long as 5 h. The electrical transport properties of YBCO are sensitive to the presence of other elements and it is therefore necessary to prevent interdiffusion of detrimental elements between the substrate and the film. Reactions that consume considerable amounts of the YBCO leaving non-superconducting products behind should also be avoided. These interactions can be divided into three main types: (1) substitutions of the substrate atoms in the superconductor lattice; (2) secondary phase formation where the formed phase acts as a passivation layer preventing further interaction; and (3) secondary phase formation without formation of a passivation layer. Limited reactions and some degree of interdiffusion can be tolerated as evidenced by films grown on, for example, YSZ, $PrGaO_3$, and $NdGaO_3$ substrates which will be addressed in further detail below.

Figure 4.9 shows the degradation in ac susceptibility (χ') of four YBCO films after annealing at 750°C for 8 h in flowing oxygen [66]. The films were of different thickness (100 to 350 nm) and their initial stoichiometry was not exactly identical. It is therefore not possible to quantify the extent of chemical interaction by the change in the ac susceptibility values. However, the results can be used to classify the substrate interactions into the three categories mentioned above. The gallate and the aluminate showed a continuous reduction in χ' with increasing annealing times indicating a type III interaction.

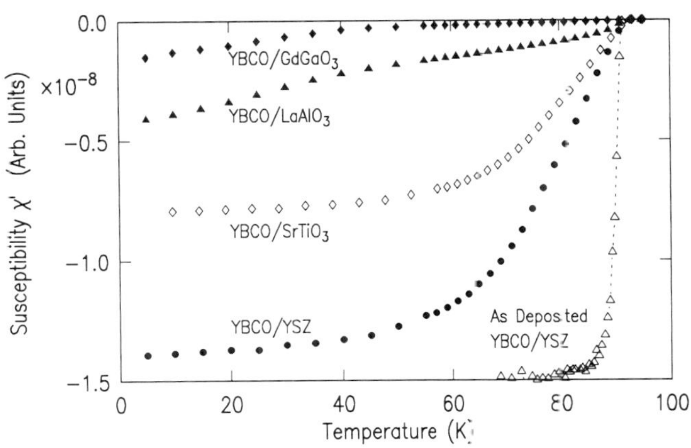

FIGURE 4.9. Degradation of ac susceptibility (χ') after annealing for 8 h at 750°C for YBCO thin films on various substrates. The measurements were carried out at 250 Hz, 6 Oe(rms) with the magnetic field parallel to the plane of the film.

The reduction in χ' was gradual in the case of $SrTiO_3$ and the change may have been related to substitution of Sr and Ti into the YBCO lattice. The results from the YSZ substrate showed clear evidence for the formation of a barrier layer. There was an initial reduction in superconducting volume fraction of the film but no further reduction was observed after prolonged annealing at 750°C. TEM investigations showed the presence of a $BaZrO_3$ layer at the interface between the YBCO and the YSZ suggesting that the intermediate $BaZrO_3$ layer may act as a barrier.

4.4.1 *Substrates Without Buffer Layers*

$SrTiO_3$, CeO_2, Y_2O_3 [67], and MgO are examples of substrate materials where the interdiffusion and chemical reactions are either absent or do not noticeably influence the YBCO at the relevant deposition temperatures. Buffer layers are therefore not necessary for preventing these types of interactions. There may still be a need for buffer layers, for example homoepitaxial layers, that eliminate defects or alter the surface structure in order to improve the epitaxy [39].

YSZ substrates can also be used without intermediate buffer layers and YSZ is in fact used as a buffer layer material in other systems where a diffusion barrier is required [e.g., Refs. 14, 16–19, 21]. It is therefore surprising to find that a reaction layer is often observed between the YSZ and YBCO [39, 40, 42, 68]. Transmission electron microscopy (TEM) investigations indicate that the YBCO initially nucleates on the YSZ substrate and establishes an epitaxial orientation relationship. Subsequent reactions between the YBCO and the YSZ result in a polycrystalline $BaZrO_3$ layer which exhibit a preferred orientation relationship with the YBCO and consequently also with the YSZ. CuO, $Y_2Cu_2O_5$, and Y_2BaCuO_5 are other byproducts of the reaction [69]. The $BaZrO_3$ layers are usually about 3–8 nm thick for depositions of 300-nm YBCO films at temperatures between 600 to 800°C where the layer thickness increases with increasing temperature [42]. The reaction does not have a negative influence on the film properties for films which are of the order of 200 to 300 nm thick. Care must be taken when depositing extremely thin films since the reaction either depletes the YBCO of Ba or results in the formation of secondary phases. Both alternatives influence the superconducting properties. It should also be noted that in some cases the YSZ does not react with the YBCO [17]. The reason for the difference in behavior is not clear.

The $LaAlO_3$, $NdGaO_3$, and the $PrGaO_3$ are other materials that can be used as substrates without buffer layers even though both Al and Ga have been shown to deteriorate the T_c and j_c of YBCO. Evidence for interdiffusion between the substrates and the films has been presented but provided that proper deposition conditions are chosen the interdiffusion can be kept at a low level [28, 31, 32, 70]. The presence of grain boundaries provide diffusion paths where the diffusion rate is enhanced. The effect of temperature on the interaction between $PrGaO_3$ and YBCO was studied as a function of temperature and it was found that the reaction layer was about 10 nm at 760°C [70]. At 785°C the interaction was severe and a 230-nm YBCO layer was destroyed due to the interaction between the layers.

4.4.2 *Substrates that Require Buffer Layers*

Si reacts with YBCO even at temperatures as low as 550°C [71] and also has a large mismatch in thermal expansion coefficient, see Figure 4.3(b). Despite these significant disadvantages the interest in Si as a substrate is intense due to the advanced process techniques that are already available for Si and also due to the fact that much of today's semiconducting technology is based on Si substrates. The interfacial diffusion and reactions result in the formation of metal-silica compounds and severely degrade YBCO films The high reactivity of the Si substrates makes it necessary to separate the YBCO films from these substrates using a buffer layer. The buffer layer should grow epitaxially with one single orientation relationship on the substrate in order to allow subsequent epitaxial growth of the YBCO film. Grain boundaries should preferably be avoided in the buffer layer since they provide rapid diffusion paths that reduce the effectiveness of the diffusion barrier. The layer should also have good matching of the thermal expansion coefficient. YSZ buffer layers produced under proper conditions have been shown to fulfill these requirements better than other investigated candidate materials [13]. It has also been found that the surface smoothness of the buffer layer plays an important role for the subsequent deposition of the YBCO [72]. Another advantage of the YSZ is that the thermal strain in YBCO that arises due to the mismatch in thermal expansion between YBCO and Si is reduced by the use of YSZ buffer layers. The critical thickness above which severe cracking of the film ensues is therefore increased. It should be noted that the Si surface most often has a native silicon oxide that needs to be removed before depositing any subsequent epitaxial layer in order to obtain epitaxy between the Si and the buffer layer. Even when this native oxide is removed amorphous silicon oxide is most often observed at the interface between Si and YSZ [13, 72]. This oxide has been suggested to form after epitaxy has been established through oxygen diffusion from the YSZ into Si. The oxide may be beneficial for relieving the thermal stresses that occur during cooling.

$CoSi_2$ [73], MgO [74], CeO_2 [75], RuO_2 [76], indium–tin–oxide [77], CaF_2, BaF_2, MgF_2 and SrF_2 [78] have also been tried as buffer layers on Si with limited success. There were difficulties in obtaining epitaxial YBCO films with a well-defined epitaxial relationship and in addition the T_c and j_c were usually low in comparison to the results from YSZ buffer layers. CaF_2 did in fact react with the YBCO forming an insulating film. The interest in RuO_2 and $CoSi_2$ arose since they can provide conducting layers contacting the YBCO.

The growth of YBCO films on GaAs introduces one more complication as compared to Si substrates since the GaAs surface needs to be encapsulated by a buffer layer at lower temperatures ($<500°C$) preventing the Ga from agglomerating into liquid droplets. The droplets cause thermal etching of the substrate interface and give rise to pronounced surface roughness. The surface roughness subsequently results in poor quality of the YBCO. The need to deposit the buffer layers at low temperatures has hampered the development of efficient buffer layers. Indium–tin–oxide [79] was used as buffer layers but did not result in acceptable film quality. Better results were obtained with double-buffer layers of Si_3N_4 and YSZ [80]. The films were c-axis oriented but with no alignment in the plane of the film. Further improvement was instead obtained by depositing epitaxial MgO [81, 82].

Sapphire is another technologically important substrate due to its low cost, high mechanical strength, low dielectric constant, low tangent loss, and high thermal conductivity. The disadvantage with Al_2O_3 is that interdiffusion reduces the quality of the YBCO films. YSZ can provide sufficient buffer layers on silicon-on-sapphire [83] as well as on Al_2O_3 substrates [19]. Experiments with amorphous YSZ buffer layers on Al_2O_3 also showed the role of the thickness of the buffer layer by a variation of T_c with the buffer layer thickness. The YBCO reacted with the amorphous YSZ forming $BaZrO_3$. The resulting $BaZrO_3$ layer was not continuous for YSZ layer thicknesses below 6 nm and Al was thus allowed to diffuse directly into the YBCO lowering the T_c. Above 6 nm the $BaZrO_3$ layer was continuous and the T_c gradually increased and reached a maximum value for layer thicknesses of 14 nm. The surprising observation was that the T_c decreased again for even thicker layers. TEM showed that for all thicknesses below 14 nm the YBCO films were c-axis oriented. This c-axis orientation was disrupted as the buffer layer thickness increased and the T_c decreased. It was mentioned above that as YBCO reacted with YSZ CuO, $Y_2Cu_2O_5$, and possibly Y_2BaCuO_5 would form in addition to $BaZrO_3$ [69]. Particles of $Y_2Cu_2O_5$ were indeed observed and there were also amorphous phases present in the YBCO grain boundaries. It thus appears as though the degraded properties were associated with secondary phases present at the YBCO grain boundaries and also the disruption of the c-axis orientation. It should be noted that the amorphous phases could be expected to have a higher reactivity than crystalline YSZ.

4.5 Concluding Remarks

The initial stages of in situ growth of YBCO thin films determine the orientation with respect to the substrate. This has been used as a tool to introduce 45° bi-epitaxial grain-boundary junctions at predetermined locations to be used for superconducting quantum interference devices (SQUIDs) [10–12]. The 45° boundary was obtained by knowledge of the different orientation relationships induced by the lattice match, see Figure 4.3, and using thin seed layers to alter the orientation in restricted areas. The YBCO thus nucleated in different orientations and a 45° boundary formed at the interface between areas with and without seed layer [12]. It was found that the performance of the grain boundary depended upon the residual stresses in the film [11]. This illustrates the need for detailed knowledge of the initial stage interactions as well as of the subsequent mechanical interactions.

The development of the bi-epitaxial grain-boundary junctions also show that buffer layers can have multiple functions. They can act as diffusion barriers preventing detrimental reactions and interdiffusion to occur but they can also provide seed layer altering the orientation relationships. A third effect is that the reproducibility of the film deposition may improve by the use of buffer layers since they may beneficially alter the interface structure.

Finally, future applications and studies of the fundamental properties of these high-T_c superconductors will require that they are integrated in multilayer structures together with metals, semiconductors, and insulators. Reproducibility and the understanding of the behavior of the multilayers require control of the interfacial interactions between the different layers. It is also essential to obtain smooth films for the ultrathin multilayers. It is yet not clear what factors are instrumental in determining the surface smoothness. The presence of unit-cell steps and screw dislocations has been shown using TEM and scanning tunneling microscopy [24, 84–86] with substantial surface roughness of films on both MgO and $SrTiO_3$. There are indications that the details of the deposition technique will decide whether the growth will be two-dimensional (layer-by-layer) or whether an island mechanism ensues immediately.

Acknowledgments. Eva Olsson would like to thank J.A. Alarco, Yu. Boikov, G. Brorsson, P. Chaudhari, T. Claeson, R. Gross, A. Gupta, Z.G. Ivanov, P.Å. Nilsson, J. Ramos, J.M. Rowell, E.A. Stepantsov, M.D. Thouless, A.Ya. Tzalenchuk, S.Z. Wang, and D. Winkler for stimulating discussions on related issues. Financial support from the Swedish National Board for Industrial and Technical Development (NUTEK) is gratefully acknowledged.

References

1. P. Chaudhari, J. Mannhart, D. Dimos, C.C. Tsuei, J. Chi, M.M. Oprysko, and M. Scheuermann, Phys. Rev. Lett. **60**, 1653 (1988); D. Dimos, P. Chaudhari, J. Mannhart, and F.K. Legoues, Phys. Rev. Lett. **61**, 219 (1988); Rev. Lett. **61**, 2476 (1988); J. Mannhart, R. Gross, K. Hipler, R.P. Huebener, C.C. Tsuei, D. Dimos, and P. Chaudhari, Science **245**, 839 (1989); R. Gross, P. Chaudhari, M. Kawasaki, and A. Gupta, Phys. Rev. B **42**, 10, 735 (1990).
2. S.E. Russek, D.K. Lathrop, B.H. Moeckley, R.A. Buhrman, D.H. Shin, and J. Silcox, Appl. Phys. Lett. **57**, 1155 (1990); D.K. Lathrop, B.H. Moeckley, S.E. Russek, and R.A. Buhrman, Appl. Phys. Lett. **58**, 1096 (1991).

3. Z.G. Ivanov, P.Å. Nilsson, D. Winkler, J.A. Alarco, T. Claeson, E.A. Stepantsov, and A.Ya. Tzalenchuk, Appl. Phys. Lett. **59**, 3030 (1991).
4. R. Gross, P. Chaudhari, M. Kawasaki, M.B. Ketchen, and A. Gupta, Appl. Phys. Lett. **57**, 727 (1990).
5. G. Friedl, B. Roas, M. Römheld, L. Schultz, and W. Jutzi, Appl. Phys. Lett. **59**, 2751 (1991).
6. Z.G. Ivanov, J.A. Alarco, T. Claeson, P.Å. Nilsson, E. Olsson, H.K. Olsson, E.A. Stepantsov, A.Ya. Tzalenchuk, and D. Winkler, Beijing International Conference on High-T_c Superconductivity (BHTSC'92) May 25–29, 1992 Beijing, China.
7. J. Ramos, Z.G. Ivanov, E. Olsson, and T. Claeson, Beijing International Conference on High-T_c Superconductivity (BHTSC'92) May 25–29, 1992 Beijing, China.
8. J.A. Edwards, J.S. Satchell, N.G. Chew, R.G. Heumphreys, M.N. Keene, and O.D. Dosser, Appl. Phys. Lett. **60**, 2433 (1992).
9. C.L. Jia, B. Kabius, K. Urban, K. Herrman, G.J. Cui, J. Schubert, W. Zander, A.I. Braginski, and C. Heiden, Physica C **175**, 545 (1991); C.L. Jia, B. Kabius, K. Urban, K. Herrman, J. Schubert, W. Zander, and A.I. Braginski, Physica C **196**, 211 (1992).
10. K. Char, M.S. Colclough, S.M. Garrison, N. Newman, and G. Zaharchuk, Appl. Phys. Lett. **59**, 733 (1991).
11. K. Char, M.S. Colclough, L.P. Lee, and G. Zaharchuk, Appl. Phys. Lett. **59**, 2177 (1991).
12. S.J. Rosner, K. Char, and G. Zaharchuk, Appl. Phys. Lett. **60**, 1010 (1992).
13. D.K. Fork, D.B. Fenner, R.W. Barton, J.M. Phillips, G.A.N. Connell, J.B. Boyce, and T.H. Geballe, Appl. Phys. Lett. **57**, 1161 (1990).
14. E. Olsson, A. Gupta, M.D. Thouless, A. Segmüller, and D.R. Clarke, Appl. Phys. Lett. **58**, 1682 (1991).
15. T.K. Worthington, W.J. Gallagher, and T.R. Dinger, Phys. Rev. Lett. **59**, 1160 (1987).
16. E. Narumi, L.W. Song, F. Yang, S. Patel, Y.H. Kao, and D.T. Shaw, Appl. Phys. Lett. **56**, 2684 (1990).
17. R.P. Reade, X.L. Mao, and R.E. Russo, Appl. Phys. Lett. **59**, 739 (1991).
18. Y. Iijima, N. Tanabe, O. Kohno, and Y. Ikeno, Appl. Phys. Lett. **60**, 769 (1992).
19. Yu. Boikov, Z.G. Ivanov, E. Olsson, J.A. Alarco, G. Brorsson, and T. Claeson, J. Appl. Phys. **72**, 199 (1992).
20. D.P. Norton, D.H. Lowndes, J.D. Budai, D.K. Christen, E.C. Jones, K.W. Lay, and J.E. Tkaczyk, Appl. Phys. Lett. **57**, 1164 (1990).
21. K.S. Harshavardhan, R. Ramesh, T.S. Rvi, S. Sampere, A. Inam, C.C. Chang, G. Hull, M. Rajeswari, T. Sands, T. Venkatesan, M. Reeves, J.E. Tkaczyk, and K.W. Lay, Appl. Phys. Lett. **59**, 1638 (1991).
22. C.W. Nih, L. Anthony, J.Y. Josefowicz, and F.G. Krajenbrink, Appl. Phys. Lett. **56**, 2138 (1990).
23. K. Kamigaki, H. Terauchi, T. Terashima, Y. Bando, K. Iijima, K. Yamamoto, K. Hirata, H. Hayashi, I. Nakagawa, and Y. Tomii, J. Appl. Phys. **69**, 3653 (1991).
24. S.K. Streiffer, B.M. Lairson, C.B. Eom, B.M. Clemens, C.J. Bravman, and T.H. Geballe, Phys. Rev. B **43**, 13,007 (1991).
25. Thermophysical Properties of Matter, edited by Y.S. Touloukian and C.Y. Ho, The TRCP Data Series (IFI/Plenum, New York, 1975), Vol. 13.
26. E.D. Specht, C.J. Sparks, A.G. Dhere, J. Brynestad, O.B. Cavin, D.M. Kroeger, and H.A. Oye, Phys. Rev. B **37**, 7426 (1988).

27. E.A. Giess, R.L. Sandstrom, W.J. Gallagher, A. Gupta, S.L. Shinde, R.F. Cook, E.I. Cooper, E.J.M. O'Sullivan, J.M. Roldan, A. Segmüller, and J.Angilello, IBM J. Res. Devel. **34**, 1946 (1990).
28. M. Sasaura, M. Mukaida, and S. Miyazawa, Appl. Phys. Lett. **57**, 2728 (1990).
29. H.M. O'Bryan, P.K. Gallagher, G.W. Berkstresser, and C.D. Brandle, J. Mater. Res. **5**, 183 (1990).
30. A.H. Carim, S.N. Basu, and R.E. Muenchausen, Appl. Phys. Lett. **58**, 871 (1991).
31. R.L. Sandstrom, E.A. Giess, W.A. Gallagher, A. Segmüller, E.I. Cooper, M.F. Chisholm, A. Gupta, S. Shinde, and R.B. Laibowitz, Appl. Phys. Lett. **53**, 1874 (1988).
32. G. Koren, A. Gupta, E.A. Giess, A. Segmüller, and R.B. Laibowitz, Appl. Phys. Lett. **54**, 1054 (1989).
33. M. Sasaura, S. Miyazawa, and M. Mukaida, J. Appl. Phys. **68**, 3643 (1990).
34. B.H. Moeckley, S.E. Russek, D.K. Lathrop, R.A. Buhrman, J. Li, and J.W. Mayer, Appl. Phys. Lett. **57**, 1687 (1990).
35. D.M. Hwang, T.S. Ravi, R.Ramesh, Siu-Wai Chan, C.Y. Chen, L. Nazar, X.D. Wu, A. Inam, and T. Venkatesan, Appl. Phys. Lett. **57**, 1690 (1990).
36. S.K. Streiffer, B.M. Lairson, and J.C. Bravman, Appl. Phys. Lett. **57**, 2501 (1990).
37. T.E. Mitchell, S. Basu, N.A. Nastasi, and T. Roy, Mater. Res. Soc. Symp. Proc. **183**, 357 (1990).
38. M.G. Norton, S.R. Summerfelt, and C.B. Carter, Appl. Phys. Lett. **56**, 2246 (1990).
39. D.K. Fork, S.M. Garrison, M. Hawley, and T.H. Geballe, J. Mater. Res. to be published July 1992.
40. L.A. Tietz, C.B. Carter, D.K. Lathrop, S.E. Russek, R.A. Buhrman, and J.R. Michael, J. Mater. Res. **4**, 1072 (1989).
41. S.M. Garrison, N. Newman, B.F. Cole, K. Char, and R.W. Barton, Appl. Phys. Lett. **58**, 2168 (1991).
42. J.A. Alarco, G. Brorsson, Z.G. Ivanov, P.Å. Nilsson, E. Olsson, and M. Löfgren, Appl. Phys. Lett. **61**, 723 (1992).
43. R.B. Laibowitz, R.H. Koch, A. Gupta, G. Koren, W.J. Gallagher, V. Foglietti, B. Oh, and J.M. Viggiano, Appl. Phys. Lett. **56**, 686 (1990).
44. A. Inam, C.T. Rogers, R. Ramesh, K. Remschnig, L. Farrow, D. Hart, T. Venkatesan, and B. Wilkens, Appl. Phys. Lett. **57**, 2484 (1990).
45. T. Hashimoto, M. Sagoi, Y. Mizutani, J. Yoshida, and K. Mizushima, Appl. Phys. Lett. **60**, 1756 (1992).
46. J.B. Barner, C.T. Rogers, A. Inam, R. Ramesh, and S. Bersey, Appl. Phys. Lett. **59**, 742 (1991).
47. C.B. Eom, A.F. Marshall, S.S. Laderman, R.D. Jacowitz, and T.H. Geballe, Science **249**, 1549 (1990).
48. K.H. Young and J.S. Sun, Appl. Phys. Lett. **59**, 2448 (1991).
49. J. Zhao, Y.Q. Li, C.S. Chern, P. Lu, P. Norris, B. Gallois, B. Kear, F. Cosandey, X.D. Wu, R.E. Muenchausen, and S.M. Garrison, Appl Phys. Lett. **59**, 1254 (1991).
50. R. Ramesh, A. Inam, D.L. Hart, and C.T. Rogers, Physica C **170**, 325 (1990).
51. M.E. López-Morales, D. Rios-Jara, J. Tagüena, R. Escudero, S. La Placa, A. Bezinge, V.Y. Lee, E.M. Engler, and P.M. Grant, Phys. Rev. B **41**, 6655 (1990).
52. Y. Enomoto, T. Murakami, M. Suzuki, and K. Moriwaki, Jpn. J. Appl. Phys. **26**, L1248 (1987).

53. T. Terashima, Y. Bando, K. Iijima, K. Yamamoto, and K. Hirata, Appl. Phys. Lett. **53**, 2232 (1988).
54. G. Linker, X.X. Xi, O. Meyer, Q. Li, and J. Geerk, Solid State Commun. **69**, 249 (1989).
55. A. Inam, C.T. Rogers, R. Ramesh, K. Remschnig, L. Farrow, D. Hart, T. Venkatesan, and B. Wilkens, Appl. Phys. Lett. **57**, 2484 (1990).
56. J. Kwo, R.M. Fleming, H.L. Kao, D.J. Werder, and C.H. Chen, Appl. Phys. Lett. **60**, 1905 (1992).
57. J.H. van der Merwe, J. Appl. Phys. **34**, 123 (1963).
58. J.W. Matthews and A.E. Blakeslee, J. Cryst. Growth **29**, 118 (1974); **27**, 273 (1975).
59. W.D. Nix, Metall. Trans. A **20**, 2217 (1989).
60. A. Gupta, R. Gross, E. Olsson, A. Segmüller, G. Koren, and C.C. Tsuei, Phys. Rev. Lett. **64**, 3191 (1990).
61. R. Gross, A. Gupta, E. Olsson, A. Segmüller, and G. Koren, Appl. Phys. Lett. **57**, 203 (1990).
62. L.B. Freund, J. Appl. Mech. **54**, 553 (1987).
63. A. Catana, R.F. Broom, J.G. Bednorz, J. Mannhart, and D.G. Schlom, Appl. Phys. Lett. **60**, 1016 (1992).
64. H. Ledbetter and M. Lei, J. Mater. Res. **6**, 2253 (1991).
65. M.D. Thouless, E. Olsson, and A. Gupta, Acta Metall. Mater. **40**, 1287 (1992).
66. S.L. Shindé et al., to be published.
67. A. Oishi, H. Teshima, K. Ohata, H. Izumi, S. Kawamoto, T. Morishita, and S. Tanaka, Appl. Phys. Lett. **59**, 1902 (1991).
68. D.M. Hwang, Q.Y. Ying, and H.S. Kwok, Appl. Phys. Lett. **58**, 2429 (1991).
69. M.J. Cima, J.S. Schneider, S.C. Peterson, and W. Coblenz, Appl. Phys. Lett. **53**, 710 (1988).
70. G. Brorsson, P.Å. Nilsson, E. Olsson, S.Z. Wang, T. Claeson, and M. Löfgren, Appl. Phys. Lett. **61**, 486 (1992).
71. D.B. Fenner, A.M. Viano, D.K. Fork, G.A.N. Connell, J.B. Boyce, F.A. Ponce, and J.C. Tramontana, J. Appl. Phys. **69**, 2176 (1991).
72. A. Lubig, Ch. Buchal, J. Schubert, C. Copetti, D. Guggi, C.L. Jia, and B. Stritzker, J. Appl. Phys. **71**, 5560 (1992).
73. A. Kumar and J. Narayan, Appl. Phys. Lett. **59**, 1785 (1991).
74. D.K. Fork, F.A. Ponce, J.C. Tramontana, and T.H. Geballe, Appl. Phys. Lett. **58**, 2294 (1991).
75. T. Inoue, Y. Yamamoto, S. Koyama, S. Suzuki, and Y. Ueda, Appl. Phys. Lett. **56**, 1332 (1990).
76. Q.X. Jia and W.A. Andersoon, Appl. Phys. Lett. **57**, 304 (1990).
77. B.J. Kellett, J.H. James, A. Gauzzi, B. Dwir, D. Pavuna, and F.K. Reinhart, Appl. Phys. Lett. **57**, 1146 (1990).
78. S.-W. Chan, E.W. Chase, B.J. Wilkens, and D.L. Hart, Appl. Phys. Lett. **54**, 2032 (1989).
79. B. Kellet, A. Gauzzi, J.H. James, B. Dwir, D. Pavuna, and F.K. Reinhart, Appl. Phys. Lett. **57**, 2588 (1990).
80. Q.X. Jia, S.Y. Lee, W.A. Anderson, and D.T. Shaw, Appl. Phys. Lett. **69**, 1120 (1991).
81. D.K. Fork, K. Nashimoto, and T.H. Geballe, Appl. Phys. Lett. **60**, 1621 (1992).
82. L.D. Chang, M.Z. Tseng, and E.L. Hu, Appl. Phys. Lett. **60**, 1753 (1992).

83. D.K. Fork, F.A. Ponce, J.C. Tramontana, N. Newman, J.M. Phillips, and T.H. Geballe, Appl. Phys. Lett. **58**, 2432 (1991).
84. M. Hawley, I.D. Raistrick, J.G. Beery, and R.J. Houlton, Science **251**, 1587 (1991).
85. H.U. Krebs, Ch. Krauns, X. Yang, and U. Geyer, Appl. Phys. Lett. **59**, 2180 (1991).
86. D.G. Schlom, D. Anselmetti, J.G. Bednorz, R.F. Broom, A. Catana, T. Frey, Ch. Gerber, H.-J. Güntherrodt, H.P. Lang, and J. Mannhart, Z. Phys. B **86**, 163 (1992).

5
The Twin-Boundary Structure in $YBa_2Cu_3O_{7-\delta}$ Superconductors

YIMEI ZHU and MASAKI SUENAGA

5.1 Introduction

Since the discovery of high-T_c superconductivity in $YBa_2Cu_3O_{7-\delta}$ the relationship between the twin boundaries and the oxide's superconducting properties, such as critical temperature T_c, and the critical current density J_c has been explored extensively. For example, Fang et al. [1] discussed the possibility that the boundary region is an area of higher T_c than the matrix. However, Deutscher and Müller [2] proposed that the boundary layer is weakly superconducting, and that the magnetic flux lines flow freely along the boundary. On the other hand, Chaudhari [3] and Kes [4] argued that the twin boundaries act as strong pinning sites for the flux lines. Evidence for preferential pinning of the vortex lattice along the twin boundaries was observed in experiments on flux lattice decoration in $YBa_2Cu_3O_{7-\delta}$ crystals [5, 6]. More recently, magnetic and transport measurements have shown that the twin boundaries can pin flux lines if the magnetic field is applied accurately parallel to the boundary planes [7–9]. Otherwise, the boundaries do not act as effective pinning centers. Furthermore, the effectiveness of boundary pinning is observable only in cases where no other strong pinning centers are present. If other strong pinning centers coexist with the boundaries, as in thin films, the boundaries will add very little to the pinning strength of that material [10]. Thus, now it is believed that the twin boundaries do not play a primary role in determining the critical current densities in $YBa_2Cu_3O_7$. However, the boundaries can be used as a model system to study the nature of the interaction between the flux lines and a planar defect in $YBa_2Cu_3O_{7-\delta}$, and can provide useful insight in flux pinning mechanisms. In addition, Cu and O play an important part in the superconductivity of the $YBa_2Cu_3O_{7-\delta}$ system. Replacement of Cu by two valent (Zn, Ni) and three valent (Fe, Al) transition elements was shown to change the structures of the boundaries and may affect the flux pinning mechanisms. Thus, detailed inves-

Materials Science Division, Brookhaven National Laboratory, Upton, NY 11973.

tigations of the structure of the twin boundaries, including the effects of substitution on the boundary structure, are of interest and have been extensively studied.

In characterizing the microstructure of the $YBa_2Cu_3O_{7-\delta}$ system, conventional transmission electron microscopy (TEM) [11–16] and high-resolution electron microscopy (HREM) [17–22] have been widely used to study the twinning and twin-boundary structure although systematic studies as a function of the stoichiometry of samples are not plentiful. Using various electron microscopy techniques, we have systematically studied the structure of twin boundary in carefully prepared $YBa_2Cu_3O_{7-\delta}$ ($\delta \approx 0.0$–0.4) and $YBa_2(Cu_{0.98}M_{0.02})_3O_{7-\delta}$ (M = Fe, Al, Ni, or Zn) samples. Here, we review our studies on the twin-boundary structure, based on TEM and HREM as well as electron diffraction observations [23–26]. Our experimental procedures are described briefly in Sec. 5.2. In Sec. 5.3, we discuss the gross features of the twin boundary of $YBa_2Cu_3O_{7-\delta}$ and cation substitution effects using two-beam, selected-area diffraction (SAD) and HREM techniques. In Sec. 5.4, we focus our attention on the observed displacement at twin boundaries by HREM in situ experiments, on analyses of fringe contrast and diffuse scattering of the twin boundaries; and we present evidence of the variable nature of the twin boundaries. In Sec. 5.5, we illustrate the geometry of the twin boundary and twin-boundary dislocation using coincidence-site-lattice (CSL) and displacement-shift-complete-lattice (DSCL) theories. The results are summarized in Sec. 5.6.

5.2 Experimental Procedure

Bulk samples of Y–Ba–Cu–O oxide were produced by grinding together appropriate amounts of Y_2O_3, $BaCO_3$, CuO, Fe_2O_3, $NiCO_3$, ZnO, and Al_2O_3 powders in an agate mortar, and subsequently firing them. After calcining two times at 900°C, the mixtures were heated in oxygen flow at 960°C for 40 h, then at 650°C for 8 h, followed by furnace cooling. The bulk samples produced were very uniform and of high quality, as characterized by electron microprobe [27], x-ray [27–29], and neutron diffraction (lattice parameters measurements) [28, 29], x-ray absorption (extended x-ray absorption fine-structure spectroscopy, x-ray absorption near-edge structure) [30], superconductivity measurements (resistivity, susceptibility, and superconducting critical temperature) [27, 28, 30], and accurate mass measurements to determine the oxygen content [27–29]. The specimens for HREM were produced by crushing bulk samples into fine fragments under acetone, and then putting a few drops of the suspension on a carbon-coated holely film spanning the openings of a Cu grid. The specimens were immediately placed into the microscope to prevent degradation in air. The specimens for TEM were prepared by ion-milling dimpled, thin (~ 50 μm) disks of the oxides in an ion beam with incident angle 8° to 10° at 4 kV. (The final stage of milling was done at 3 kV.) Electron microscopy was carried out in a JEOL JEM-2000 FX (side-entry, $C_s = 2.3$ mm, $C_c = 2.2$ mm), a JEM-2000 EX (top-entry, $C_s = 0.7$ mm, $C_c = 1.3$ mm), and a JEM-2010 (side-entry, $C_s = 0.5$ mm, $C_c = 1.0$ mm) all operating at 200 kV.

5.3 Gross Features of the Twin Boundary

5.3.1 *Conventional TEM Observations*

Twin boundaries are the most commonly observed defects in $YBa_2Cu_3O_{7-\delta}$. It is now well-known that in $YBa_2Cu_3O_{7-\delta}$ crystals, twins are formed in two equivalent orientations ([110] and [−110]) to accommodate the strain energy of the tetragonal to orthorhombic phase transformation as increasing oxygen is absorbed by the oxide when it is cooled from high temperature in an oxygen atmosphere. Each twinning forms two twin variants. Visually, the morphology of the twinning consists of alternating lamellae contrast from the twin variants, as shown in Figure 5.1(a). The width of the lamellae of $YBa_2Cu_3O_{7-\delta}$ usually depends on the grain size, which is controlled by processing parameters. The difference in contrast between neighboring lamellae

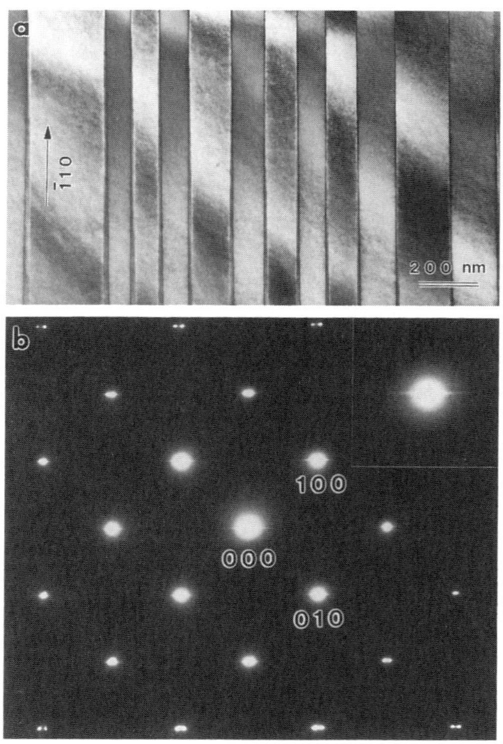

FIGURE 5.1. (a) A typical two-beam image of the twinning morphology in Y–Ba–Cu–O systems showing alternative lamellae. The contrast between neighboring lamellae is twin boundaries. (b) A well-aligned (001) selected-area diffraction pattern covering several twin boundaries from a $YBa_2Cu_3O_7$ sample. Note that sharp streaks superimpose on all the diffraction spots in a direction perpendicular to the twin boundary. The inset shows an enlarged reciprocal spot.

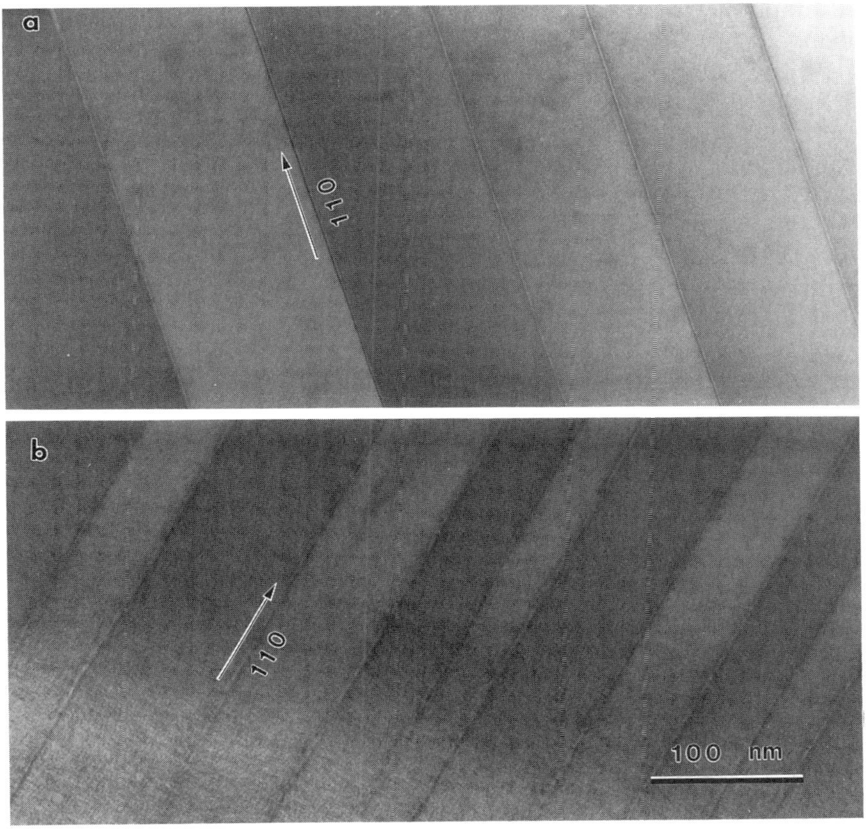

FIGURE 5.2. Multi-beam images of twin boundaries viewed edge-on along the [001] direction from a $YBa_2Cu_3O_7$ sample (a), and a $YBa_2(Cu_{0.98}Al_{0.02})_3O_7$ sample (b). The twin-boundary image from $YBa_2(Cu_{0.98}Al_{0.02})_3O_7$ is wider and fuzzier than that of $YBa_2Cu_3O_7$.

arise from their difference of deviation from the Bragg reflection. When the twin boundary is viewed inclined, it shows fringe contrast under most two-beam conditions. A close inspection of the boundary shows that the boundary has a perceptible thickness. Figures 5.2(a) and 5.2(b) are typical multi-beam images of the twin boundaries seen edge-on along the (001) zone axis in $YBa_2Cu_3O_7$ and $YBa_2(Cu_{0.98}Al_{0.02})_3O_{7-\delta}$, respectively. Double thin lines of contrast at the twin boundary are clearly visible. Such a contrast suggests that the crystal structure at the boundary may be differ from the orthorhombic twin matrix. The thin layer between the lines represents a transition region between the two neighboring twin variants. The width of the twin boundary ranges from 7 to 39 Å, depending on stoichiometry (cation substitution) of the samples, level of oxygenation, and heat treatment. The observation of the twin boundary as a transitional region, rather than as a monoatomic plane, is consistent with the x-ray studies of Laue patterns, x-ray oscillating patterns, and angular scanning topograms [31].

Figure 5.1(b) shows a well-aligned (001) selected-area diffraction (SAD) pattern obtained by using a nearly parallel electron beam, which extends over an area of 0.3 μm in diameter covering several (110) twin boundaries, such as that shown in Figure 5.1(a). Because of twinning, all diffraction spots are split, except for the (110) systematic row in which the reflection of the two twin variants is common. However, the main feature to be noted in Figure 5.1(b) is that sharp streaks are superimposed on all reciprocal spots including the origin in a direction perpendicular to the boundaries. The streaks are straight and narrow, the latter feature suggesting a coherence parallel to the twin boundaries in excess of 500 Å. Although the intensity of the streaks is not strong, they can easily be observed by having the second condenser lens of the electron microscope well overfocused.

To explore the origin of the streaks, a near parallel-beam dark-field image technique was used allowing half of a streak to pass through the objective aperture (marked by a circle on the inset of Figure 5.3) to form an image. The enhanced contrast of the twin boundary unambiguously suggests that these streaks result from the boundaries rather than from the shape of the twin lamellae. Such streaks can be due to a thin boundary layer with a different structure, and/or a phase shift at the boundary. These two possibilities can be distinguished by comparing the intensity distribution at different reflections in reciprocal space [32]. However, to analyze such a diffraction pattern in a highly symmetric orientation from a thick specimen , one must bear in mind that there are strong multiple-scattering effects, as commonly observed in

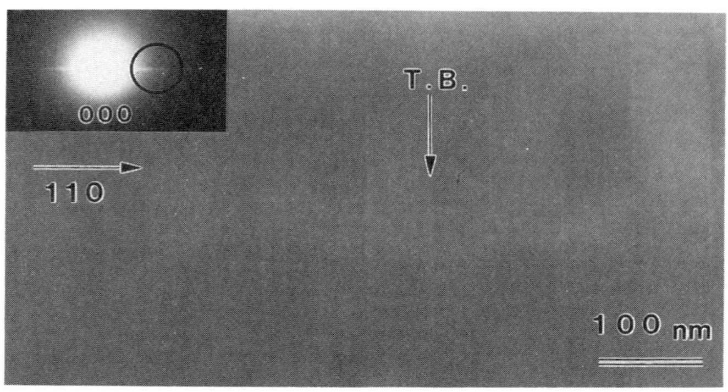

FIGURE 5.3. A dark-field image obtained by allowing half of a streak, marked by a circle on the inset, to pass through the objective aperture. The enhanced contrast of the twin boundaries unambiguously demonstrates that the streaks result from the boundaries.

electron diffraction when many Bragg beams are excited simultaneously; this complicates the interpretation.

5.3.2 *HREM Observations*

Figure 5.4 is an example of the structural image of a twin boundary of a fully oxidized $YBa_2Cu_3O_{7-\delta}$ ($\delta \approx 0.0$) sample viewed along the [001] direction with a point-to-point resolution better than 1.9 Å. The bright dots arranged in a near-square cell correspond to the projection of the fundamental perovskite unit cell. The approximate 3.82 × 3.88-Å periodic dot patterns correspond, respectively, to the *a* and *b* lattice parameters of the crystal. Computer-image simulations under our imaging conditions [33] sug-

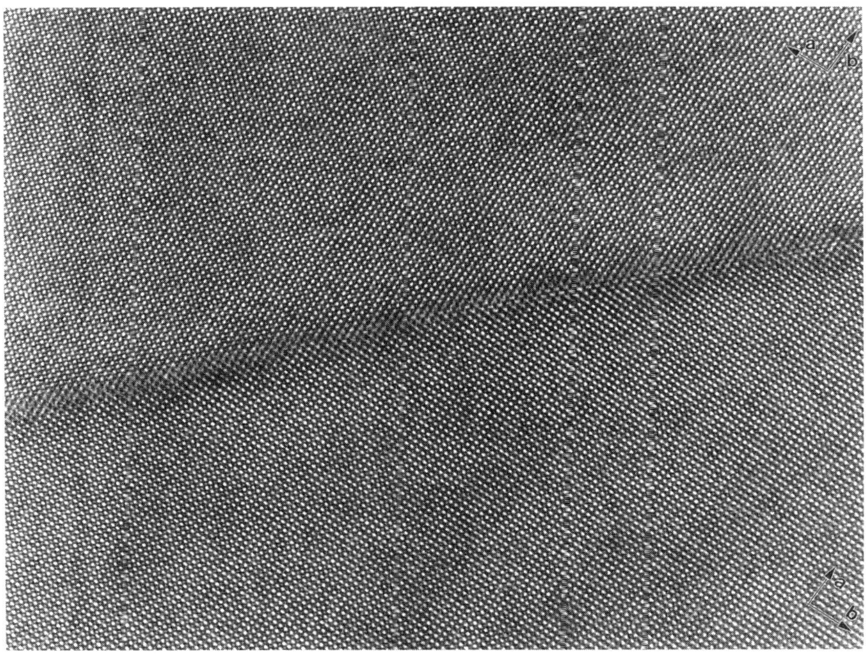

FIGURE 5.4. A structural image of a twin boundary of $YBa_2Cu_3O_7$ viewed along the [001] direction. The bright dots arranged in a near-square cell correspond to the projection of the fundamental perovskite unit cell. Computer-image simulations under the imaging conditions suggest that the atom of the Ba(Y) columns appears in strong white contrast, and the Cu(O) columns in weak white contrast. The oxygen columns are shown between the Ba(Y) columns. The sharp, uniform atomic image (except for the twin-boundary region) implies that the columns of atoms are accurately aligned in the direction of the incident beam on both sides of the twin boundary.

gest that the heavy atom of the Ba(Y) columns appears in strong white contrast, and the Cu(O) columns in weak white contrast. The plane-oxygen (chain-oxygen) columns are shown between the Ba(Y) columns. The sharp, uniform atomic image (except for the twin-boundary region) implies that the columns of atoms are accurately aligned in the direction of the incident beam on both sides of the twin boundary.

Several observations can be made about the structural image of the twin boundaries in $YBa_2Cu_3O_7$: (1) The twin boundary consists of several severely distorted atomic layers (10 to 30 Å), in which the exact positions of the atoms are difficult to define; (2) some twin boundaries show steps perpendicular to the boundary plane; and (3) there is a lattice discontinuity at the boundary, or a lattice shift along the boundary. The interchange of the *a* and *b* planes across the twin boundary is accomplished not only by a rotation of the planes in one twin variant with respect to the other by an angle of $90^\circ - \Theta$, where $\Theta = 2(b - a)/(a + b)$, but also by a translation of the lattice plane by approximately $(1/3\text{–}1/2) \cdot 2d_{110}$ along the twin boundary. The lattice shift is most clearly seen by tilting the micrograph and observing the series of the dots along the [100] or [010] direction across the boundary, as demonstrated in Figure 5.5(a) which was taken by placing an optical camera at a

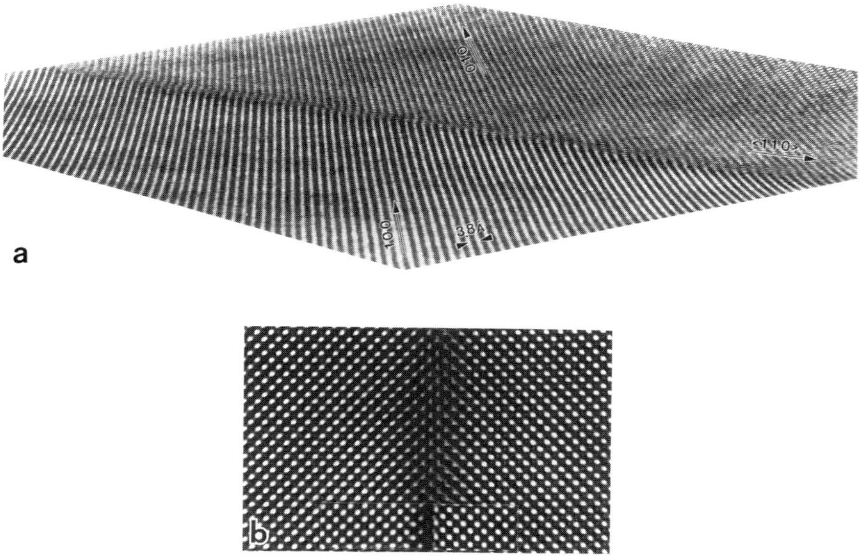

FIGURE 5.5. (a) A photograph taken from near glancing angle of the twin-boundary image shown in Figure 5.4. It can be clearly seen that the *a* and *b* lattice planes interchanged at the twin boundary involve a shifting along the boundary as well as a rotation across the boundary. (b) A calculated twin boundary with a lattice shift at the boundary matches well with the experimental HREM image of $YBa_2Cu_3O_7$ (after digitizing).

near-glancing angle along the [100] direction from a micrograph of a twin boundary shown in Figure 5.4. Unless the twin boundary had been damaged (including the loss and/or disordering of oxygen) by electron-beam irradiation, such a rigid-body translation at the twin boundary is readily observable in $YBa_2Cu_3O_{7-\delta}$ ($\delta \approx 0.0$), except for oxygen-reduced cases ($\delta \approx 0.4$). A preliminary computer-simulated image, which focused on the observed lattice shift rather than the exact atomic position at the boundary, is shown in Figure 5.5(b) [33]. The twin structure was simulated by a large supercell (81.72 × 21.72 × 11.70 Å, 1492 atoms) with a lattice translation along the boundary. The good match of the image on both sides of the twin boundary between the experimental (after digitizing) and the calculated HREM image [inset of Figure 5.5(b)] confirmed the existence of a $(1/3–1/2)\cdot 2d_{110}$ lattice translation.

5.3.3 *The Effects of Cation Substitution*

To study the effect on the structure of the twin boundary of replacing Cu in $YBa_2Cu_3O_{7-\delta}$ by another element, we studied $YBa_2(Cu_{0.98}M_{0.02})_3O_{7-\delta}$ for $\delta \approx 0.0$ and M = Ni, Zn, Fe, and Al. Examples of the HREM images of these twin boundaries viewed along the [001] orientation are shown in Figures 5.6

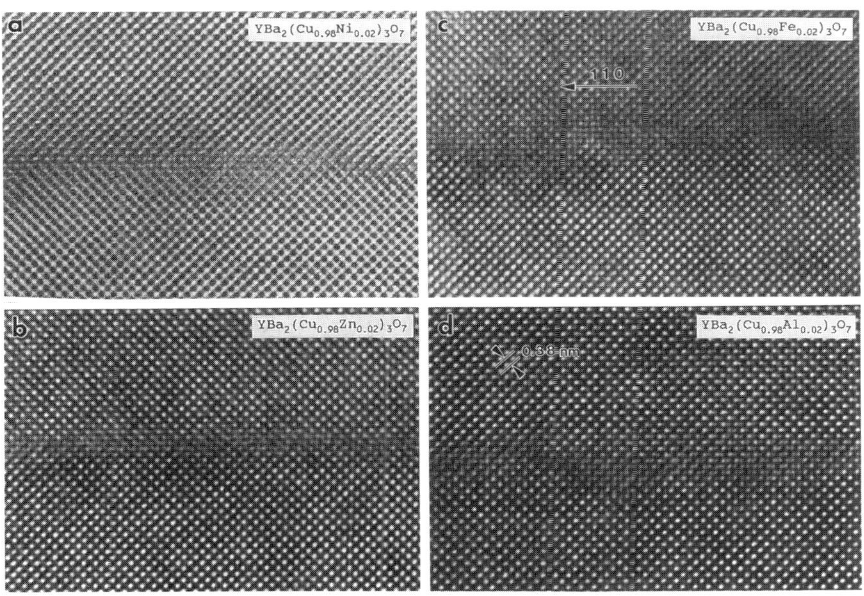

FIGURE 5.6. High resolution structure images of twin boundaries observed along the [001] direction for (a) $YBa_2(Cu_{0.98}Ni_{0.02})_3O_7$, (b) $YBa_2(Cu_{0.98}Zn_{0.02})_3O_7$, (c) $YBa_2(Cu_{0.98}Fe_{0.02})_3O_7$, and (d) $YBa_2(Cu_{0.98}Al_{0.02})_3O_7$. The horizontal distorted region in the middle of each micrograph is a twin boundary.

(a), 5.6(b), 5.6(c), and 5.6(d), respectively. The horizontal distorted areas are the twin-boundary regions. It is generally agreed that, at least at low concentrations, Fe and Al substitute predominately for Cu on chain sites. In the case of M = Zn and Ni the sites are more controversial. We believe that Zn substitutes on plane sites and Ni occupies both sites [27]. Compared with pure $YBa_2Cu_3O_7$, the widths of the twin boundaries of $YBa_2(Cu_{0.98}Ni_{0.02})_3O_7$ [Figure 5.6(a)] and $YBa_2(Cu_{0.98}Zn_{0.02})_3O_7$ [Figure 5.6(b)] are essentially unchanged, but are widened considerably with Fe [Figure 5.6(c)] and Al [Figure 5.6(d)] substitutions. We also noted that the image of the boundary in Zn- and Ni-doped $YBa_2Cu_3O_{7-\delta}$ is very similar to that of the pure $YBa_2Cu_3O_7$, that is, straight and narrow. On the other hand, the addition of Fe and Al causes the boundary not only to widen but also to become diffuse and wobbly. The twin-boundary thickness varies from $\sim$1 nm for the pure and the Zn- and Ni-substituted $YBa_2Cu_3O_{7-\delta}$ ($\delta \approx 0.0$), to 2.5 to 3 nm for the Fe- and Al-substituted oxides and for oxygen-deficient $YBa_2Cu_3O_{7-\delta}$. The samples with wider twin boundaries usually have lower orthorhombicity than those with narrow ones as measured by x-ray and electron diffraction. Table 5.1 lists the values of twin boundary thickness (as measured by the width of the distorted boundary region in HREM images and the length of the streaks in diffraction pattern) together with the values of twin spacing, orthorhombicity, and critical temperature T_c (as measured by an ac technique) [34].

It is notable that the twin-boundary width changes with cation substitution. It has been speculated that some dopants (such as Fe, Al, or Co, which form cross-links) tend to segregate at the twin boundary [35, 36]. However, after carefully examining a large number of high-resolution twin-boundary images in the Y–Ba–Cu–O system, we could reach no conclusions about the clustering of the dopants at the boundaries. Instead, we found that the general feature of these twin boundaries is the existence of a lattice translation [see Figures 5.6(a)–5.6(d)]. The magnitude of the translation may vary with the amount and type of substitution. The equilibrium width of the twin

TABLE 5.1. Examples of twin-boundary thickness, orthorhombicity, and superconducting critical temperature in $YBa_2(Cu_{0.98}M_{0.02})_3O_{7-\delta}$ where M = Cu($\delta \approx 0.0$, 0.45), Zn, Ni, Fe, and Al($\delta \approx 0.0$).

	Thickness of twin boundary	Orthorhombicity		T_c(K) midpoint
		X-ray	SAD	
$YBa_2Cu_3O_7$	1.0 to 1.5 nm	0.0171	0.0170	90.5
$YBa_2Cu_3O_{6.55}$	2.0 to 3.0 nm	0.0060	0.0100	40.0
$YBa_2(Cu_{0.98}Zn_{0.02})_3O_7$	0.9 to 1.4 nm	0.0173	0.0166	67.0
$YBa_2(Cu_{0.98}Ni_{0.02})_3O_7$	0.7 to 1.0 nm	0.01765	0.0169	82.5
$YBa_2(Cu_{0.98}Fe_{0.02})_3O_7$	2.0 to 3.0 nm	0.0117	0.0115	86.5
$YBa_2(Cu_{0.98}Al_{0.02})_3O_7$	2.6 to 3.9 nm	0.0119	0.0114	88.0

boundary appears to be determined by a balance among the chemical potential of the O–Cu–O chain, the internal stress present at the twin boundary, and the Coulomb repulsion force between like-atoms on opposite sides of the twin boundary, during the twin formation. A clear indication that chain-oxygen dominates the nature of the distortion is seen from the effects on the width of the boundary of the substituting elements Ni, Zn, Fe, and Al. Zn (at a level of 2%) is thought to substitute for Cu in the CuO_2 plane and, thus, is not likely to influence the location of the chain-oxygen. On the other hand, for Fe and Al, additional oxygen is incorporated in the chain region. This extra oxygen disorders the O–Cu–O chains at the boundary region and creates a diffuse and wider boundary. A similar broadening in the boundary was observed in oxygen-reduced specimens, $YBa_2Cu_3O_{7-\delta}$ ($\delta \approx 0.3$–0.4).

5.3.4 *The Effects on Superconducting Properties*

The observation of the finite width (10 to 30 Å) of the distorted, anti-phase-like, twin-boundary region is of significant interest because it may influence the superconducting properties of the bulk material. For instance, there might be a depression in the superconducting energy gap at the boundary. As pointed out by Deutscher and Müller [2], such a depression would form a weakly coupled junction at the boundary, and magnetic flux lines could move freely along the boundary. Thus, for a sharp boundary, such as fully oxidized $YBa_2Cu_3O_{7-\delta}$ or one formed with the addition of Ni or Zn, the gap should not be degraded drastically since the width of the boundary is ~ 10 Å, compared to the coherence length in the a–b plane of 15 to 30 Å [37, 38]. Conversely, specimens with Fe or Al substitutions or with insufficient oxygen are expected to have a large degradation of the gap at the boundary because the boundary width is approximately equal to or greater than the value of the coherence length. Thus, the critical current density J_c in these specimens is likely to be lower than in those with a sharp boundary, if the above theory applies. On the other hand, Kes [4], Chaudhari [3], and Matsushita et al. [39] have argued that pinning by twin boundaries is strong, and high-critical currents, which are observed in single crystals and in $YBa_2Cu_3O_{7-\delta}$ thin films, are caused by flux pinning by the boundaries. For his argument, Kes used a mechanism of pinning based on electron scattering by the boundary, while Chaudhari pointed out a possible pinning of the flux lines through core pinning at the boundary. In either case, both the width and sharpness of the boundary will determine the pinning strength. Thus, the variations in the twin-boundary thicknesses should also influence the pinning strength. A preliminary study of intragrain critical currents, measured by the magnetization of oriented powders [40, 41] suggested that those alloying elements which widen the boundary layer decrease J_c by nearly an order of magnitude. However, substitutions which narrow the boundary tend not to decrease J_c in comparison to that for pure $YBa_2Cu_3O_{7-\delta}$ ($\delta \approx 0.0$). This result indicates that an increase in the boundary width reduces J_c. However, Wördenweber et al. [42] pointed out that J_c is also determined by other factors, such as H_{c2}, in addition to the size of the defects. Further study is needed to elucidate the actual relationship between the pinning strength of the boundary and its width in this high-T_c material.

5.4 Studies of the Displacement at the Twin Boundary

5.4.1 *Evolution of the Twin-Boundary Structure During Electron-Beam Irradiation*

Observations by TEM in situ experiments provide an useful insight into understanding the structural changes in a crystal. Electron-beam irradiation and/or beam heating in a microscope has been used to study the evolution of twinning in $YBa_2Cu_3O_{7-\delta}$ [43–46]. During the orthorhombic-tetragonal phase transformation under beam irradiation, the crystal evolves from a twinned structure through an ortho-II phase (a partially ordered structure) to a tweed structure [43]. Twin contrast, fading irreversibly or reappearing reversibly, was also observed under beam irradiation [47, 48]. These structural changes are attributed to oxygen reduction and to oxygen disordering through a knock-on [45] or an ionization process [49]. The aim of our HREM in situ experiment was to elucidate the evolution of the twin-boundary structure on an atomic scale.

Figure 5.7 is a series of micrographs showing the structural changes in a twin boundary during irradiation by 200-kV electrons with a beam diameter ~ 600 Å and current density ~ 30 A/cm^2. The HREM in situ observations start with a $YBa_2Cu_3O_7$ sample, viewed along the c axis. As is also seen in Figures 5.7(a) and 5.7(b), the twin boundary of $YBa_2Cu_3O_7$ shows a characteristic lattice shift at the boundary. Two minutes after the start of irradiation, the lattice shift has disappeared preferentially [comparing the reference lines of the $(100)_{t1}/(010)_{t2}$ lattice in Figure 5.7(b) and 5.7(a)] at a relatively thin area (left-hand side of the micrograph) without destroying the twinning ($\sim 89.1°$ lattice rotation). After the area was irradiated for about 10 min, the lattice shift at the twin boundary became completely invisible [Figure 5.7(c)] over the whole area available for structural image observations. Despite the fading of boundary contrast and the enhancement of lattice distortion, the $\sim 0.9°$ rotation of the $(100)_{t1}$ and $(010)_{t2}$ lattices (or $\sim 89.1°$ rotation of the O–Cu–O chain) caused by twinning was still visible across the boundary, suggesting that, at an early stage in electron irradiation, only the $(1/3–1/2)\cdot 2d_{110}$ lattice shift at the boundary is eliminated. The evolution of twin structure during irradiation occurs preferentially from the twin boundary. This finding is in good agreement with electron diffraction observations using a special convergent-beam technique, which also shows the orthorhombic-tetragonal transformation occurring first at the twin bound-

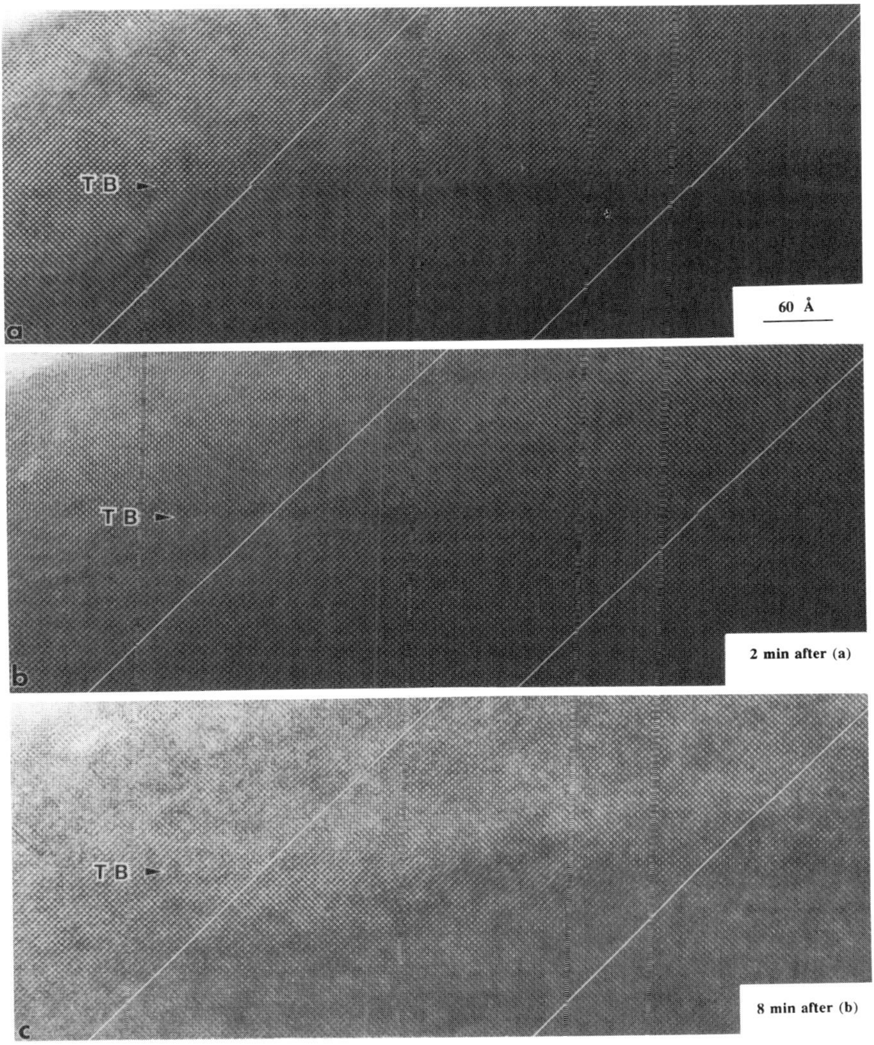

FIGURE 5.7. HREM in situ observations of structural changes of a twin boundary during electron-beam irradiation. (a) Original structural image of a $YBa_2Cu_3O_7$ twin boundary. The two reference lines indicate the existence a lattice shift across the boundary. (b) Two minutes after the start of the electron irradiation (30 A/cm^2, 200 kV). (c) Ten minutes after the start of the irradiation. Note that there is no lattice shift across the boundary (see the reference lines), however, the lattice rotation due to the twinning is still visible.

ary [50]. The structural changes in the twin boundary during irradiation appear to result from a gradual loss of oxygen at the boundary.

5.4.2 *Evidence of the Variability of the Twin Boundaries*

The above in situ experiments suggest that the twin boundary in $YBa_2Cu_3O_{7-\delta}$ exhibits variability (with or without a lattice shift). Additional evidence of such variability was obtained by HREM observations of samples with different oxygen stoichiometries. Figure 5.8 compares the structural images of $YBa_2Cu_3O_7$ [Figure 5.8(a)] and $YBa_2Cu_3O_{6.6}$ [Figure 5.8(b)]. Disregarding the details of the twin-boundary region, we consistently observed an important difference between pure samples of $YBa_2Cu_3O_{7-\delta}$, with $\delta \approx 0.0$ and an oxygen-deficient one with $\delta \approx 0.4$, as indicated schematically in Figure 5.9. For $\delta \approx 0.0$, the (100) and (010) planes are shifted $\sim d_{200}$ and $\sim d_{020}$ ($a/2$ or $b/2$) across the twin boundaries, and, by measuring the distance normal to the twin boundary between the corresponding rows of dots in the interior of the twins, we found it to be $(n + 1/2)2d_{110}$ (n is an integer) for $\delta \approx 0.0$, while, for $\delta \approx 0.4$, the shift is not present and this distance is $n2d_{110}$. As also discussed below (Sec. 5.4.4), these observations are consistent with the twin boundaries being centered at the (110) planes through the oxygen atoms in CuO planes for $\delta \approx 0.0$ [Figure 5.9(a)] and at the cations for $\delta \approx 0.4$ [Figure 5.9(b)]. Both types of boundaries have same crystallographic symmetry; in other words, in both twin boundaries the neighboring twins are related by a mirror operation across the twin boundary. However, because the O atoms in the CuO chains contribute negligibly to the electrostatic potential, other atoms—mainly the cations—define the correspondence between the structure and the image. Thus, the structure of Figure 5.9(a) appears to be an anti-phase boundary.

Previous HREM observations and discussions of the structure of the twin boundary were limited to the twinning symmetry, strain contrast, and possible lack of oxygen at the boundary region [16, 19, 51]. Although a model of an oxygen-centered twin boundary was proposed before, the earlier HREM observations of the twin boundary supported the Cu-centered boundary model [52]. The rigid-body translation at the twin boundary was first observed by us [23], then confirmed by other groups [53–55]. The observations were puzzling, since such a lattice translation would likely be a high-energy process. A recent HREM study [53] suggests that the displacement at the twin boundary may be a consequence of small differences in the orientation of crystal that is not ideally oriented. However, for a reflection twin boundary, the regions away from a twin boundary should give a similar image on either side of the boundary (except for a rotation around the [001] axis) when the thickness and the orientation of the crystal relative to the electron beam are the same. This is what we saw in HREM when there was proper alignment; the lattice shift was repeatedly observed at the twin boundary in perfectly aligned twinned crystals in $YBa_2(Cu_{0.98}M_{0.02})_3O_7$ (M = Cu, Ni, and

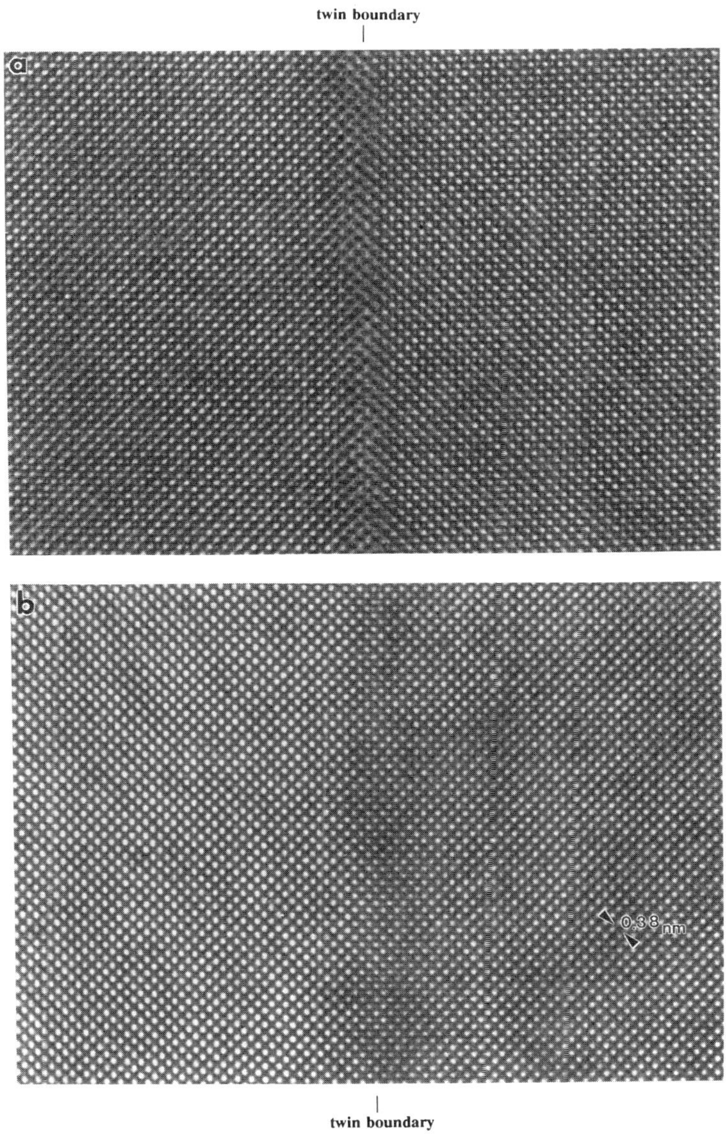

FIGURE 5.8. A comparison of structural images of twin boundaries of (a) $YBa_2Cu_3O_7$ and (b) $YBa_2Cu_3O_{6.6}$.

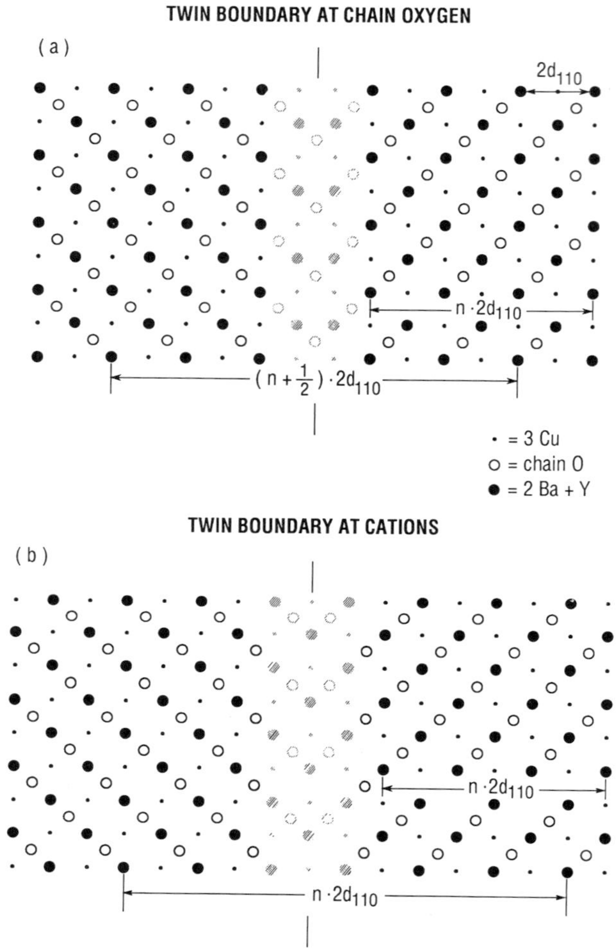

FIGURE 5.9. Idealized drawing of the twin-boundary region consistent with the observations of Figure 5.8(a) and Figure 5.8(b), respectively. Most of the oxygen atoms are omitted, except for chain-oxygen atoms.

Zn). The inclination of the boundary plane normal, and misalignment of the microscope, would not generate a lattice shift across the boundary. Therefore, we believe that the displacement is an intrinsic feature of the twin boundary in the $YBa_2Cu_3O_7$ system. The discrepancies, which exist in the literature on the detailed structure of the boundary, may be due to oxygen loss or surface relaxations at the boundary region, as we also observe the twin boundary without lattice shift in a very thin area. To avoid any misleading imaging artifacts from HREM observations, complementary methods to study the displacement from a thick area are desirable.

5.4.3 *Analyses of Fringe Contrast of the Twin Boundary*

Fringe contrast analysis has been very successful in studying the nature of boundaries or interfaces. The symmetry of the intensity variation, or rocking curve, from a tilted boundary provides important information about the crystallographic features of respective crystals across the boundary. Figures 5.10(a) and 5.10(b) show the fringe contrasts of an inclined twin boundary of a $YBa_2Cu_3O_{6.6}$, obtained with the [-110] reflection being strongly excited. The outer fringes (the fringes from the top and bottom surfaces of the foil) of the boundaries in a bright-field (BF) are asymmetric [Figure 5.10(a)], that is, dark–bright and bright–dark, while in a dark-field (DF) they are symmetric [Figure 5.10(b)], that is, bright–bright and dark–dark. For a boundary with pure twinning characteristics, these are the results we expect to observe [56].

In contrast, the symmetry of the twin-boundary fringes from $YBa_2(Cu_{0.98}M_{0.02})_3O_{7-\delta}$ (M = Cu, Zn, Ni, and $\delta \approx 0.0$) samples, which exhibit a lattice displacement seen in HREM, is different. In a BF image of the boundaries [Figure 5.11(a)], the outer fringes of the boundaries are symmetric. In this particular case, $S_1 = -S_2$ (where S_1 and S_2 are the deviation from the Bragg reflection of the adjacent twin domains), the intensity of the background and the fringes in each image of the boundaries are the same. On the other hand, in a DF image in Figure 5.11(b), the outer fringes of the boundaries are alternatively symmetric and asymmetric for the series of the boundaries. Such contrast is not caused by overlapping boundaries because the spacing between the boundaries [see Figure 5.11(c) where the twin bound-

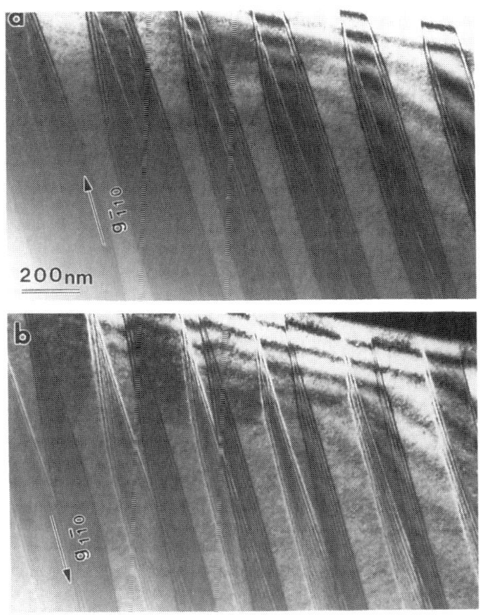

FIGURE 5.10. Fringe contrasts of an inclined twin boundary of $YBa_2Cu_3O_{6.6}$, obtained with the [-110] reflection being strongly excited. The outer fringes of the boundaries in a bright-field image are asymmetric (a), while, in a dark-field image, they are symmetric (b).

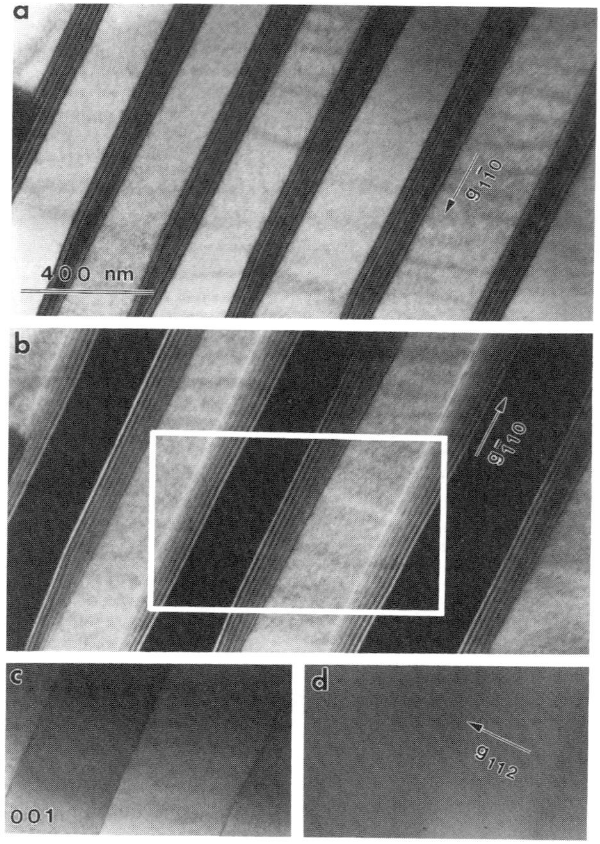

FIGURE 5.11. Twin boundaries observed in $YBa_2(Cu_{0.98}Zn_{0.02})_3O_7$. (a) and (b) are of the same region of the specimen, and (c) and (d) are from the region in (b) which is enclosed by a white border. (a), (b), and (d) were imaged in an orientation tilted about 15° away from the [001] zone (c). Note that the outer fringes of the inclined boundaries are symmetric in bright-field (a), and alternatively symmetric and asymmetric in dark-field (b). Also, when a **g** perpendicular to the twin boundary is excited, the boundary fringes are out of contrast (d).

ary is edge-on] is substantially larger than the image widths of the tilted boundaries. Similar alternative fringes was observed in Nb_2O by van Landuyt et al. [57].

Gevers et al. [58] and Amelinckx and van Landuyt [56] subdivided the types of planar interfaces into three categories, depending on the nature of the interface: the (1) α interface ($\alpha = 2\pi \mathbf{g} \cdot \mathbf{R}$, where **g** and **R** are a diffraction and a displacement vector); the contrast of the interface is due to the phase difference, that is, a displacement, or to a small misfit at the boundary, such as a stacking fault or an anti-phase boundary; the (2) δ interface ($\delta = \xi_{g1}S_1 - \xi_{g2}S_2$); the contrast of the interface is due to a slight misorientation, that is, the different deviation S from the Bragg reflection of the adjacent crystals, together with a possible difference in extinction distance ξ_g, such as in ordered domains, or a twin boundary; and (3) the α–δ interface; a mixture of (1) and (2). These authors also calculated the fringe intensity profile from these three interfaces in a thick foil in terms of anomalous absorption and two-beam dynamical theory. They found that the outer fringe for an α interface is symmetric in BF but asymmetric in DF, while the δ interface is asymmetric in BF but symmetric in DF. For a α–δ mixed interface, the outer fringe has an intermediate characteristic between that of pure α and pure δ fringes, and depends on the relative contributions from each of these types. According to the above criteria, the boundary fringes from $YBa_2Cu_3O_{6.6}$ are a pure δ fringe, while the fringes observed in Figures 5.11(a) and (b) are not a pure α or a pure δ type, but have both α and δ characters. The δ character could arise from the twinning, whereas the α character arises from the displacement at the boundary.

The conventional $\mathbf{g}\cdot\mathbf{R} = 0$ invisibility criterion can be used to determine the nature of the observed displacement. Figures 5. 11(c) and 5.11(d) are the same as the marked area of Figure 5.11(b). Figure 5.11(c) is imaged in an exact [001] zone axis, which was tilted about 15° from Figures 5.11(a) and 11(b). The fringes from the inclined twin boundary do not always show contrast. We tilted these boundaries $\pm 35°$ from [001] through [−1−11] to [−3−32] zone axes. When $\mathbf{g}$ = [110], [112], and [113] in these zone axes (i.e., $\mathbf{g}$ perpendicular to the boundary), the fringes are out of contrast, as shown in Figure 5.11(d) for $\mathbf{g}$ = [112]. This finding implies that the phase angle $\alpha = 2\pi\mathbf{g}\cdot\mathbf{R}$ is either 0 or $2n\pi$, where n is an integer. Then, unless the component of $\mathbf{R}$, which is perpendicular to the twin boundary, is $n/2\cdot[110]$, this two-beam contrast study suggests that the displacement at the twin boundary is along the boundary. However, this does not completely rule out the possibility of a displacement perpendicular to the twin boundary. If such $\mathbf{R}$ exists, it has to be a multiple of d_{110}. To examine whether there is such volume expansion at the boundary, the average d_{110} spacings along the [110] direction were determined in two regions: an area containing a boundary, and the undistorted matrix, as shown in Figure 5.9(a). There was no difference between the measured d_{110} spacing in these two regions. All these observations suggest that the displacement is [−110] type, which is parallel to the twin boundary, and this is consistent with the HREM observation discussed above.

Recent work [55] based on the use of the symmetry properties of the α and δ character in diffraction contrast images of pure and Fe-doped $YBa_2Cu_3O_{7-\delta}$ has confirmed that the twin boundary is an α–δ interface (the α component is $\mathbf{R}$ = 1/2[110]). The observed displacement is also in good agreement with our calculations from diffuse scattering and electron diffraction observations (see Sec. 5.4.4).

5.4.4 *Analyses of the Streaks of Diffuse Scattering*

The sharp streaks seen in the SAD pattern [Figure 5.1(b)] reflect the important features of the twin-boundary structure, as shown in a dark-field image produced by a streak (Figure 5.3). These streaks of low intensity are longest and most clearly visible for $YBa_2(Cu_{0.98}M_{0.02})_3O_{7-\delta}$ for $\delta \approx 0.0$ and M = Cu, Zn, Ni, where the streak lengths are typically 1/4 of $\mathbf{g}_{110}$. Much shorter streaks are observed for pure specimens with an oxygen content of $\delta \approx 0.4$, and for fully oxidized specimens with M = Al, Fe. The length is then typically 1/10 of $\mathbf{g}_{110}$. Streaks observed in reciprocal space can be primarily attributed to a crystallographic shape effect (such as a thin layer) and/or a strain effect (such as lattice displacement) in real space. From a kinematic point of view, the main basis for the distinction between these two follows from the fact that the extension of the streaks due to the shape effect extend equally on all diffraction spots, including the origin. For a lattice with displacement vector **R**, streaks in the diffraction pattern will be visible through reflection **g** for which $\mathbf{g} \cdot \mathbf{R} \neq n\ (n = 0, 1, \cdots)$ and invisible for $\mathbf{g} \cdot \mathbf{R} = 0$. Originally, the streaks were interpreted as evidence for the existence of a thin non-orthorhombic layer. The shortening of the streaks were attributed to the broadening of the distorted boundary region, as confirmed by HREM.

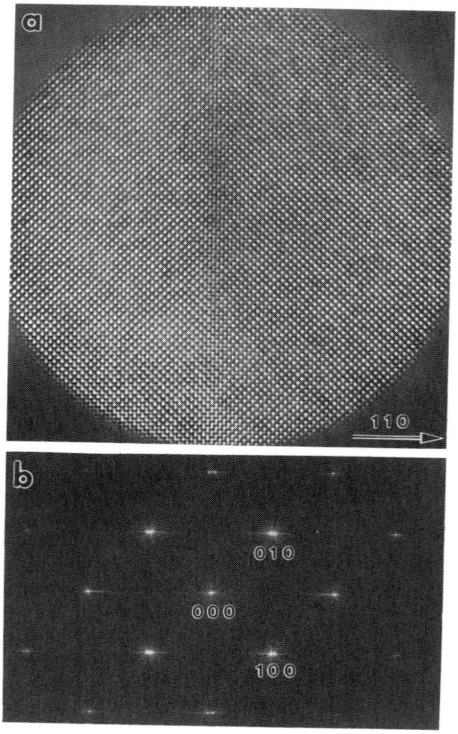

FIGURE 5.12. (a) A digitized HREM image of a twin boundary (with a lattice shift) from a thin area with less multiple-scattering effects. (b) A diffractogram Fourier transformed from (a) showing sharp streaks. Note the difference in intensity distribution of the streaks at different reciprocal spots.

Similar streaks were also observed in a diffractogram produced by Fourier transforming a digitized HREM micrograph of a single twin boundary (Figure 5.12). The diffractograms, obtained from an HREM image from a ~100-Å thin area (having less multiple scattering), show that the streaks at different reflections have a different intensity distribution. To minimize the multiple-scattering effects in SAD observations it is desirable to reduce the number of reflections. By tilting the specimens slightly about the axis normal to the twin boundaries, we found that the streaks at the direct beam and the ($hh0$) reflections disappear, suggesting that they are caused by multiple scattering, because the ($hh0$) line of reciprocal space was also observed after tilting. From careful observations of these streaks for different incident-beam directions, it appears that the strongest streaks are through the (100)- and (010)-type reflections. The diffraction pattern is schematically shown in Figure 5.13(a).

As we discussed earlier, a twin boundary with a cation lattice shift acts like an anti-phase boundary for electrons. Anti-phase boundaries result in a splitting of reflections for which $\alpha = 2\pi\mathbf{g}\cdot\mathbf{R} \neq 2\pi n$ ($n = 0, 1, \cdots$) when a small electron probe is used or the domain sizes are small [59, 60]. When the domain size increases, a streak normal to the anti-phase-like boundary is more pronounced than the splitting. Two reflections which are most strongly perturbed by the anti-phase-like boundary are those for which $\alpha = 2\pi\mathbf{g}\cdot\mathbf{R} = 2\pi(n + 1/2)$, for example, the (100)- and (010)-type reflections when $\mathbf{R} = 1/2[1-10]$, which is the displacement for the cations lattice in Figure 5. 9(a).

To support our arguments made from diffraction observations, we calculated the intensity of the diffuse scattering of twin boundaries using a periodic boundary condition. With a kinetic treatment, the intensity of diffuse scattering from the periodic twin boundaries of $YBa_2Cu_3O_7$ can be expressed by

$$I \approx \left\{ \frac{1}{2} + \sum_{n=1}^{\frac{D}{2}} \cos 2\pi(nh + n\phi' k) + \sum_{n=\frac{D}{2}+1}^{D-1} \cos 2\pi[(nh + (D\phi'/2 - n\phi' + \alpha')k] \right\}^2, \quad (1)$$

where ϕ' is the orthorhombicity, α' is the lattice displacement ($\alpha' = 0.5$ for $|\mathbf{R}| = d_{110}$), and D is the twin spacing. The calculated intensities of (1–10), (100), and (110) reflections are shown in Figure 5.13(b) for $\mathbf{R} = 0$ and Figure 5.13(c) for $\mathbf{R} = 1/2[1-10]$; note the difference in streak length for (100) reflection in Figure 5.13(b) and 5.13(c). The intensity profile for $\mathbf{R} = 1/2[1-10]$ is consistent with the SAD observation of $YBa_2Cu_3O_7$ when the multiple-scattering effect is eliminated. Similar observations were made by laser simulation using optical masks. For a pure twin-boundary configuration (only a ~89.1° rotation about the [001] axis without lattice distortion at the boundary), no streaks are seen in the diffractogram, although streaks were observed for a twin with a lattice shift at the boundary. Thus, the diffuse scattering experiments support the interpretation of the HREM observations and the proposed models shown in Figure 5.9(a) and 5.9(b).

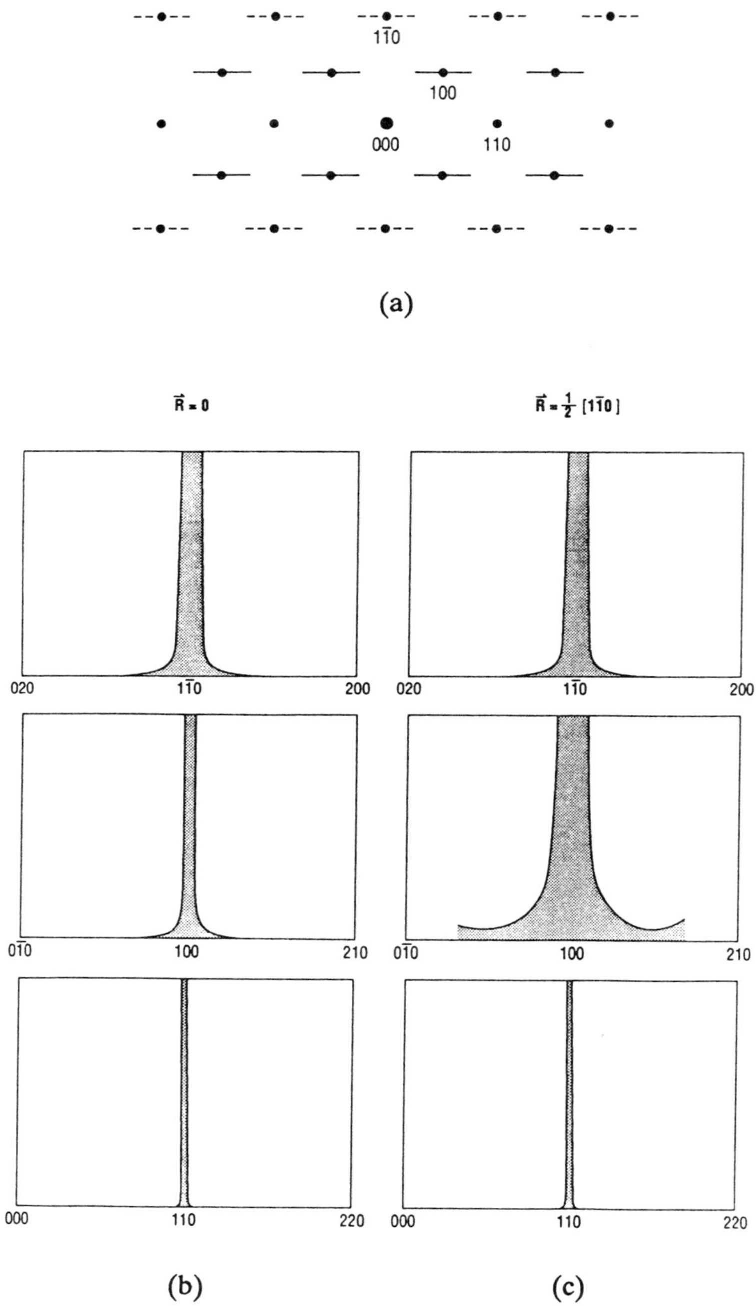

FIGURE 5.13. (a) Schematics of the observed streaks in the (001) diffraction patterns of $YBa_2(Cu_{0.98}M_{0.02})_3O_{7-\delta}$ for $\delta \approx 0.0$ and M = Ni, Zn, and Cu. The dashed streaks may be caused by multiple scattering. (b) and (c) are calculated intensity profiles along the streaks using a twin spacing of 270 Å and periodic boundary conditions for two different displacement vectors **R**. (b) **R** = 0; (c) **R** = 1/2[−110]. Note the difference in streak length (intensity profile) for the (100) reflection. The difference in the intensity profile width of (110) and (1−10) reflection for **R** = 0 is caused by twinning.

The analyses of the diffuse scattering results above indicates that a major fraction of the intensity of the streaks is also consistent with a shift of the cation lattice parallel to the twin boundary. Then, considering the effect of multiple scattering on the intensities of the streaks, and weak streaks observed in oxygen-deficient samples (without a lattice shift), it appears that the streaks observed in SAD [Figure 5.1(b)] result from a combination of both effects, the lattice displacement at the twin boundary and the distorted boundary with a limited width. The increased length of the streak observed in M = Zn, Ni, and Cu appears to be caused by the abrupt lattice shift occurring within a ~10 Å thin boundary region.

5.4.5 *Interfacial Energy of the Boundaries*

To interpret the lattice displacement at the twin boundary and the difference between fully oxygenated and oxygen-deficient $YBa_2Cu_3O_{7-\delta}$, we proposed two twin-boundary models, illustrated in Figure 5.9(a) and Figure 5.9(b). The twin boundary centered at oxygen would appear to be energetically unfavorable because the cations and the O atoms in the CuO_2 and BaO_2 layers would only be separated by 2.72 Å for an ideal twin boundary rather than 3.82 Å in the matrix region. Nevertheless, the alternative of a 90° bend in the O–Cu–O chains at the Cu atoms which accompanies a cation-centered twin boundary is also thought to be very energetically unfavorable, as was pointed out by de Fontaine et al. [61] from considerations of effective pair interactions. In addition, strong experimental evidence against 90° bends is provided by the observed short-range order diffuse scattering from oxygen-deficient $YBa_2Cu_3O_{7-\delta}$, where narrow rods of diffuse scattering are observed, suggesting long unbroken O–Cu–O chains [62, 63]. This type of oxygen-centered boundary may require considerable local displacement of the atoms at the boundary to reduce the energy. Thus, the HREM images suggest that the shift of the corresponding features on one side of the twin boundary relative to the other side of the twin boundary may not be exactly as shown in Figure 5.9(a) (which gives a shift of d_{110} along the twin boundary); rather, the observed shift is $(0.80 \pm 0.15)\, d_{110}$ (while it is zero when the twin boundary is at the cations). The local structure disorder and compositional variation at the twin-boundary regions were also confirmed by field-ion microscopy observations [64].

In contrast, for oxygen-deficient samples, the number of O–Cu–O chains is reduced, thus reducing the strain energy and also the net energy penalty for the 90° bends in the chains. Hence, forming an oxygen-centered twin boundary becomes unnecessary, consistent with our observation that the twin boundary occurs at the cations for lower oxygen content [Figure 5.9(b)]. The energy associated with the formation of the twin boundary through the cations may be further reduced if the twin-boundary region has a lower oxygen content than the interior of the twins. It was observed that, during electron irradiation, $YBa_2Cu_3O_{7-\delta}$ becomes tetragonal first at the twin boundaries, suggesting that the energy associated with the reduction of oxygen content is lowest at the twin boundaries. When Fe or Al is added, the presence of these elements causes the formation of oxygen cross links and disrupts the linear Cu–O chain structure [65]. Hence, these elements are likely to eliminate the chain-bending energy penalty and favor the formation of cation-centered twin boundaries. Thus, our observations suggest that these modifications reduce the energy associated with the formation of the twin at the cations below that required for forming the boundary at the O atoms in the CuO plane. The twin boundary forms at cations below a certain oxygen content or above a certain Al or Fe content.

The transformation from tetragonal to orthorhombic is a disorder-to-order transition which occurs when oxygen is localized to (0 1/2 0) sites, leaving the (1/2 0 0) sites vacant. During the transformation, if no constraint is present, such as in single crystals [66], the lattice is free to relax and no twinning results. With constraints, the shear strains associated with the transformation can be relieved by twinning at the cost of twin-boundary energy, which mainly comes from the Coulomb repulsion. The spacing of the twin is determined by minimizing the total energy of the system associated with the elastic strain energy and twin-boundary interfacial energy [67]. The relationship between twin spacing D, the twin-boundary energy γ, the elastic modulus M, the grain size g, and the orthorhombicity, $\phi = 2(b - a)/(a + b)$, is given by [68]

$$D \approx (g\gamma/CM\phi^2)^{1/2}, \tag{2}$$

where C is a constant ~ 1. The validity of this relationship was shown for bulk [69–71] and also for thin films, in which case the critical length is the thickness of the film rather than the grain size [72]. Using $M = 2 \times 10^{12}$ dyns/cm^2 as reported [73, 74] and measuring the twin-boundary spacing, we find $\gamma \approx 80$ ergs/cm^2 for $YBa_2Cu_3O_{7-\delta}$. It is interesting to note that substitution with Al or Fe at a high oxygen content or reducing the oxygen content in pure $YBa_2Cu_3O_7$ both substantially reduce the average twin-boundary spacing D [24, 27, 69, 75] and the twin-boundary energy γ was found to decrease by an order of magnitude [69]. (Interestingly, these values of γ are in the range of the twin-boundary energies which are found in metals and alloys.) This indicates that the Cu-centered boundaries are lower in energy than the O-centered ones, if the boundary regions can be broadened by the deficiency or the excess of oxygen, by low oxygen contents, or by substitution with Fe, Al, etc.

5.5 Crystallographic Analyses of the Twin Boundary [76]

5.5.1 *A Σ64 Coincidence Boundary*

An elegant way to elucidate a boundary geometry is by using a coincidence-site-lattice (CSL) model. CSL can be produced by rotating two crystals rela-

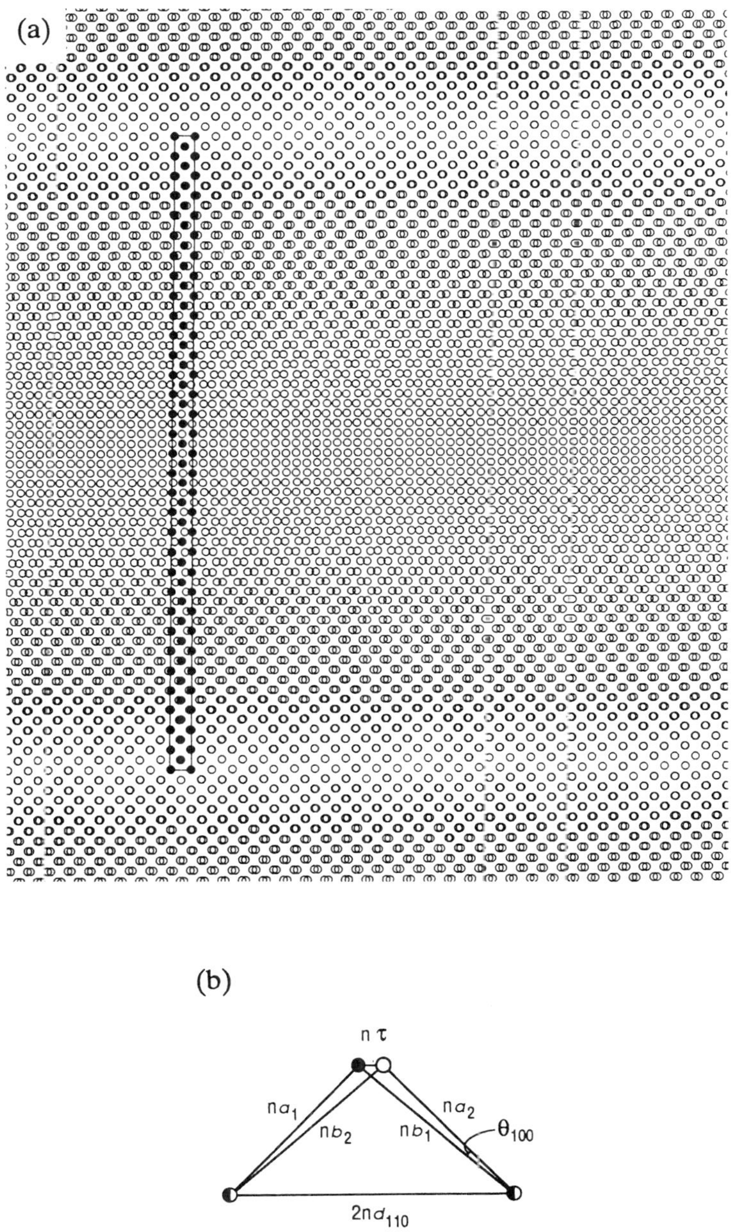

FIGURE 5.14. (a) A superposition of two neighboring twins of $YBa_2Cu_3O_{7-\delta}$ ($a = 3.82$ Å, $b = 3.88$ Å), or a coincidence-site lattice of the twin boundary, produced by rotating 89.1° of the lattice one to another along the [001] axis. The CSL unit cell ($32 \cdot 2d_{110} \times 2d_{110}$ in size and 64 lattice sites) is also depicted. (b) Atom rearrangement under the twinning shear.

tive to one another about a common axis, using a lattice site as the origin. The reciprocal density of CSL (or the ratio of the unit-cell volume of CSL to that of the crystal) is usually denoted by Σ. Considerable evidence shows that low-energy interfaces are associated with a high density, short-period coincidence lattice, that is, a boundary with low Σ value [77–80]. Furthermore, based on CSL theories, boundary dislocations have been successfully used to explain the boundary structure [81–83]. To shed light on this geometry, crystallographic analysis of the fully oxygenated twin boundary ($YBa_2Cu_3O_7$) was carried out. Figure 5.14(a) shows the superposition of the neighboring twin crystals, produced by a rotation of 89.1° of the near-square lattice ($a = 3.82$ Å and $b = 3.88$ Å)—one to another—along the [001] axis. The perfect-match positions, or coincidence sites, form a systematic row, and can be considered as the twin boundary. The lattice mismatch increases with increasing the distance from the coincidence sites and changes periodically. According to the definition, we have

$$\mathbf{\Sigma} = [\mathbf{C}_1 \cdot (\mathbf{C}_2 \times \mathbf{C}_3)]/[\mathbf{a} \cdot (\mathbf{b} \times \mathbf{c})] = 64, \tag{3}$$

where **a**, **b**, and **c** are lattice vectors of a $YBa_2Cu_3O_7$ crystal, and $\mathbf{C}_1$, $\mathbf{C}_2$, and $\mathbf{C}_3$ are three primitive vectors of the CSL unit cell, as also indicated in Figure 5.14(a). Thus, the twin boundary can be described by a CSL system of **Σ**64, 89.1°/[001].

The relative lattice shift due to the twinning shear $\boldsymbol{\tau}$ is schematically shown in Figure 5.14(b), where Θ_{100} is the rotation angle between $\mathbf{a}_1$ and $\mathbf{b}_2$ or $\mathbf{a}_2$ and $\mathbf{b}_1$. By examining the geometry of the twinning, the minimum lattice-twinning shear $\boldsymbol{\tau}$ can be given by

$$\boldsymbol{\tau} = \mathbf{a} - \mathbf{b}. \tag{4}$$

The magnitude of the twinning shear $|\boldsymbol{\tau}|$ is then

$$\begin{aligned} |\boldsymbol{\tau}| &= |\mathbf{a} - \mathbf{b}| \\ &= (a^2 + b^2 - 2ab\cos\Theta_{100})^{1/2} \\ &= \{a^2 + b^2 - 2ab\cos[\pi/2 - 2\tan^{-1}(a/b)]\}^{1/2}. \end{aligned} \tag{5}$$

(Note $|\mathbf{a} - \mathbf{b}| \neq |a - b|$) For $a = 3.82$ Å and $b = 3.88$ Å, $|\mathbf{a} - \mathbf{b}| = 0.085$ Å, which is the shear displacement at the distance of d_{110} from the twin boundary. At a distance of $64 \cdot d_{110}$ from the boundary, the lattices reach a 2π phase shift (0.085 Å × 64 = 5.44 Å = $2d_{110}$), which gives a period of 64 (110) lattice planes in the direction of the boundary plane normal.

5.5.2 *Twin-Boundary Steps and Twinning Dislocations*

Careful inspections of the twin boundaries reveal that some have steps (ledges), where the center of the twin boundary leaves one (110) lattice plane and moves to a neighboring (110) lattice plane (Figure 5.15). This observation

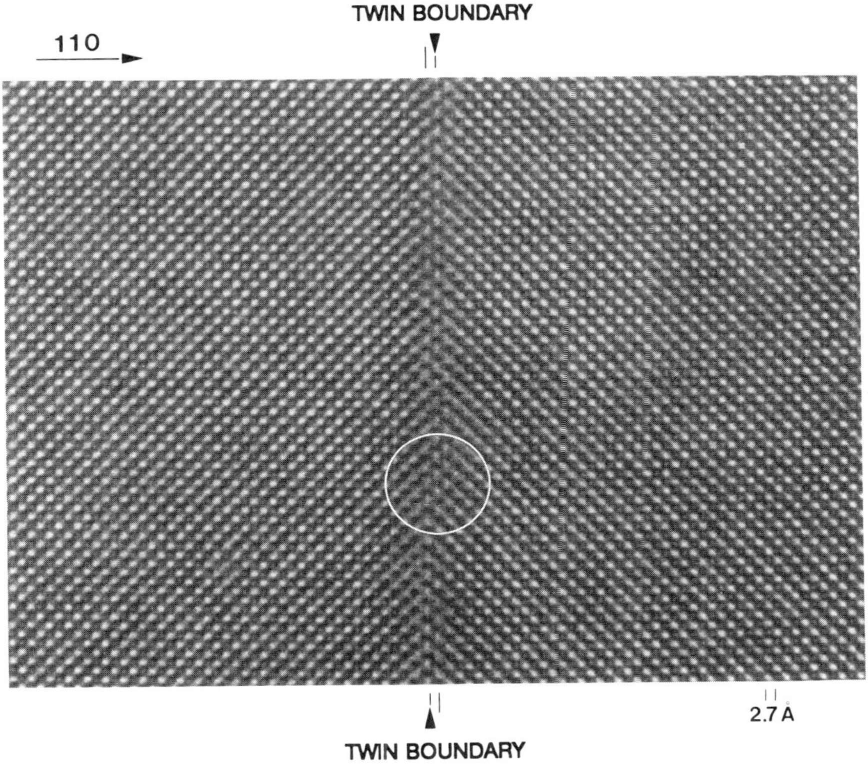

FIGURE 5.15. High-resolution image of a (110) twin boundary associated with a d_{110} twinning step, indicated by a circle, in $YBa_2Cu_3O_7$ view along the [001] axis. Note that the center of the twin boundary (marked by an arrow) shifts to a neighboring (110) plane at the step. The step is most visible by looking the twin boundary under grazing incidence along the (110) lattices.

of a twinning step is believed to be the first in $YBa_2Cu_3O_7$, although a twinning-step model was proposed to explain the broadening of the twin boundary by an oxygen-diffusion mechanism [53]. The twinning step runs perpendicular to the micrograph with a d_{110} step parallel to the twin-boundary plane normal. The step is easily visible by looking at the twin boundary under grazing incidence along the ($\bar{1}$10) lattice planes. As the step propagates, atoms near the step rearrange themselves in the manner of a dislocation motion. The dislocation associated with such a step is called a twinning dislocation, since it exists only in a twin boundary. A twinning dislocation can glide in the twin-boundary plane, and, as it does so, the amount of one twin domain grows at the expense of the other. The Burger vector of a twinning dislocation is not a lattice vector of the twinned lattices;

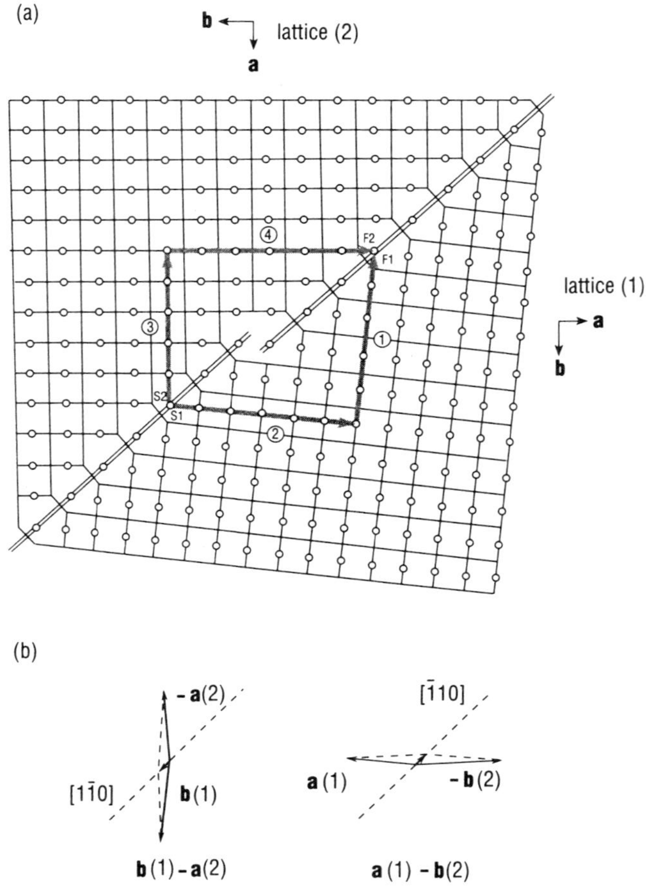

FIGURE 5.16. (a) Schematic drawing of a twinning step in $YBa_2Cu_3O_7$. The rectangular lattice represents the Cu–O sublattice in the basal plane. The open circles represent chain-oxygen atoms, and the copper atoms are located at the corners. The thick shade lines are the Frank circuit. (b) Determination of the resultant direction from the Frank circuit. Note the differences between the *a* and *b* lattices and the rotation angle are exaggerated to emphasize the characteristic feature of the boundary.

its magnitude should be equal to the distance moved by each lattice plane due to the twinning shear and be proportional to the height of the twinning step.

A twinning dislocation can be defined rigorously by a line integral along the so-called Frank circuit, which is equivalent to the Burger circuit for a lattice line-defect, surrounding the interfacial dislocation. Figure 5.16(a) is a scheme of a (110) twin boundary associated with a twinning step (step height

2.7 Å) in $YBa_2Cu_3O_7$. The rectangular lattice represents the Cu–O sublattice in the CuO plane and the open circles represent chain-oxygen atoms; the copper atoms are located at the corners. The differences between a and b lattices and the rotation angle are exaggerated to emphasize the characteristic features of the boundary. The d_{110} lattice shift can be seen from the Cu–O sublattice across the boundary.

To characterize the Burger vector of the twinning dislocation, the circuit must begin and end on the interface, as suggested by Frank [84] and Read [85]. First, consider a reference twin boundary without steps. The Frank circuit is closed whether or not the boundary has a d_{110} cation lattice shift. Then, we construct the same circuit in the real system containing a step [shown as a circuit of thick dash lines in Figure 5.16(a)]. The local Burger vector of the twinning dislocation **B**, defined by an SF/RH convention, can be given by the closure failure of the circuit. Let routes 1 and 2 be positive and 3 and 4 be negative; then we have

$$\begin{aligned} \mathbf{B} &= \oint_1 (\partial \mathbf{u}/\partial \mathbf{l})d\mathbf{l} \\ &= [5\mathbf{b}(1) - 5\mathbf{a}(2)] + [6\mathbf{a}(1) - 6\mathbf{b}(2)] \\ &= \mathbf{a}(1) - \mathbf{b}(2), \end{aligned} \tag{6}$$

where **u** is the elastic displacement around the dislocation, **l** is taken in a right-handed sense relative to the dislocation line direction ξ, and (1) and (2) denote the twins 1 and 2, respectively. As shown in Figure 5.16(b), the directions of both $\mathbf{b}(1) - \mathbf{a}(2)$ and $\mathbf{a}(1) - \mathbf{b}(2)$ vectors are along the twin boundary with the same magnitude, but with an opposite sense. Therefore, the direction of the resultant vector **B** also lies along the boundary. Reversing the sense ξ causes **B** to reverse its direction. Thus, the geometry of the observed twinning step at a (110) twin boundary can be associated with a perfect edge dislocation, with a line direction of [001], and a Burger vector of $|\mathbf{a} - \mathbf{b}|[-110]$.

5.5.3 *A Displacement-Shift-Complete-Lattice (DSCL) Treatment*

The magnitude of the Burger vector of the observed twinning dislocation is less than 1.6% of the primitive lattice vector of a $YBa_2Cu_3O_{7-\delta}$ unit cell. Such a small interfacial dislocation is usually called a secondary dislocation which can be characterized in a more graceful but equivalent circuit construction using a DSCL [86]. The DSCL defines all the vector displacements of the two crystals relative to one another under conditions where the overall pattern of lattice points produced by the two interpenetrating lattices in the coincidence orientation remains unchanged. Therefore, this DSCL defines all the possible Burger vectors of interfacial dislocations.

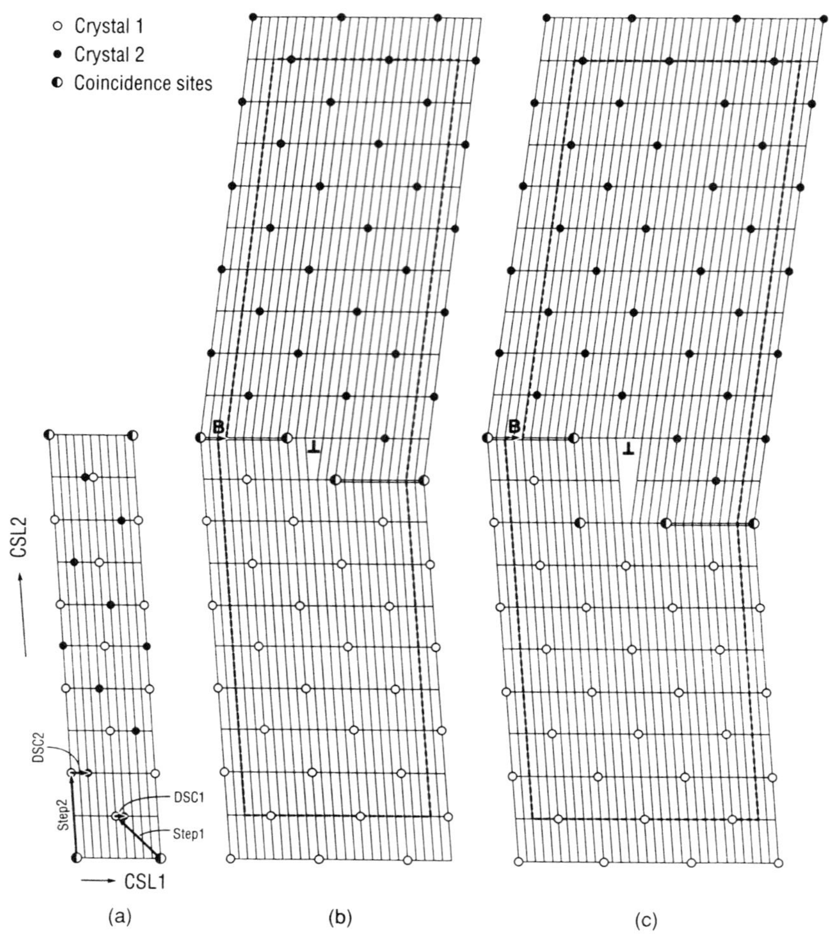

FIGURE 5.17. A demonstration of a twinning dislocation associated with a twinning step described by a DSC lattice. (a) A CSL unit cell of the twin boundary in $YBa_2Cu_3O_7$. Note that the difference of the scale between CSL1 and CSL2 is understated and the real lattice sites and CSL unit cell is shown in Figure 5.14. (b) A twinning dislocation ($\mathbf{B}$ = DSC1 = $|\mathbf{a} - \mathbf{b}|[-110]$) associated with a d_{110} step ($\mathbf{s} = [010]$, $h = d_{110}$). (c) A twinning dislocation ($\mathbf{B}$ = DSC2 = $2|\mathbf{a} - \mathbf{b}|[-110]$) associated with a $2d_{110}$ step ($\mathbf{s} = [110]$, $h = 2d_{110}$).

Figure 5.17(a) illustrates the DSCL of a CSL unit cell ($32 \cdot 2d_{110} \times 2d_{110}$, 64 lattice sites) of the twin boundary. The two primitive DSCL vectors can be geometrically derived as $|\mathbf{a} - \mathbf{b}|[-110]$ and $[110]$, as shown in Figure 5.17(a) with different scales. With the DSCL, the twinning dislocation can be demonstrated by extra DSCL planes on one side of the twin boundary, which is lacking in Figure 5.15(a). For a step at the twin boundary in $YBa_2Cu_3O_{7-\delta}$, the Burgers vector of a twinning dislocation can be expressed by

$$\mathbf{B} = h(\mathbf{a} - \mathbf{b})/d_{110}, \tag{7}$$

where h is the step height measured by the number of (110) lattice planes. The equation suggests that the magnitude of the Burger vector, or the number of the extra half-plane of DSCL is proportional to the height of the step, regardless of the lattice shift at the boundary. For a twinning step with a step height of $h = d_{110}$, as seen in Figure 5.15, the Burger vector of the twinning dislocation is one primitive DSC vector, that is, $|\mathbf{a} - \mathbf{b}|[-110]$ [Figure 5.17(b)], which is consistent with the result from a Frank construction using a crystal lattice [see Eq. (6)]. The step vector is then [010], as shown in Figure 5.17(a). A Frank circuit using DSCL for a twinning dislocation with a Burger vector of $2|\mathbf{a} - \mathbf{b}|[-110]$, associated with a $2d_{110}$ twinning step (step vector $\mathbf{s} =$ [110], step height $h = 2d_{110}$) is shown in Figures 5.17(c) and 5.17(a). However, one would not expect to see such an infinitesimal dislocation by transmission electron microscopy.

5.5.4 *Structure of Mixed Twin Boundaries*

Our observations suggest that the twin boundaries form at the (110) planes through the oxygen atoms in the CuO planes for fully oxygenated samples, and at the cations for oxygen-deficient ones. The atomic configuration of the transition from one to the other, seen in in situ experiments (Figure 5.7), may be understood by mixing the two types of twin boundaries. Figure 5.18 depicts the mixture [combining Figure 5.9(a) and 5.9(b)] of the twin boundary centered at a chain-oxygen plane (lower part) and at a cation plane (upper part). Here, we focus our attention only on the lattice shift and the position of the twin boundary. The change of the boundary center produces a $1/4 \cdot 2d_{110}$ twinning step ($h = 1.35$ Å). The step can also be related to a perfect edge dislocation at the (110) twin boundary, as shown in the middle of Figure 5.18. Constructing a Frank circuit results in a closure failure along the boundary. The Burger vector of such a dislocation is determined to be $\mathbf{B} = 1/2\langle 110\rangle + 6\mathbf{b}(1) + 5.5\mathbf{a}(1) - 6\mathbf{a}(2) - 5.5\mathbf{b}(2) = 1/2\langle 110\rangle + \boldsymbol{\delta}$, where $\boldsymbol{\delta} = 1/2|\mathbf{a} - \mathbf{b}|\langle 110\rangle$. In the real case, because of the lattice distortion at the twin boundary and the limit of the spatial resolution of the microscope, a clear 1.35-Å step may not be visible. However, by counting the number of the (110) planes on both sides of the twin boundary between the regions with and without a lattice shift (seen in Figure 5.19, an enlarged micrograph of Figure 5.7), we did find an extra (110) lattice plane in the lower part of the twin matrix, as indicated by the numbers shown in Figure 5.19. This finding strongly supports our dislocation model for the twin-boundary structure and suggests that the dislocation is essential for the transition between the two types of twin boundaries in $YBa_2Cu_3O_{7-\delta}$. A dislocation ($\mathbf{B} = 1/2\langle 110\rangle = 1/2|\mathbf{a} - \mathbf{b}|\langle 110\rangle$) passing through a coherent twin boundary, would produce a $\sim d_{110}$ lattice shift at the boundary.

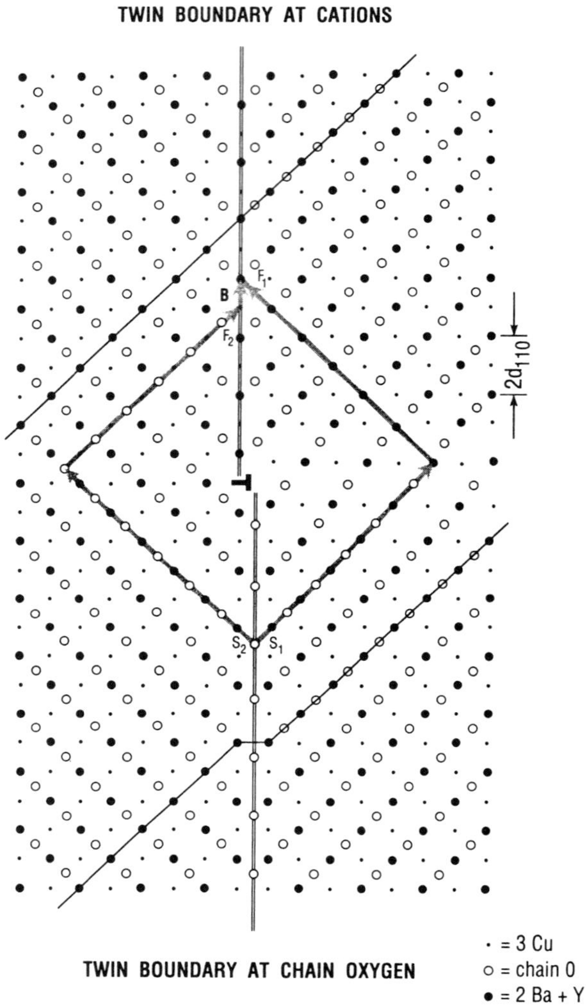

FIGURE 5.18. Schematic representation of a mixed twin boundary: a twin boundary through cation atoms with a cation sublattice shift (upper part) and through oxygen atoms without cation sublattice shift. Note the $1/4 \cdot 2d_{110}$ step and a perfect dislocation at the boundary.

The early stage of the evolution of the twin boundary in $YBa_2Cu_3O_7$ is found to be associated with switching the twin boundary centered from oxygen atoms to cation atoms, as suggested by HREM in situ experiments. The kinetics of such a transition may be explained by the motion of twin-boundary dislocations. The observed transformation-twinning dislocations and the crystallographic analyses of the twin boundary using CSL and DSCL

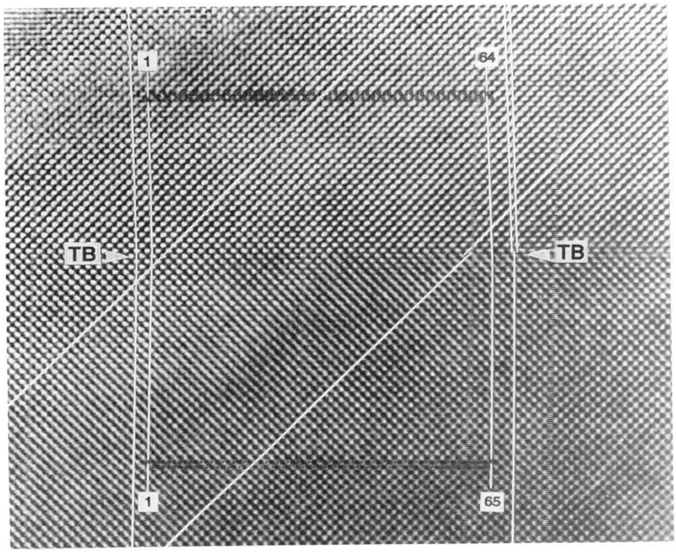

FIGURE 5.19. An enlarged micrograph of Figure 5.7 shows the mixed type of a twin boundary as proposed in Figure 5.9. The reference lines indicate that there is no lattice shift on the left-hand side of the boundary, but a lattice shift on the right-hand side. The difference in the number of the (110) planes across the boundary suggests the existence of an edge dislocation at the twin boundary.

concepts imply that dislocations play an important role in the growth of the twin. (The spiral growth of a $YBa_2Cu_3O_{7-\delta}$ thin film was shown to be related to the presence of screw dislocations by scanning tunneling microscopy studies [87, 88]. Under irradiation, the strain energy can be reduced by the loss and/or disordering of the chain-oxygen. The dislocations, associated with twinning steps moving through the (110) twin-boundary plane, would eliminate the cation lattice shift. This would create a cation-centered twin boundary, as shown in Figure 5.18, resulting in a lower boundary energy. For a transition from a cation-centered twin boundary to an oxygen-centered twin boundary during oxidation, a similar mechanism may apply. However, the origin of the formation energy of the dislocation will be different. Since such twin-boundary dislocations have been determined to be edge dislocations with a Burgers vector of $\mathbf{B} = 1/2[-110] - 1/2|\mathbf{a} - \mathbf{b}|[-110]$, the (110) twin-boundary plane is the dislocation glide plane, which contains both the dislocation line and its Burger vector. Therefore, the dislocation is glissile, and its motion is conservative. The motion of a glissile dislocation is a rapid process and can take place without any atomic diffusion. The formation of such dislocations may require a certain amount of energy. Once formed, however, the dislocation can move very fast even under the action of small internal shear stress.

5.6 Summary

(1) The twin boundary in $YBa_2(Cu_{1-x}M_x)_3O_{7-\delta}$ ($\delta \approx 0.0$) consists of a distorted region with a width of several atomic layers. The width depends on substitution and on oxygen content; it varies from ~1.0 nm for M = Cu, Zn, Ni, $x = 0.02$, to ~3.0 nm for M = Fe, Al, $x = 0.02$.

(2) The twin boundary in $YBa_2Cu_3O_7$ involves a twinning rotation of the respective lattice across the boundary, and also a translation along the boundary. The $(1/3\text{–}1/2) \cdot 2d_{110}$ lattice shift can be easily detected in samples for M = Cu, Zn, and Ni by HREM. Computer simulations confirmed such observations.

(3) The narrow streaks of the diffuse scattering observed in the SAD pattern arise from the twin boundaries. Calculation of the intensity of the diffuse scattering from periodic twin boundaries with a lattice translation **R** = 1/2[1−10] shows consistent intensity profiles with the observed streaks.

(4) Fringe contrast analysis of the twin boundary suggests that the boundary is an α–δ interface. The δ characteristic arises from the twinning, whereas the α character arises from the lattice displacement.

(5) HREM observations of twin boundaries with different oxygen stoichiometries (including in situ experiments) suggest that there are two types of twin boundaries—one with, and the other without, a lattice shift. For the former, the twin boundary is probably centered through the O atoms in the Cu–O layer, while, for the latter, it is through cation atoms.

(6) Crystallographic analyses reveal that the twin boundary in the $YBa_2Cu_3O_{7-\delta}$ system is a $\Sigma 64$ interface with a twinning shear of **a** − **b**. The observations of twinning steps and twinning dislocations suggest that dislocations may play an important role in forming such a twin boundary with a cation-lattice shift.

Acknowledgments. The authors would like to thank J. Tafto for discussion and assistance in calculating the intensity of diffuse scattering. Useful discussions with D. Welch and Z. Cai are also acknowledged. This research was performed under the auspices of the U.S. Department of Energy, Division of Materials Sciences, Office of Basic Energy Sciences, under Contract No. DE-AC02-76CH00016.

References

1. M.M. Fang, V.G. Kogan, D.K. Finnemore, J.R. Clem, L.S. Chumbley, and D.E. Farrell, Phys. Rev. B **37**, 2334 (1988).
2. G. Deutscher and K.K. Müller, Phys. Rev. Lett. **59**, 1745 (1987).
3. P. Chaudhari, Jpn. J. Appl. Phys. **26-3**, 2023 (1987).
4. P.H. Kes, Physica C **153-5**, 1121 (1988).
5. P.L. Gammel, et al., Phys. Rev. Lett. **59**, 2592 (1987).

6. L.Y. Vinnikov, L.A. Gurevich, G.A. Yemelchenko, and Y.A. Ossipyan, Solid State Commun. **67**, 421 (1988).
7. E.M. Gyorgy, R.B. van Dover, K.A. Jackson, L.F. Schneemeyer, and J.V. Waszczak, Appl. Phys. Lett. **55**, 283 (1989).
8. E.M. Gyorgy, R.B. van Dover, L.F. Schneemeyer, A.E. White, H.M. O'Bryan, R.J. Felder, J.V. Waszczak, and W.W. Rhodes, Appl. Phys. Lett. **56**, 2465 (1990).
9. W.K. Kwok, U. Welp, G.W. Crabtree, K.G. Vandervoort, R. Hulscher, and J.Z. Liu, Phys. Rev. Lett. **64**, 966 (1990).
10. B.M. Lairson, S.K. Streiffer, and J.C. Brarman, Phys. Rev. B **42** 10,067 (1990).
11. R. Bayer, G. Lim, E.M. Engler, R.J. Savoy, T.M. Shaw, T.R. Dinger, W.J. Gallagher, and R.L. Sandstrom, Appl. Phys. Lett. **50**, 1918 (1987).
12. S. Iijima, T. Ichihashi, Y. Kubo, and J. Tabuchi, Jn. J. Appl. Phys. **26**, L1478 (1987).
13. C.S. Pande, A.K. Singh, L.E. Toth, D.U. Gubser, and S.A. Wolf, Phys. Rev. B **36**, 5669 (1987).
14. S.X. Dou, A.J. Bourdillon, C.C. Sorrell, S.P. Ringer, K.E. Easterling, N. Savvides, J.B. Dounlop, and R.B. Roberts, Appl. Phys. Lett. **51**, 535 (1987).
15. M. Sarikaya, R. Kikuchi and I.A. Aksay, Physica C **152**, 161 (1988).
16. T.M. Shaw, S.L. Shimde, D. Dimos, R.F. Cook, P.R. Duncombe, and C. Kroll, J. Mater. Res. **4**, 248 (1989).
17. Y. Syono, M. Kikuchi, K. Oh-ishi, K. Hiraga, H. Arai, Y. Matsui, N. Kobayashi, T. Sasaoka, and Y. Muto, Jpn. J. Appl. Phys. **26**, L498 (1987).
18. G. Van Tendeloo, H.W. Zanbergen and S. Amelinckx, Solid State Commun. **63**, 389 (1987). H.W. Zandbergen, G. Vantendeloo, T. Okabe, and S. Amelinckx, Phys. Status Soidi A **103**, 45 (1987).
19. J.C. Barry, J. Electron Microsc. Technol. **8**, 325 (1988).
20. Z. Hiroi, M. Takano, and Y. Bando, Physica C **158**, 269 (1989).
21. K. Hiraga, D. Shindo, M. Hirabayashi, M. Kikuchi, and Y. Syono, J. Electron Microsc. **36**, 261 (1987).
22. R.A. Camps, J.E. Evetts, B.A. Glowacki, S.B. Newcomb, and W.M. Stobbs, J. Mater. Res. **2**, 750 (1987).
23. Y. Zhu, M. Suenaga, and Y. Xu, Philos. Mag. Lett. **60**, 51 (1989).
24. Y. Zhu, M. Suenaga, Y. Xu, R.L. Sabatini, and A.R. Moodenbaugh, Appl. Phys. Lett. **54**, 374 (1989).
25. Y. Zhu, M. Suenaga, and Y. Xu, J. Mater. Res. **5**, 1380 (1990).
26. Y. Zhu, M. Suenaga, J. Tafto, and D.O. Welch, Phys. Rev. B. **44**, 2871 (1991).
27. Y. Xu, R.L. Subatini, A.R. Moodenbaugh, Y. Zhu, S.G. Shyu, M. Suenaga, K.W. Dennis, and R.W. McCallum, Physica C **169**, 205 (1990)
28. Y. Xu, M. Suenaga, J. Tafto, R.L. Sabatini, A.R. Moodenbaugh, and P. Zolliker, Phys. Rev. B **39**, 6667 (1989).
29. P. Zolliker, D.E. Cox, J.M. Tranquada, and G. Shirane, Phys. Rev. B **38**, 6575 (1988).
30. C.Y. Yang, S.M. Heald, J.M. Tranquada, D.O. Welch, and M. Suenaga, Phys. Rev. B **39**, 6681 (1989).
31. Y. Ossipyan, V.S. Shekhtman, and I.M. Shmyt'ko, Physica C **153–155**, 970 (1988).
32. Y. Zhu, M. Suenaga, and J. Tafto, Philos. Mag. Lett. **64**, 29 (1991).
33. Y. Zhu, M. Pan, Z.G. Li, M. Suenaga, and D.O. Welch, *Proc. 12th International Congress for Electron Microscopy, Seattle, August 1990*, edited by L.D. Peachey and D.B. Williams (1990), Vol. 4, p. 18.

34. Y. Zhu, M. Suenaga, Y. Xu, and M. Kawasaki, Mater. Res. Soc. Symp. Proc. **159**, 413 (1990).
35. P. Bordet, J.L. Hodeau, P. Strobel, M. Marezio, and A. Santoro, Solid State Commun. **66**, 435 (1988).
36. Z. Hiroi, M. Takano, Y. Takeda, R. Kanno, and Y. Bando, Jpn. J. Appl. Phys. Lett. **27**, 580 (1988).
37. A. Umezawa, G.W. Grabtree, J.Z. Liu, T.J. Moran, S.K. Malik, L.H. Zunez, W.L. Kwok, and C.H. Sowers, Phys. Rev. B **38**, 2843 (1988).
38. B. Oh, K. Char, A.D. Kent, M. Naito, M.R. Beasley, T.H. Geballe, R.H. Hammond, and A. Kapitulnik, Phys. Rev. B **37**, 7861 (1988).
39. T. Matsushita, K. Fumaki, M. Takeo, and K. Yamafuji, Jpn. J. Appl. Phys. **26**, L1524 (1987).
40. D.K. Finnemore, K. Athreya, S. Sanders, Y. Xu, and M. Suenaga (unpublished).
41. R.N. Shelton, J.L. Peng, Y. Xu, A.R. Moodenbaugh, and M. Suenaga (unpublished).
42. R. Wördenweber, G.V.S. Sastry, K. Heinemann, and H.C. Freyhardt, J. App. Phys. **65**, 1649 (1989).
43. S. Iijima, T. Ichihashi, Y. Kubo and J. Tabuchi, Jpn. J. Appl. Phys. **26**, L1790 (1987).
44. T.E. Mitchell, T. Roy, R.B. Schwarz, J.F. Smith, and D. Wohlleben, J. Electron Microsc. Technol. **8**, 317 (1988).
45. T. Roy, K. Sickafus, F.W. Clinard, and T.E. Mitchell, G.W. Bailey, Eds., *Proc. 47th Annual Meeting of the Electron Microscopy Society of America* (1989), p. 184.
46. K. Sasaki, K. Kuroda, H. Saka, and T. Imura, J. Electron Microsc. **36**, 232 (1987).
47. J.F. Smith and D. Wohlleben, Z. Phys. B **72**, 323 (1988).
48. J. Zou, D.J.H. Cockayne, G.J. Auchterlonie, D.R. McKenzie, S.X. Dou, A.J. Bourdillon, C.C. Sorrell, K.E. Easterling, and A.W.S. Johnson, Philos. Mag. Lett. **57**, 157 (1988).
49. D.J. Eaglesham, C.J. Humphreys, N.M. Alford, W.J. Clegg, M.A. Harmer, and J.D. Birchall, Appl. Phys. Lett. **51**, 457 (1988).
50. P.A. Midgley, R. Vincent, and D. Cherns (preprint).
51. C.J. Jou and J. Washburn, J. Mater. Res. **4**, 795 (1989).
52. E.A. Hewat, M. Dupuy, A. Bourret, J.J. Capponi, and M. Marezio, Solid State Commun. **64**, 517 (1987).
53. G. van Tendeloo, D. Brodden, H.W. Zandbergen, and S. Amelinckx, Physica C **167**, 627 (1990).
54. J. Kulik, Z.J. Huang, J. Bechtolf, Y.Y. Xue, and P.H. Hor, and G.W. Bailey, Eds., *Proc. of 49th Annual Meeting of the Electron Microscopy Society of America* (San Francisco Press, San Francisco, 1991), p. 1084.
55. S.C. Cheng, S.S. Sheinin, T. Jung, M.K. Yu, and J.P. Frank, Phys. C (submitted).
56. S. Amelinckx and J. van Landuyt, *in Electron Microscopy in Mineralogy*, edited by J.M. Christie, J.M. Colley, A.H. Hener, G. Thomas, and N.J. Tighe (1976), p. 68.
57. J. van Landuyt, R. Gevers, and S. Amelinckx, Phys. Status Solidi **13**, 467 (1966).
58. R. Gevers, J. van Landuyt, and S. Amelinckx, Phys. Status Solidi **11**, 689 (1965).
59. J. Zhu and J.M. Cowley, Acta Crystallogr. Sect. A **38**, 718 (1982).
60. D. van Dyck, G. van Tendeloo, and S. Amelinckx, Ultramicroscopy **15**, 357 (1984).
61. D. de Fontaine, L.T. Wille, and S.C. Moss, Phys. Rev. B **36**, 5709 (1987).

62. D. J. Werder, C.H. Chen, R.J. Cava, and B. Batlogg, Phys. Rev. B **38**, 5130 (1988).
63. Y. Zhu, A.R. Moodenbaugh, M. Suenaga, and J. Tafto, Physica C **167**, 363 (1990).
64. G.P.E.M. Van Bakel, P.A. Hof, J.P.M. Van Engelen, P.M. Bronsveld, and J.Th.M. De Hosson, Phys. Rev. B **41**, 9502 (1990).
65. Z. Cai and S. D. Mahanti, Phys. Rev. B **40**, 6558 (1989).
66. C. Thomsen, M. Cardona, R. Liu, B. Gegenheimer, and A. Simon, Physica C **153–155**, 1756 (1988).
67. A.G. Khachturyan, *Theory of Structural Transformations in Solids* (Wiley, New York, 1983); A.L. Roitburd, Sov. Phys. Solid Sate **10**, 2870 (1969).
68. D. O. Welch (unpublished).
69. N. Chandrasekar, M. Suenaga, and D.O. Welch (unpublished).
70. L.S. Chumbley, J.D. Verhoeven, M.R. Kim, A.L. Cornelius, and M.J. Kramer, IEEE Trans. Magn. **25**, 2337 (1989).
71. T. Roy and T.E. Mitchell, Philos. Mag. A **63**, 225 (1991).
72. S.K. Streiffer, E.M. Zielinski, B.M. Lairson, and J.C. Bravman, Appl. Phys. Lett. **58**, 2171 (1991).
73. R.C. Baetzold, Phys. Rev. B **38**, 304 (1988).
74. P. Baumgart, S. Blumenröder, A. Erlr, B. Hillebrands, P. Splittgerber, G. Güntherodt, and H. Schmidt, Physica C **162–164**, 1073 (1989).
75. R. Wördenweber, G.V.S. Sastry, K. Heinemann, and H. C. Freyhardt, J. Appl. Phys. **65**, 1648 (1989).
76. Y. Zhu (unpublished).
77. C.P. Sun and R.W. Balluffi, Philos. Mag. A **46**, 49 (1982).
78. C. d'Anterroches and A. Bourret, Philos. Mag. A **49**, 738 (1984).
79. S.E. Babcock and R.W. Balluffi, Philos. Mag. A **55**, 643 (1987).
80. F.R. Chen and A.H. King, Philos. Mag. A **57**, 431 (1988).
81. See, for example, *Grain Boundary Structure and Kinetics*, edited by R.W. Bulluffi (American Society for Metals, Metals Park, OH, 1979).
82. C.B. Carter, Acta Metall. **36**, 2753 (1988).
83. R.W. Balluffi, A. Brokman, and A.H. King, Acta Metall. **30**, 1453 (1982).
84. F.C. Frank, Philos. Mag. **42**, 809 (1951).
85. W.T. Read, Jr., *Dislocations in Crystals* (McGraw-Hill, New York, 1953).
86. J.P. Hirth and R.W. Baluffi, ACTA Metall. **21**, 929 (1973).
87. M. Hawley, I.D. Raistrick, J.G. Beery, and R.J. Houlton, Science, **251**, 1587 (1991).
88. S. Jin, G.W. Kammlott, S. Nakahara, T.H. Tiefel, and J.E. Graebner, Science **253**, 427 (1991).

6
Grain-Boundary Josephson Junctions in the High-Temperature Superconductors

RUDOLF GROSS

6.1 Introduction

Defects can determine the electrical transport properties of solids. This statement also holds true for the high-temperature superconductors. In these materials defects play an important role concerning two aspects. On the one hand, high transport critical current densities in these materials can only be achieved by the presence of a high density of defects providing pinning centers for the magnetic flux lines. It is well-known that ideal defects for flux pinning should have a diameter equal to about the coherence length. Furthermore, their density should be high enough to provide sufficient flux pinning at high magnetic fields. Due to the short coherence length of the cuprate superconductors their order parameter is depressed considerably by local defects. In this way point defects can be effective pinning cites for magnetic flux lines increasing the transport critical current density in single crystalline materials. On the other hand, extended defects in the cuprate superconductors in most cases form weak links reducing the transport critical current density. Extended defects in these materials seem to be organized in planes as, for example, stacking faults or small and large-angle grain boundaries. This shows that defects can both increase and decrease the transport critical current density depending on their detailed nature. Grain boundaries are probably the most important defects in the cuprate superconductors. On the one hand, they are known to cause disappointingly low transport critical current density in polycrystalline materials thereby limiting high current density applications. However, on the other hand, their weak-link nature enables the fabrication of high-T_c Josephson elements, which are useful for cryoelectronic devices and circuits. That is, the detailed study of the nature of grain boundaries in the high-temperature superconductors is important both with respect to high and low current density applications.

Lehrstuhl Experimentalphysik II, Universität Tübingen, Morgenstelle 14, D-72076 Tübingen, Germany.

Certainly, the anisotropic nature of the cuprate superconductors constrains the direction of current flow. High critical currents are observed when the current is flowing parallel to the copper oxide planes. However, aligning the CuO_2 planes is not sufficient. It is found from studies of single grain boundaries [1–9] in $YBa_2Cu_3O_{7-\delta}$ that the grain-boundary critical current density is quite sensitive to the misorientation angle [2, 4], even if the CuO_2 planes of the adjacent grains are essentially parallel. Although the vast majority of the work has been carried out on the $YBa_2Cu_3O_{7-\delta}$ superconductor, bicrystals of the bismuth superconductor [10] and, more recently, on bicrystal films of the neodimium, bismuth, and thallium superconductor show similar characteristics. The superconducting transport properties depend strongly on the misorientation angle between the adjoining grains. Below a critical angle the grain boundaries show a critical current density limited by the motion of Abrikosov vortices along the grain boundary and above by Josephson coupling. The critical angle at which the crossover between the two regimes occurs depends on the superconducting system [11].

Although the weak-link nature of grain boundaries in the high-temperature superconductors is disappointing with respect to the use of polycrystalline materials in applications requiring high critical current densities, the same weak-link nature allows the development of grain-boundary Josephson junctions for microelectronics applications. Artificially generated, so-called engineered grain-boundary Josephson junctions (GBJs) have been successfully used to fabricate useful superconducting quantum interference devices (SQUIDs) [12–26]. Understanding the mechanisms responsible for the weak-link nature of grain boundaries and other extended planar defects in the high-T_c oxides is crucial. For high current density applications this will probably allow one to develop remedies to enhance the superconducting coupling across grain boundaries what, in turn, would significantly hasten the use of the high-temperature superconductors for many applications. For those applications making use of the weak-link nature of grain boundaries the knowledge of the mechanisms determining the weak-link properties is required for the development of controllable high-T_c Josephson devices in which single grain-boundaries are used. Up to now different types of artificially generated grain-boundary Josephson junctions have been fabricated and shown to be useful for numerous applications [12–26].

At present, we do not unequivocally understand the precise microstructural origin of the weak-link behavior of grain boundaries in the cuprate superconductors. However, as we will discuss in this review, recent experiments narrow the range of possible mechanisms. At present, most valuable information is obtained from the study of well controllable, individual GBJs. In this article we summarize the characteristics of individual grain-boundary junctions in the high-temperature superconductors based on the investigation of a large number of samples with different misorientation geometries and angles. We will discuss the current-voltage characteristics (IVCs), the resistive transition, the dependence of the grain-boundary critical current J_c

on the misorientation angle and temperature, the magnetic field dependence of the critical current, the temperature and voltage dependence of the normal resistance times area, ρ_N, and the scaling behavior of the $J_c\rho_N$ products. Furthermore, we will discuss the noise characteristics of the GBJs. Based on reproducible experimental data from a large number of samples theoretical models describing the current transport across grain boundaries in the high-temperature superconductors are discussed.

Up to now the majority of work on the study of individual grain boundaries was carried out on $YBa_2Cu_3O_{7-\delta}$. This is caused by the fact that high-quality epitaxial films are required for the fabrication of artificial thin-film GBJs (see Sec. 6.2). Discussing the transport and noise characteristics of GBJs we therefore will concentrate only on $YBa_2Cu_3O_{7-\delta}$ GBJs. Preliminary results on GBJs fabricated from the bismuth, thallium, or neodimiun superconductors [10, 11] indicate that the behavior observed for the $YBa_2Cu_3O_{7-\delta}$ GBJs may be general for all high-T_c cuprates. Considering that the discovery of the high-temperature superconductors is less than five years old, the progress with respect to the understanding of the nature of grain boundaries is very rapid. Therefore, much of what we say here should be viewed as a snapshot of our current knowledge.

6.2 Fabrication of Thin-Film Grain-Boundary Junctions

Thin-film grain-boundary junctions (GBJs) can be obtained by different techniques as shown schematically in Figure 6.1. According to their fabrication technique we can distinguish between bicrystal GBJs, biepitaxial GBJs, and step edge GBJs. Bicrystal GBJs are obtained by the epitaxial growth of a high-T_c film on a bicrystalline substrate. In general, every bicrystalline substrate allowing the epitaxial growth of a high-T_c film on top of it is suitable for this technique. Up to now $SrTiO_3$ [2, 4] or Y–ZrO_2 [18] bicrystals have been used as substrates. In this technique, a single crystalline substrate is cut and fused together to form a bicrystal with a misorientation that can be chosen arbitrarily. By the epitaxial growth of a $YBa_2Cu_3O_{7-\delta}$ film on this substrate a thin-film bicrystal with the same misorientation as the substrate is obtained. A basic advantage of this technique is the possibility of fabricating GBJs with different misorientation geometries, such as [001] and [100] tilt or [100] twist grain boundaries (see Ref. [4]), and arbitrary misorientation angles. In Figure 6.1 a [001] tilt GBJs is shown. The bicrystal technique is not very suitable when a high number of junctions is needed at different positions on the substrate.

Biepitaxial GBJs are fabricated by controlling the in-plane epitaxy of the deposited high-T_c film using seed and buffer layers [19–21]. A prerequisite of this technique is the availability of a seed layer which both grows epitaxially on a substrate with a defined angle of rotation of major symmetry axes and serves as a substrate for the epitaxial growth of a high-T_c film or buffer layer

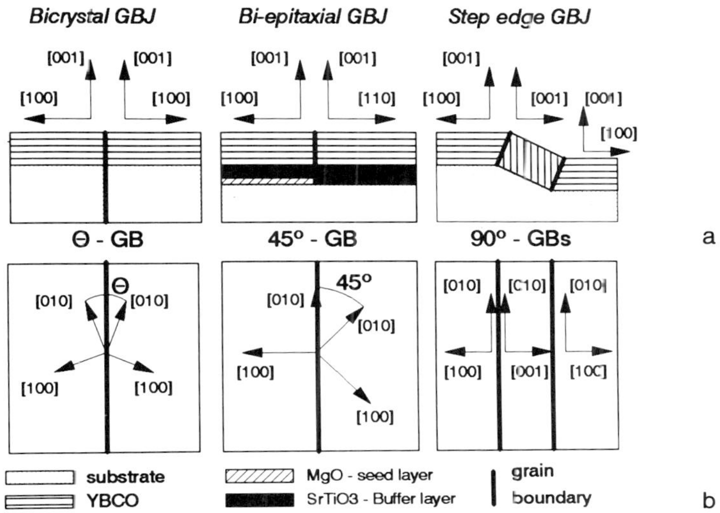

FIGURE 6.1. Configuration of different types of GBJs: bicrystal, biepitaxial, and step edge GBJ. (a) and (b) show a cross-sectional and planar view, respectively.

without rotation. Furthermore, the substrate itself must be suitable for the epitaxial growth of the high-T_c film or buffer layer. Of course, the opposite situation, that is, the seed layer grows without rotation but causes rotation of the high-T_c film or buffer layer is also possible. Using this technique asymmetric 45° grain boundaries could be produced using a MgO seed and a $SrTiO_3$ buffer layer on an *r*-plane sapphire substrate. The use of a MgO seed and a CeO_2 buffer layer on a $SrTiO_3$ substrate is also possible [20]. An important advantage of this technique is the possibility of photolithographically defining the grain boundary at an arbitrary position. However, it is difficult to fabricate GBJs with different misorientation geometries and arbitrary misorientation angles. Up to now, only 45° GBJs have been fabricated by this technique.

Step edge GBJs are fabricated using a steep step in a single crystalline substrate, which must be suitable for the epitaxial growth of a high-T_c film [22–25]. At a sufficiently steep step angle ($\alpha > 45°$) two 90° grain boundaries are generated in the high-T_c film covering the step. The position of the step can be chosen at an arbitrary position on the substrate. However, in this technique it is difficult to vary the misorientation angle.

Thin-film grain-boundary junctions, have also been fabricated in $YBa_2Cu_3O_{7-\delta}$ films grown on single crystalline MgO substrates. [27, 28]. Depending on the preparation of the MgO substrate and the film deposition parameters, $YBa_2Cu_3O_{7-\delta}$ films can be produced that are fully *c*-axis oriented, but contain a variable amount of large-angle grain boundaries. The preferred film orientation is with the *a* and *b* axis aligned with the MgO cubic axis, however, there are also grains rotated by 45° about the c axis [27, 28].

Individual GBJs are obtained by patterning a microbridge straddling the grain boundary in the bicrystalline high-T_c film. Typically, the width W and the length L of the microbridge are a few microns, the thickness d of the film is a few 100 nm. The microbridge is patterned using laser ablation or photolithography and ion-beam milling. With respect to the application of GBJs in cryoelectronic circuits we note that both bicrystal and biepitaxial GBJs can be fabricated in a controlled way and thermally cycled between 77 K and room temperature many times without observing any change in their characteristics. For the step edge GBJs a precise control and good stability of the junction parameters might be more difficult to achieve. Furthermore, in the step edge technique two GBJs connected in series are always obtained. All types of GBJs can be fabricated with a high yield.

6.3 Structural Properties

Reports on impurity segregation [29, 30] at grain boundaries and variations of the chemical stoichiometry [31, 32] have led to the view that deviations from the ideal composition at the grain boundary are responsible for their weak-link behavior. However, recent images of $YBa_2Cu_3O_{7-\delta}$ grain boundaries obtained by a scanning transmission electron microscope in the Z-contrast mode [33] show that chemical segregation does not necessarily occur at these boundaries. However, GBJs without impurity phases and the correct chemical stoichiometry also show weak-link behavior. Unfortunately, the STEM analysis is quite insensitive to oxygen because of its relatively low atomic number. Therefore, it cannot provide information on the exact oxygen stoichiometry at the grain boundary, which may be crucial for the electric transport characteristics of GBJs. In general, the structural analysis of artificial grain boundaries generated as described above show that these boundaries are clean without any second phases and with the lattice distortions confined to within one to two lattice spacings [4].

The bicrystal grain boundaries consist of an array of uniformly spaced edge or screw dislocation produced to accommodate the mismatch [34]. For the biepitaxial GBJs the situation is expected to be very similar. A simple model of the strain associated with the grain-boundary dislocations can provide a simple physical explanation of the reduced critical current density. Assuming that the superconducting order parameter is depressed within the strained region associated with the grain-boundary dislocations and that these strained regions have a well-defined radius r_m in the plane of the boundary, the critical current density across the grain boundary should decrease for small misorientation angles ($\theta < 10°$) as [33]

$$J_c(\theta)/J_c(0) = 1 - \frac{2r_m}{|\mathbf{b}|}\theta. \tag{1}$$

Here, **b** is the Burgers vector of the dislocation. In this model, the decrease of J_c with increasing θ is simply caused by a reduction of the effective area of the grain-boundary plane. Equation (1) describes the almost linear decrease of the grain-boundary critical current density found for small-angle $YBa_2Cu_3O_{7-\delta}$ bicrystal GBJs [4]. For larger misorientation angles ($\theta > 10°$) the boundary seems to lose the structural order required for superconductivity and becomes a weak link. That is, strongly coupled regions are no longer present in these grain boundaries. Strain also plays an important role for the higher angle grain boundaries. This is caused by the strong sensitivity of the superconducting properties of the high-T_c cuprates to strain. For $YBa_2Cu_3O_{7-\delta}$, for example, superconductivity is produced by the addition of charge to the CuO_2 planes. Disruption of the charge reservoir layers by disordering the CuO chain layer can destroy superconductivity. This may lead to a significant volume of non-superconducting, most likely insulating material surrounding each dislocation core. Weak-link behavior is obtained if these non-superconducting regions overlap. For $YBa_2Cu_3O_{7-\delta}$ GBJs this seems to be the case for $\theta > 10°$.

Strain cannot arise only due to dislocations but also due to a large difference of the thermal expansion coefficients of the substrate material and the high-T_c film. This is very important for the biepitaxial GBJs, which have a complicated layer structure. It is likely that the different transport characteristics observed for junctions, which were fabricated with varying layer structures [20], are caused by strain effects resulting from a mismatch of the thermal expansion coefficients. For the 90° grain boundaries in step edge GBJs the dislocation array should be widely spaced and it is difficult to explain their weak-link behavior by the dislocation core idea. Although the detailed structure of these grain boundaries is not well studied up to now, it seems that the strain fields at the boundary, rather than the dislocations themselves, may be responsible for the suppression of superconductivity at these boundaries.

6.4 Transport Characteristics of $YBa_2Cu_3O_{7-\delta}$ Grain-Boundary Junctions

6.4.1 *Resistive Transition*

A typical resistive transition of a $YBa_2Cu_3O_{7-\delta}$ bicrystal GBJ is shown in Figure 6.2. The sharp drop in resistance at about 90 K originates from the resistive transition of the grain material forming the microbridge parts adjoining the grain boundary. The foot structure at the bottom of the resistive transition, which is shown more clearly in the inset, is caused by thermally activated phase slippage (TAPS) [6, 37]. For an overdamped Josephson junction TAPS results in a finite resistance $R_P(T) = R_N\{I_0[\gamma(T)/2]\}^{-2}$, that is, in a finite slope of the IVCs, even for $I \rightarrow 0$ [6]. Here, I_0 is the modified Bessel function, R_N is the normal resistance of the GBJ, and the dimension-

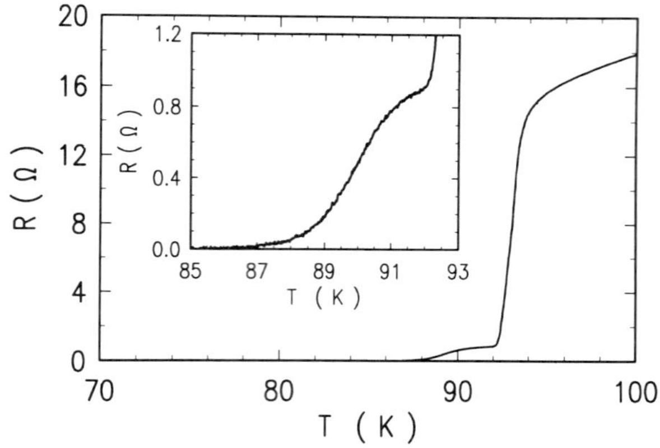

FIGURE 6.2. Resistance versus temperature curve of a $YBa_2Cu_3O_{7-\delta}$ grain-boundary junction. The inset shows part of the transition curve on an enlarged scale.

less parameter

$$\gamma(T) = \frac{2E_J}{k_B T} = \frac{\hbar I_c(T)}{ek_B T} = \frac{\Phi_0 I_c(T)}{\pi k_B T} \tag{2}$$

is given by the ratio of twice the Josephson coupling energy E_J of the GBJ and the thermal energy $k_B T$. The dimensionless parameter $\gamma(T)$ represents a normalized barrier height for the thermally activated slippage of the phase difference across the grain boundary. $\gamma(T)$ increases with decreasing temperature due to the increase of I_c and the decrease of the thermal energy. Note that $\gamma(T)$ is not determined by the critical current density but by the critical current $I_c(T) = J_c(T)A$, where A is the area of the GBJ. That is, a small γ is obtained both by a small J_c and a small area of the GBJ. Decreasing the temperature a measurable resistance $R_P(I \to 0)$ is observable down to a temperature T^* well below the critical temperature T_c of the grain material. T^* is given by the temperature at which the coupling energy of the junction E_J is sufficiently large to make TAPS and, hence, $R_P(I \to 0)$ unmeasurably small. The value of T^* is determined by the magnitude of E_J, that is, for a fixed value of J_c by the area A of the GBJ, and by the threshold criterion used for the resistance measurement. Fabricating, for example, GBJs with varying area, that is, with varying γ, on the same substrate results in different temperatures T^* for each GBJ, where T^* is decreasing with decreasing junction area. This shows that T^* has no physical meaning and is to be distinguished from the critical temperature T_c of the grain material. There is considerable confusion in literature on this point.

In Figure 6.3 the normalized resistive transition of a bicrystal GBJ is plotted on a logarithmic scale together with R_P/R_N dependences calculated according to the resistively shunted junction (RSJ) model [6, 35, 37] includ-

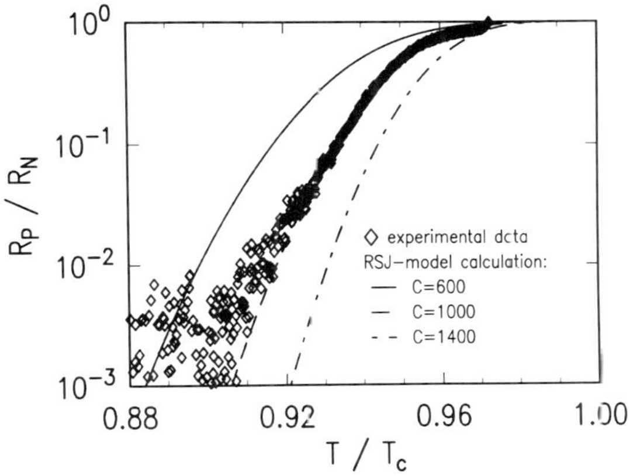

FIGURE 6.3. Experimental R_P/R_N dependence (diamonds) of a $YBa_2Cu_3O_{7-\delta}$ GBJ and curves calculated according to the resistively shunted junction (RSJ) model including thermally activated phase slippage. The best fit is obtained for $\gamma(T) = C(1 - T/T_c)^2$ with $C = 1350$.

ing TAPS. Figure 6.3 demonstrates that the broadened resistive transition of the GBJ can be fitted well by the RSJ model including TAPS. Fitting the data yields the temperature dependence of γ, that is, of I_c. For the $YBa_2Cu_3O_{7-\delta}$ GBJ the best fit is usually obtained for $\gamma(T) \propto (1 - T/T_c)^2$, that is, we have $I_c(T) \propto (1 - T/T_c)^2$ close to T_c.

The broadened resistive transition shown in Figures 6.2 and 6.3 should also be observed for step edge and biepitaxial GBJs. However, for these junctions the ideal Bessel function dependence $R_P(T) = R_N\{I_0[\gamma(T)/2]\}^{-2}$ often is not observed. This indicates that these GBJs may deviate considerably from an ideal RSJ-type junction. As will be discussed below such deviations are also observed with respect to the current-voltage characteristics and the magnetic field dependence of the critical current. It is possible that the deviations are caused by the presence of strongly coupled regions at the grain boundary. Such regions will both yield resistive transitions and IVCs deviating from the RSJ model prediction and are expected to cause a strongly inhomogenous current density distribution along the grain boundary, which in turn results in unusual $I_c(B)$ dependences.

6.4.2 *Current-Voltage Characteristics*

In order to characterize the current-voltage characteristics (IVCs) of the GBJs one has to distinguish between narrow and wide GBJs. Here, GBJs with a width $W \leq 4\lambda_J$ are denoted narrow and those with $W \geq 4\lambda_J$ wide GBJs. The characteristic length $\lambda_J = (\hbar/2e\mu_0 J_c t)^{1/2}$ is the Josephson penetration depth [35], where t is about two times the London penetration depth λ_L

of the grain material. Using $t \simeq 2\lambda_L \simeq 300$ nm, we obtain $\lambda_J \simeq 300\ \mu\text{m}/(J_c)^{1/2}$ taking J_c in A/cm^2. That is, $\lambda_J \simeq 1\ \mu$m for $J_c = 1 \times 10^5$ A/cm^2. Note that the bicrystal, biepitaxial, and step edge GBJs usually have an overlap geometry [35]. For such junctions the critical current does not saturate when the width of the junction becomes larger than about $4\lambda_J$ but increases proportional to the width [35].

The IVCs of narrow GBJs with misorientation angles larger than about 10° are close to those calculated according to the RSJ model [6, 8, 18–20, 25, 27, 35]. The differential resistance $R_d = dV/dI$ decreases with increasing voltage approaching a constant value $R_d = R_N$ at higher voltages ($V \geq I_c R_N$). Here, in accordance with the RSJ model, R_N is denoted as the normal resistance of the GBJ. For the bicrystal, biepitaxial, and the step edge GBJs the normal resistance of the GBJs is found to be independent of temperature between 4.2 K and T_c [6, 8, 20, 25]. From the critical current density J_c, the normal resistance times area, ρ_N, and the specific capacitance C of the GBJs one obtains their plasma frequency [35]

$$\omega_p = \left(\frac{2eJ_c}{\hbar C}\right)^{1/2}, \tag{3}$$

their characteristic frequency [35]

$$\omega_c = \frac{2eJ_c\rho_N}{\hbar}, \tag{4}$$

and their McCumber parameter [36]

$$\beta_c = \frac{2eJ_c\rho_N^2 C}{\hbar}. \tag{5}$$

For satisfactory performance in most applications sufficiently high ω_c, say between about 10^{12} and 10^{13} s^{-1} at 77 K, and a high plasma frequency, that is, a high J_c and a low junction capacitance is required. Whereas high J_c values ($>10^4$ A/cm^2 at 77 K) and, hence, high plasma frequencies can be achieved, the small $J_c\rho_N$ products of GBJs result in small values for ω_c (typically $<10^{12}$ s^{-1} at 77 K). The McCumber parameter of the GBJs is usually smaller than 1 for temperatures ranging between 4.2 K and T_c resulting in nonhysteretic IVCs [35, 36]. Only some GBJs on $SrTiO_3$ bicrystals show hysteretic IVCs ($\beta_c \leq 5$) for temperatures smaller than about 30 K [8]. The reason for that is the relatively large specific capacitance of GBJs on $SrTiO_3$ bicrystals, which is determined by the considerable stray capacitance caused by the high dielectric constant of $SrTiO_3$ at low temperatures [8].

For T close to T_c the IVCs are rounded by thermally activated phase slippage (TAPS) [6, 37]. TAPS results in a finite resistance $R_P(T) = R_n \{I_0[\gamma(T)/2]\}^{-2}$, that is, in a finite slope of the IVCs, even for $I \to 0$ [6]. The temperature regime ($T^* < T < T_c$) where the rounding of the IVCs is observable is determined by the coupling energy of the GBJ. The rounded IVCs can be closely modeled by the RSJ model if TAPS is included [6].

The IVCs of wide GBJs usually deviate from these predicted by the RSJ model, which, of course, is applicable only for Josephson junctions with dimensions smaller than the Josephson penetration depth. Wide GBJs show a finite so-called excess current I_{exc} that is, at higher voltages their IVCs approach an ohmic line $V = R_N(I - I_{exc})$ in contrast to the narrow GBJs, where $I_{exc} = 0$. The behavior of the wide GBJs may be explained by the onset of flux motion along the grain boundary. The critical current of a wide GBJ corresponds to the current where flux lines begin to enter the junction. These flux lines are not usual Abrikosov vortices with a normal core. What we call flux lines here are really only regions in which the phase difference across the junction changes by 2π. It has been shown [38] by extending the RSJ model to the case of larger junctions, that the detailed dynamics of the flux line motion results in a finite excess current for $W/\lambda_J \geq 4$ with $I_{exc} \rightarrow I_c$ for $W/\lambda_J \rightarrow \infty$. This qualitatively agrees with our experimental results. Due to the higher operation temperatures of the oxide superconductors also thermally activated motion of the flux lines has to be taken into account. This may explain the smooth voltage onset observed in the IVCs of some wide GBJs in a similar way as thermally activated motion of Abrikosov vortices explains the smooth voltage onset in the IVCs of the fully epitaxial lines.

Current-voltage characteristics deviating from the RSJ model curves are also observed for bicrystal GBJs with small misorientation angles ($\theta < 10°$). These GBJs show flux-flow-type IVCs. Here, in contrast to the RSJ-type IVCs, the differential resistance increases with increasing current and does not approach at a constant value at $I \gg I_c$. This can be understood in terms of the microstructure of small-angle GBJs. Since the regions of depressed superconducting order parameter around the dislocation cores probably do not overlap (see Sec. 6.3), strongly coupled regions may be present along the grain boundary. Therefore, flux-flow-type behavior instead of weak coupling RSJ-type behavior is expected. Flux-flow-type IVCs are also observed for some step edge GBJs. The reason why some step edge GBJs show flux-flow and some RSJ-type IVCs is not understood up to now. However, it is likely that the presence of strongly coupled regions is responsible for this behavior.

6.4.3 *Dependence of the Critical Current Density on the Misorientation Angle*

The dependence of the critical current density on the misorientation angle θ between the adjoining grains has been studied for [001] and [100] tilt and [100] twist $YBa_2Cu_3O_{7-\delta}$ grain boundaries fabricated on $SrTiO_3$ [1–4, 12, 49] and Y–ZrO_2 bicrystals [18]. Such a study is difficult to perform with biepitaxial or even step edge GBJs, since their fabrication technique does not allow to vary the misorientation angle in an arbitrary way. The measurements on individual bicrystal GBJs have provided direct evidence that grain boundaries are weak links. As discussed above, for small misorientation angles the weak-link behavior is dominated by flux motion along the grain boundary. However, above a critical misorientation angle, about 10° for $YBa_2Cu_3O_{7-\delta}$, the weak-link behavior changes from a flux-flow to a Josephson coupling-type behavior.

The critical current density J_c across the grain boundaries is found, within the experimental error to be a function only of the misorientation angle, if the $YBa_2Cu_3O_{7-\delta}$ films are prepared under the same conditions. For each type of misorientation geometry, J_c decreases monotonously with increasing θ. The amount by which J_c decreases with increasing θ is about the same for different misorientation geometries and for different substrate materials. This is shown in Figure 6.4 for $YBa_2Cu_3O_{7-\delta}$ GBJs fabricated on $SrTiO_3$ and

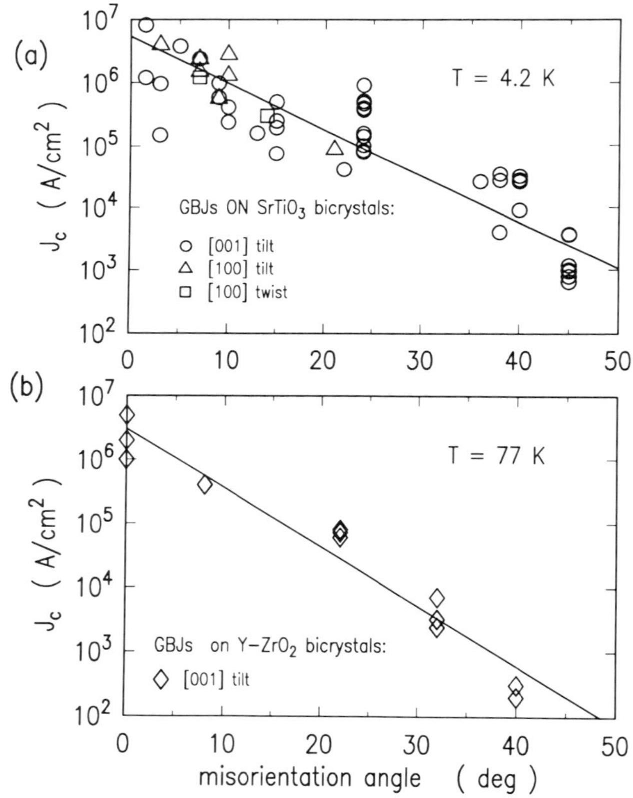

FIGURE 6.4. Variation of the critical current density of [100] twist and [001] and [100] tilt GBJs on $SrTiO_3$ bicrystals with the misorientation angle at 4.2 K (a). In (b) the same dependence is shown for [001] tilt GBJs on $Y\text{–}ZrO_2$ bicrystals at 77 K. The lines represent least-square fits to the data.

Y–ZrO_2 bicrystals. For small θ ($< 10°$) the decrease of J_c can be understood by the decreasing distance of the dislocations along the grain boundary. However, for larger misorientation angles ($\theta \geq 10°$) the regions of depressed order parameter should overlap and the observed further reduction of J_c with increasing θ is not evident. According to the GBJ models discussed below the decrease of J_c should be related to an increasing thickness of the perturbed interface layer with increasing misorientation angle. This interpretation is reasonable since strain and disorder at the grain boundary are expected to increase with increasing θ.

In Figure 6.4 we have plotted the absolute value of J_c versus the misorientation angle and not the J_c value normalized to the critical current density J_c^G of the adjoining grain material. Since J_c^G is determined by the flux pinning strength in the grain material, whereas J_c is determined by the Josephson coupling between the grains, the ratio J_c/J_c^G contains information both on the transport properties of the grain boundary and the grains. Therefore, the true angle dependence of the grain-boundary critical current density can be masked by variations of the grain properties. For example, a smaller value of J_c^G causes a larger ratio J_c/J_c^G even if the absolute magnitude of J_c stays constant. On the other hand, a reduced value of J_c/J_c^G can originate simply from a higher value of J_c^G, caused, for example, by a better flux pinning in the grain material. For bicrystal GBJs the absolute magnitude of J_c is found to be similar for samples with different J_c^G values [12, 49]. This gives evidence that J_c only depends on the weak-link properties of the grain boundary. Obviously, there is no relation between J_c and J_c^G.

As shown in Figure 6.4(a), J_c decreases monotonously with increasing misorientation angle. However, for the same misorientation angle J_c can be varied by more than one order of magnitude. This variation is achieved by different film deposition conditions and/or by post-deposition oxygen or ozone annealing [7, 9, 17]. Using the same film deposition conditions the spread of J_c for the same misorientation angle is small [6–9, 12]. The wide range of achievable current density values of bicrystal GBJs and their good reproducibility is important for device applications of these GBJs. The strong angle dependence of J_c shows that large misorientation angles should be avoided in polycrystalline materials, if high current densities are required.

6.4.4 *Scaling Behavior of the Characteristic Voltage*

Beyond the dependence of the critical current density on the misorientation angle there is a second general feature observed for $YBa_2Cu_3O_{7-\delta}$ GBJs. It relates their characteristic voltage, $V_c \equiv J_c\rho_N$, to their critical current density, J_c, or equivalently, to the inverse of the normal resistivity times area, $1/\rho_N$. There are independent sets of measurements for GBJs fabricated on $SrTiO_3$ bicrystals [7] and on MgO single crystals [27], and for step edge GBJs [25], which show that for a variety of grain boundaries in the $YBa_2Cu_3O_{7-\delta}$ superconductor the characteristic voltage scales as

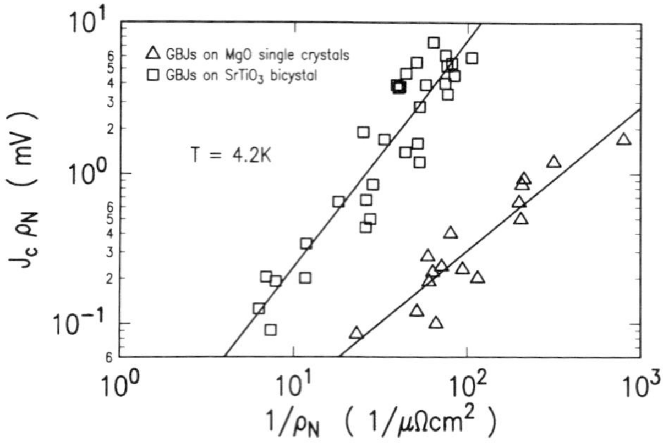

FIGURE 6.5. Variation of the characteristic voltage $V_c = J_c\rho_N$ of GBJs fabricated on single crystalline MgO (data from Ref. [27]) and bicrystalline $SrTiO_3$ substrates with their normal conductance per unit area, $1/\rho_N$, at 4.2 K. The lines represent least-square fits to the data.

$$J_c\rho_N = I_cR_N \propto (1/\rho_N)^q$$

or

$$J_c\rho_N = I_cR_N \propto (J_c)^{q/q+1}. \tag{6}$$

This scaling behavior is shown in Figure 6.5 for GBJs fabricated both on single crystalline MgO and bicrystalline $SrTiO_3$ substrates. Figure 6.5 shows that $J_c\rho_N = I_cR_N \propto (1/\rho_N)^q$ with $q \approx 1$ and $q \approx 1.5$ for the GBJ on single crystalline MgO and on $SrTiO_3$ bicrystals, respectively. The same scaling behavior, with q ranging between 1.0 and 1.5, is found for $YBa_2Cu_3O_{7-\delta}$ step edge GBJs [25]. At present, it is unclear whether the biepitaxial GBJs show a similar scaling behavior, since the scatter of the available experimental data is large [20]. For the latter type of GBJ the ρ_N values strongly depend on the substrate and the materials used for the seed and buffer layers. This suggests that stress effects at the grain boundary due to different thermal expansion coefficients of the materials used in the multilayer structure plays an important role. Due to the stress effects only a fraction of the junction width may act as an active region. This, of course, would prevent the observation of the scaling behavior found for the bicrystal and step edge GBJs.

The fact that the scaling behavior $V_c \propto (1/\rho_N)^q$ is observed for a large variety of different samples indicates that it is related to an intrinsic property of the GBJs. Since the parameters of bicrystal GBJs shift along the $J_c\rho_N \propto (1/\rho_N)^q$ dependence by post-deposition oxygen or ozone annealing [9, 17, 27], there is strong evidence that the observed behavior is related to the oxygen stoichiometry at the grain boundary. The question why different values of q and the absolute magnitude of the $J_c\rho_N$ products are observed for

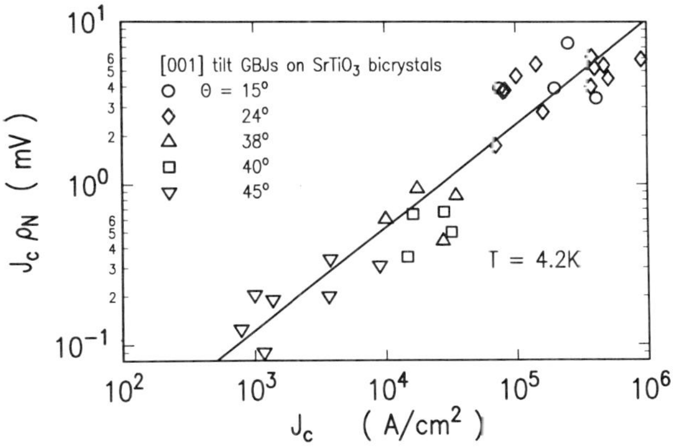

FIGURE 6.6. Characteristic voltage $V_c = J_c\rho_N$ of [001] tilt bicrystal GBJs ($\theta = 15°$, 24°, 38°, 40°, 45°) plotted versus J_c. The line represents a least-square fit to the data.

the GBJs fabricated on single crystalline MgO and on $SrTiO_3$ bicrystals cannot be answered at present. This difference may be related to the diffusion of substrate material into the grain boundary, to different film deposition conditions, or to stress caused by different thermal expansion coefficients of the substrate and the high-T_c film.

In Figure 6.6 the $J_c\rho_N$ products of [001] tilt bicrystal GBJs with five different misorientation angles are plotted versus their critical current density. Figure 6.6 shows that both $J_c\rho_N$ and J_c decrease with increasing misorientation angle. Since all biepitaxial GBJs fabricated up to now have $\theta = 45°$, they have lower $J_c\rho_N$ products and J_c values than bicrystal GBJs with smaller misorientation angles. This is important for applications requiring high characteristic voltages. The V_c values of bicrystal GBJs fabricated on $SrTiO_3$ bicrystals range between about 0.1 and 8 mV at 4.2 K [7, 8, 12, 49]. Biepitaxial GBJs and GBJs fabricated on MgO single crystals (both with $\theta = 45°$) have characteristic voltages which are typically smaller than about 1 mV at 4.2 K [20, 27]. At 77 K, the V_c values are a factor of about 20 smaller than those at 4.2 K.

For classical *SIS* (S = superconductor, I = insulator) tunnel-type Josephson junctions, the characteristic voltages $V_c = J_c\rho_N = I_cR_N$ is given by the Ambegoakar–Baratoff expression [39]

$$J_c(T)\rho_N = \frac{\pi\Delta(T)}{2e}\tanh\frac{\Delta(T)}{2k_BT}. \tag{7}$$

At low temperatures, $V_c \simeq \pi\Delta/2e$, that is, the characteristic voltage is about equal to the gap voltage Δ/e and is independent of ρ_N. This is in contrast to the $YBa_2Cu_3O_{7-\delta}$ GBJs, where the characteristic voltages are small compared to the gap voltage of $YBa_2Cu_3O_{7-\delta}$ ($\Delta/e \sim$ 15–30 mV) and scale proportionally to $(1/\rho_N)^q$. Possible reasons for this behavior are discussed in Section 6.6.

6.4.5 *Temperature Dependence of the Normal Resistance and the Critical Current*

The IVCs of most GBJs are close to the RSJ model prediction. Therefore, a well-defined normal resistance, R_N, can be attributed to these GBJs. For bicrystal, biepitaxial, and step edge GBJs, the normal resistance determined in accordance with the RSJ model is found to be independent of temperature in the entire temperature regime between 4.2 K and T_c [7, 8, 20, 25].

In the RSJ model a voltage independent normal resistance is assumed resulting in a constant conductance $dI(V)/dV$ at high voltages $V \gg V_c$. This is not observed for almost all GBJs. For most bicrystal GBJs, $dI(V)/dV$ is found to increase with increasing voltage at $V \gg V_c$. Interestingly, for some bicrystal GBJs an almost linear increase of the conductance with increasing voltage, $dI(V)/dV \propto |V|$, was found for $V \gg V_c$ [40]. Such linear $dI(V)/dV$ dependence is also reported for tunneling experiments [41]. In contrast to the bicrystal GBJs, detailed experimental data on the voltage dependence of the conductance of biepitaxial or step edge junctions are not available at present. A general problem for all types of GBJs is the difficulty to measure the $dI(V)/dV$ dependence up to high voltages in a reliable way. Due to their low ρ_N values, excessive Joule heating generates additional resistive regions in the adjoining grains already at small voltages of only a few 10 mV. This allows a reliable measurement of the GBJ conductance only at small voltages and prevents the search for structures in the conductance at higher voltages that may be related to the energy gap of the grain material.

It is well known that a white energy distribution of localized defect states in the insulating layer of metal–insulator–metal tunnel junctions results in a linear increase of the conductance with increasing voltage. That is, the presence of such states in an insulating grain-boundary layer could account for the observed linear increase of the conductance of the bicrystal GBJs. In this respect, it is important to note that some small area GBJs ($A \leq 2\ \mu m^2$) showed a large number of peaks in the $dI(V)/dV$ characteristics. These peaks may be caused by resonant tunneling via localized states. For larger area junctions such peaks in conductance probably were not resolved because they are so numerous that they effectively form a continuum. Also, the peaks disappeared at higher temperatures most likely by thermal broadening. A similar behavior has been found for the tunneling conductance of thin *a*-Si barriers [42] where resonant tunneling via localized states plays an important role.

The temperature dependence of the critical current at small reduced temperatures ($T/T_c \leq 0.9$) varies considerably for the different GBJs. Both almost linear temperature dependences (biepitaxial and step edge GBJs) and dependences, which are close to the Ambegaokar–Baratoff expression with a

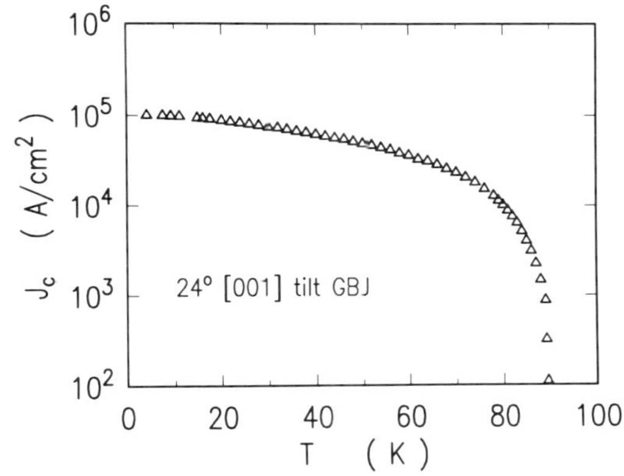

FIGURE 6.7. Temperature dependence of the critical current of a 24° [001] tilt bicrystal grain-boundary junction.

reduced energy gap Δ_i at the grain boundary (bicrystal GBJs) have been reported [3, 8, 20, 25]. A typical $J_c(T)$ dependence of a bicrystal GBJ is shown in Figure 6.7. For the bicrystal GBJs the Ambegaokar–Baratoff like $I_c(T)$ dependence is observed only for misorientation angles larger than about 10°. Low-angle GBJs show an almost linear $I_c(T)$ dependence which is close to that observed for epitaxial $YBa_2Cu_3O_{7-\delta}$ films. Although the reason responsible for the observation of different temperature dependences for different types of GBJs is not clear at present, the different $I_c(T)$ dependences observed for small- and large-angle bicrystal GBJs suggest that the presence of strongly coupled regions play an important role. Furthermore, in some cases the transition from a small ($W/\lambda_J \leq 4$) to a large ($W/\lambda_J \geq 4$) GBJ caused by the decrease of λ_J with decreasing the temperature, may have some influence on the measured $I_c(T)$ dependence.

The characteristic voltage $V_c(T) = I_c(T)R_N$ has the same temperature dependence as I_c, since R_N is independent of temperature. For bicrystal GBJs with large misorientation angles ($\theta > 10°$) the temperature dependence of the normalized characteristic voltage $V_c(T)/V_c(4.2\text{ K})$ was found to be almost identical for all GBJs. That is, the functional form of the $V_c(T)$ or equivalently $I_c(T)$ dependence does not depend on the absolute magnitude of V_c (0.1 to 8 mV at 4.2 K).

For T close to T_c, the temperature dependence of the real critical current I_c of the GBJs, that is, the critical current without any reduction due to thermal fluctuation effects, is difficult to measure due to the presence of TAPS [6, 8, 37]. As a consequence of TAPS the measured critical current I_{cm} will be zero between T_c and the temperature T^* at which TAPS causes a resistance R_P as large as the applied resistance criterion in the critical current measurement. Also below T^*, I_{cm} will be smaller than I_c. The deviation of the measured critical current I_{cm} from the real critical current I_c depends on the dimensionless parameter $\gamma(T)$ and the voltage or resistance criterion applied in the critical current measurement [8]. Thus, in general it is difficult to derive the real $I_c(T)$ dependence, which is expected to show a power-law dependence $I_c(T) \propto (1 - T/T_c)^p$ close to T_c, from the measured $I_{cm}(T)$ dependence.

Taking into account the effect of TAPS at T close to T_c, the temperature dependence of the real critical current $I_c(T)$ of the GBJs can be derived both from the resistive transition [6] and from the IVCs [8]. The $I_c(T)$ dependence obtained by such careful analysis follows:

$$I_c(T) = I_{c0}(1 - T/T_c)^2 \qquad \text{for} \quad 0.9 \leq T/T_c \leq 1.0 \tag{8}$$

for the majority of GBJs [6, 8]. Note that for T close to T_c the measured critical current I_{cm} and its temperature dependence have no physical meaning. Unfortunately, in most cases only $I_{cm}(T)$ is reported.

6.4.6 *Magnetic Field Dependence of the Critical Current*

The magnetic field dependence of the critical current yields important information on the homogeneity of the current flow across the grain boundary and on the effective area of the GBJ. For narrow GBJs with a perfectly homogeneous current density the $I_c(B)$ dependence should be a Fraunhofer diffraction pattern [35]. The effective area of the GBJ derived from the magnetic field modulation period of the diffraction pattern should be equal to $2W\lambda_L$ [35]. For narrow bicrystal GBJs, $I_c(B)$ dependences close to a Fraunhofer diffraction pattern were found [7, 18]. This demonstrates that the current flow across the grain boundary is fairly homogeneous.

Figure 6.8 shows, as an example, the modulation of the voltage across a

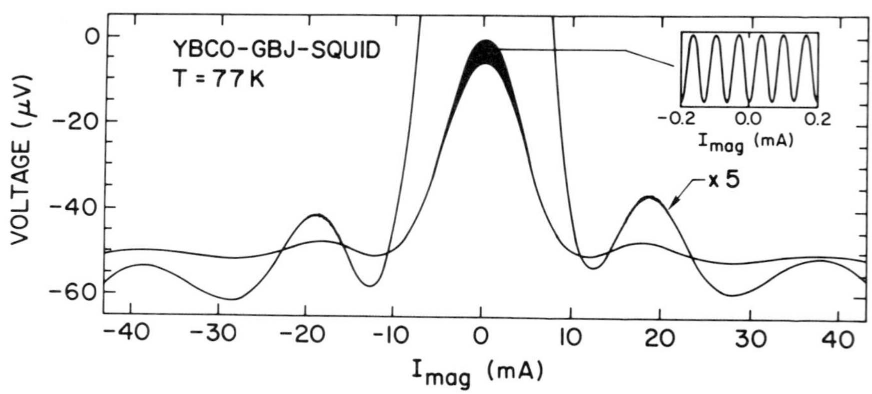

FIGURE 6.8. Modulation of the dc voltage across a current biased bicrystal GBJ dc SQUID caused by a magnetic field applied perpendicular to the plane of the SQUID loop. The inset shows the fine modulation period caused by the magnetic flux threading the SQUID loop, which is superimposed on the coarse modulation caused by the flux threading the area of the GBJs.

current-biased bicrystal GBJ dc SQUID by a magnetic field applied perpendicular to the SQUID loop. The measured dependence is close to the diffraction pattern of a two slit interference experiment [7]. The diffraction pattern clearly shows a fine and coarse modulation period, which correspond to the effective area of the SQUID loop (slit distance) and the effective area of the single GBJs (width of the single slits), respectively. The effective GBJ area derived from the coarse modulation period is found to be equal to $2W\lambda_L$ [7], with $\lambda_L \simeq 150$ nm, a value which is in close agreement with published values. This demonstrates that the current flow across the grain boundary is not restricted to some narrow filaments but flows, though distributed inhomogeneously, across the whole cross-sectional area of the grain boundary. Further evidence for this comes from the fact that the critical current of the GBJs scales proportionally to their geometrical width down to $W < 1$ μm [16]. For narrow bicrystal GBJs, $I_c(B)$ dependences deviating from the ideal Fraunhofer pattern are most likely caused by the spatial inhomogeneity of critical current density. This has been confirmed directly by imaging the spatial distribution of J_c using low-temperature scanning electron microscopy [5]. Similarly, the $I_c(B)$ dependences of wide bicrystal GBJs indicated that the current density distribution along the grain boundary is inhomogeneous [8, 9]. Note that for wide GBJs the spatial modulation of J_c by trapped magnetic flux quanta can play an important role.

The $I_c(B)$ dependences of step edge GBJs [25] usually deviate much stronger from a Fraunhofer diffraction pattern than those of bicrystal GBJs. This suggests that in step edge GBJs the spatial distribution of the current density along the grain boundary is more inhomogeneous than in bicrystal junctions. The situation is similar for bi-epitaxial GBJs. It is likely that the strong spatial variation of J_c observed for step edge and biepitaxial GBJs is related to the microstructure of these GBJs. Stress effects, a strong meandering of the photolithographically defined grain boundary, spatial variations of the step in the substrate, or variations in epitaxy are possible reasons for the observed spatial inhomogeneity of J_c.

6.4.7 *Noise Properties*

The study of the noise properties of GBJs is important both with respect to their applications in cryoelectronic circuits and the clarification of the nature of their weak-link behavior. Detailed noise measurements were performed both with single GBJs and GBJ SQUIDs [8, 9, 12–17, 23–25, 43–47, 49]. In general, all measurements showed the presence of excess low-frequency noise. Using the SQUID geometry one is able to distinguish whether the measured noise is caused by the thermally activated motion of magnetic flux quanta in the high-T_c film or whether the noise results from fluctuations of the junction parameters such as critical current or normal resistance. It has been shown that the large amount of low-frequency noise in $YBa_2Cu_3O_{7-\delta}$ GBJ dc SQUIDs is caused by parameter fluctuations of the GBJs [13–15]. Only at temperatures close to T_c the low-frequency noise is dominated by the contribution caused by thermally activated flux motion.

The white noise level of GBJs and GBJ dc SQUIDs is close to the amount expected from the Nyquist noise of the normal resistance of the GBJs. In the white noise regime, the energy resolution ε of optimized bicrystal GBJ dc SQUIDs [12, 15, 16] was found to closely follow the model prediction [48] of $\varepsilon \simeq 9k_B T L_S / R_N$. Here, k_B is the Boltzmann constant and L_S the SQUID inductance. Recently, $\varepsilon(77\ \mathrm{K}) = 3 \times 10^{-31}$ J/Hz, corresponding to about 450 times Planck's constant, was achieved for a 14pH GBJ dc SQUID [16]. This is the best energy resolution reported so far for SQUIDs operated at 77 K.

Whereas the white noise level of $YBa_2Cu_3O_{7-\delta}$ GBJ dc SQUIDs is well-understood and sufficiently low for most applications, the high amount of low-frequency noise represents a major obstacle for the use of $YBa_2Cu_3O_{7-\delta}$ GBJ in the fabrication of useful SQUID systems. For $YBa_2Cu_3O_{7-\delta}$ bicrystal GBJ dc SQUIDs the low-frequency noise, which scales proportional to about $1/f$, usually dominates already at frequencies below about 1 kHz [8, 12–17]. For $T \lesssim 83$ K, the measured $1/f$ noise level of the SQUIDs clearly could be attributed to voltage noise of the single GBJs [12–14]. Recently, it was shown that the amount of low-frequency noise in bicrystal GBJs is related to the oxygen stoichiometry at the grain boundary. Using ozone annealing, the noise level could be reduced significantly [17]. A high level of low-frequency noise has also been found for step edge GBJs. However, recent noise measurements show that the $1/f$ noise level in these GBJs is up to about one order of magnitude lower than that in the bicrystal GBJs [45–47]. The noise properties of biepitaxial GBJs have not yet been reported.

Our present knowledge of the noise properties of the different type of GBJs allows one to conclude that the excess low-frequency noise of the GBJs seems to be related to their microstructure and their chemical composition. The oxygen stoichiometry at the grain boundary seems to especially play an important role. Since the different types of GBJs are fabricated in different ways it is not surprising that they show different levels of low-frequency noise.

In order to clarify the origin of the excess low-frequency noise in bicrystal GBJs the dependence of its magnitude on bias current, temperature, and GBJ area was measured. Varying the bias current I_b at constant temperature, the voltage noise power S_v at a constant frequency is found to be proportional to R_d^2 for $I_b \simeq I_c$ [15, 17] and proportional to I_b^2 for $I_b \gg I_c$ [17]. According to small signal noise analysis one expects $S_V = R_d^2(\delta I_c)^2$ and $S_V = I_b^2(\delta R_N)^2$ for noise caused by fluctuations of the critical current and the normal resistance, respectively. Hence we can conclude that the noise of bicrystal GBJs is dominated by critical current fluctuations for $I_b \simeq I_c$ and resistance fluctuations for $I_b \gg I_c$. The overall $S_V(I_b)$ dependence could be consistently explained using a small signal noise analysis based on the RSJ model [17]. Fitting the measured noise data to the RSJ model prediction, values of the normalized fluctuations of the critical current $\delta I_c/I_c$ and the normal resistance $\delta R_N/R_N$

could be derived. The most interesting result was that the normalized fluctuations have the same ratio,

$$\frac{\left(\frac{\delta R_N}{R_N}\right)}{\left(\frac{\delta I_c}{I_c}\right)} \simeq -0.4, \tag{9}$$

for all investigated GBJs. This ratio is equal to the ratio expected from the universal correlation between the critical current and the normal resistance [Eq. (6)]. That is, the fluctuations of the critical current and the normal resistance are related by the same scaling relation observed for I_c and R_N. This gives strong evidence that the excess low-frequency noise and the observed scaling behavior have the same physical origin. Furthermore, the observation that the normalized fluctuations of the critical current and the normal resistance can be reduced by ozone annealing [17] and that the GBJ parameters can be shifted along the $J_c\rho_N \propto (1/\rho_N)^q$ dependence by such treatment, shows that the low-frequency noise and the scaling behavior is, at least in part, related to the oxygen stoichiometry at the grain boundary.

Varying the temperature and measuring the voltage noise power for $I_b \simeq I_c$, within the scatter of the data $S_V/R_d^2 \propto I_c^2$ is found [17]. According to small signal noise analysis we expect $S_V/R_d^2 = (\delta I_c)^2$ for $I_b \simeq I_c$. Hence, we can conclude that the normalized fluctuation of the critical current $\delta I_c(T)/I_c(T)$ is about constant in the temperature regime of the experiment (60 K $\leq T \leq$ 90 K). Since $I_c(T)$ increases with decreasing temperature, $S_V(T)/R_d^2$ also increases strongly [17]. For the bias current at which R_d is maximum, $S_V(T) \propto I_c^{2.6}$ was found in agreement with the model prediction [17]. At large bias currents ($I_b \gg I_c$), $S_V(T)$ is found to be independent of temperature. Since small signal noise analysis predicts $S_V = I_b^2(\delta R_N)^2$ for $I_b \gg I_c$, we can conclude that δR_N is also independent of temperature. Thus, $\delta R_N/R_N$ is temperature independent due to $R_N(T) = \text{const}$. In contrast, δI_c increases with decreasing temperature in the same way as I_c. Furthermore, we note that the ratio $(\delta R_N/R_N)/(\delta I_c/I_c)$ is independent of temperature.

Decreasing the area of bicrystal GBJs down to about 7.5×10^{-10} cm^2 the low-frequency noise from most of these junctions was found to be indistinguishable from that of junctions with larger areas. Hence, if the noise is caused by trapping centers at the grain-boundary interface, the density of these trapping cites should be much larger than about 10^{10} cm^{-2}. Alternatively, the average distance between the centers should be much less than 100 nm. For a few small area GBJs ($A \leq 2\ \mu$m^2) the noise spectra of single GBJs and dc SQUIDs exhibited considerable structures deviating from the 1/f behavior observed for the larger area samples [49]. This is shown in Figure 6.9 where the flux noise power spectral density $S_\Phi(f)$ a 50pH GBJ dc SQUID is plotted versus frequency at $T = 77$ K. At this temperature the equivalent flux noise measured for the SQUID can be attributed to the voltage noise

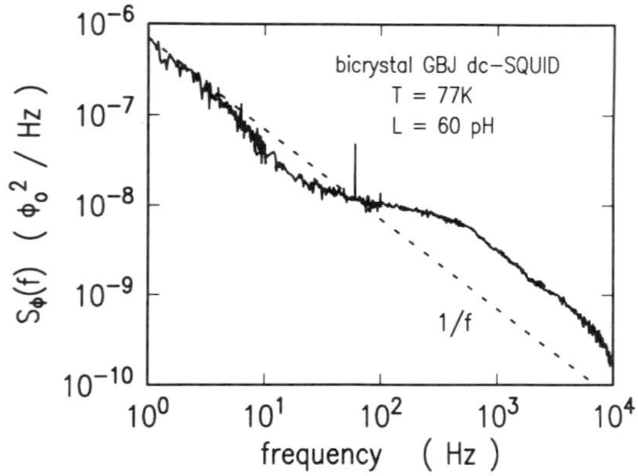

FIGURE 6.9. Flux noise power spectral density versus frequency for a 60pH GBJ dc SQUID measured at 77 K.

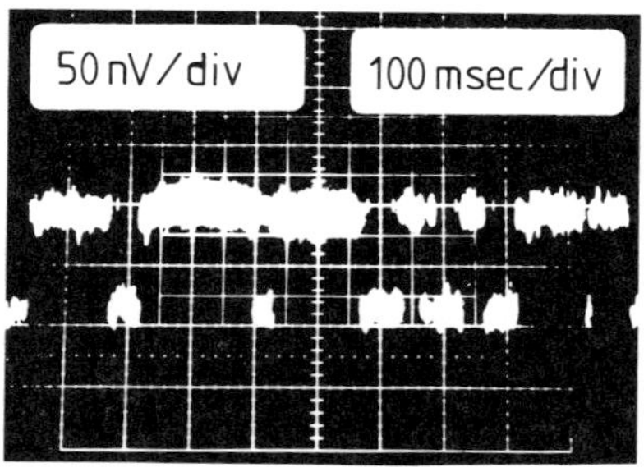

FIGURE 6.10. Real-time trace of the voltage across a 38° [001] tilt bicrystal GBJ measured at 77 K using a bias current of 1.2 times the critical current.

of the single junctions [13–15]. The most notable feature in Figure 6.9 is the presence of broad Lorentzian-like bumps in the noise spectrum. Similar bumps have been found for small area Nb–Nb_2O_5–PbBi tunnel junctions and PbIn–In_2O_3–Pb SQUIDs [50–52]. As discussed below, there is considerable evidence that the origin of the Lorentzian-like bumps are conductance fluctuations caused by the trapping and untrapping of single charge carriers into localized defect states within an insulating barrier layer at the GBJs. Similar to our observation, a number of earlier studies [50–52] have shown that the featured noise spectra turn into $1/f$ spectra when one averages over samples or increases sample size.

Figure 6.10 shows an example of the real-time voltage noise across a single GBJ. For samples for which the Lorentzian spectral features were most prominent, discrete voltage switching behavior was observed. This may be attributable to single trapping and untrapping events. Similar voltage switching has been reported in small area low-T_c tunnel junctions [50–52]. From the change in voltage across the GBJ and its dynamic resistance the fractional change of the critical current of the GBJ can be estimated to be less than 10^{-4}. Assuming that this fractional change is proportional to the effective area of a charged trap, an effective trap area of less than about $8 \times 8\ nm^2$ is obtained. A detailed analysis of the trapping kinetics has not yet been performed. Therefore, we do not know whether the traps are filled predominantly by tunneling or by thermally activated processes.

6.5 Nature of the Weak-Link Behavior

The precise origin of the weak-link behavior of grain boundaries in the cuprate superconductors is not completely understood. However, the available experimental data on the structural and electrical transport properties summarized above can be used to narrow the range of possible mechanisms. Phenomenologically, $YBa_2Cu_3O_{7-\delta}$ GBJs can be described by the RSJ model, that is, they behave as ideal overdamped Josephson junctions with a critical current I_c, which is proportional to $(1 - T/T_c)^2$ close to T_c, and a normal resistance R_N, which is independent of temperature. Furthermore, R_N is, at least for $V \leq 3 - 4I_cR_N$, almost independent of the bias voltage. The measured $I_c(B)$ dependences and the direct imaging of the current density distribution of GBJs by low-temperature scanning electron microscopy [5] demonstrate that the current flow across the grain boundaries, though spatially inhomogeneous, is not restricted to some narrow filaments. Special features of GBJs are the dependence of the critical current density on the misorientation angle and the scaling behavior of their transport and noise properties. The structural and chemical analysis of GBJs shows that dislocations, stress effects, and the oxygen stoichiometry at the grain boundary probably play a crucial role.

Discussing possible origins of the weak-link nature of GBJs we can show quite readily that the behavior of grain boundaries cannot be attributed to the possible *p*- or *d*-wave symmetry of the superconducting order parameter. In the same way it cannot be attributed to the anisotropy of the high-temperature superconductors. This follows directly from the angle dependence of the critical current of the tilt and twist boundaries described above. The critical current density J_c was found, within experimental error, to be a function only of the misorientation angle θ and not of the type of misorientation. If the symmetry of the order parameter would be important, the three types of misorientation should have yielded different angle dependences.

It has been speculated that the weak-link behavior of grain boundaries is caused by the presence of second phases. However, there is no experimental evidence for large scale composition deviations at the grain boundary (see, e.g., Ref. [33] for a summary of the transmission electron microscopy results). This strongly suggests that the weak-link behavior is not associated with large scale composition deviations. However, stress effects and the oxygen stoichiometry seem to be important. Unfortunately, transmission electron microscopy analysis does not provide valuable information on the oxygen stoichiometry at the grain-boundary interface. However, oxygen and ozone annealing experiments give direct evidence that deviations from the ideal oxygen stoichiometry at the grain boundary is, at least in part, responsible for the weak-link behavior.

We can also speculate that there is a large density of charged scattering centers at grain boundaries [53]. An obvious source of such scattering cites are dislocations, which are likely to be charged due to the strongly covalent bonding in the high-T_c material [54]. As discussed above, the density of dislocations increases with increasing misorientation angle and thus provides an obvious explanation for the observed decrease of the critical current density. Although dislocations were discussed early as a possible origin of the weak-link nature of grain boundaries there is, to date, no satisfactory explanation of their precise role. However, some aspects of their impact can be ruled out. First, GBJs with different misorientation geometries containing edge or screw dislocation show similar properties. Therefore, the type of elastic strain does not control the superconducting properties of GBJs. Second, the sign of elastic strain changes along, say, an edge dislocation. Therefore, the sign of strain also seems to play no important role. A likely explanation is that the order parameter is reduced in the strained region around the dislocation core. Once the density of dislocations is high enough that these region overlap, the order parameter is reduced at the entire grain boundary and weak-link behavior is expected to be observed. As estimated in Sec. 6.3, this is the case for a misorientation angle larger than about 10°.

With respect to the dislocation core idea one has to discuss some special grain boundaries including twin boundaries, 90° boundaries, and the set of coincidence boundaries possible for geometrical reasons. The twin boundaries, although obtained by a large angle rotation does not show weak-link behavior. The situation is similar for the 90° boundaries [55, 56]. Note that neither the twin nor the 90° grain boundaries require dislocations cores to be present at the interface. Hence, the absence of weak-link behavior is consistent with the dislocation core idea. However, there are two additional features that have to be taken into account, namely the continuity of the CuO_2 planes across the grain boundary and the orientation of the plane of the grain-boundary interface. The discontinuity of the CuO_2 planes is expected to strongly influence the superconducting properties of the boundary. Furthermore, deviations from a symmetrical grain-boundary geometry can introduce dislocations resulting in turn in weak-link behavior. This may be important for step edge GBJs.

The discussion of possible mechanisms responsible for the weak-link behavior of grain boundaries suggests that the most probable origin is the presence of dislocations and strain at the boundaries. The dislocation core idea is in accord with the existing data. However, it must be kept in mind that we arrived at this idea by a process of elimination. A direct experimental verification of this conception is still required.

6.6 Models for $YBa_2Cu_3O_{7-\delta}$ Grain-Boundary Junctions

Developing theoretical models for $YBa_2Cu_3O_{7-\delta}$ GBJs a crucial question is whether these junctions are *SIS*, *SNS*, *SNINS*, *SS'S'S*, or *SS'NS'S*-type junctions (S = superconductor, S' = superconductor with reduced order parameter, N = normal metal, I = insulator), that is, whether the interface layer at the grain boundary is a normal conducting (N) or an insulating (I) layer. Concerning this question we note the following facts: First, no normal conducting (metallic) phase has been found in the phase diagram of YBaCuO [57]. Second, transmission electron microscopy studies have shown that the grain-boundary interfaces are clean, that is, they do not contain impurity phases. Furthermore, the lattice distortion at the grain-boundary interface was found to be confined to within a few lattice constants. Third, oxygen and ozone annealing experiments [9, 17, 27] suggest that there is an oxygen deficient layer at the grain-boundary interface. Oxygen deficiency or disorder [58] is known to render YBCO insulating and not normal conducting. Fourth, the measured values of ρ_N vary between about 10^{-7} and 10^{-8} $\Omega\,cm^2$. Therefore, assuming an interface layer thickness of about 1 nm results in a resistivity of about 0.1 to 1 $\Omega\,cm$ for this layer. This value is typical for a semiconductor or bad insulator but is several orders of magnitude larger than the normal state resistivity of $YBa_2Cu_3O_{7-\delta}$. Finally, we note that oxygen disorder in the Cu–O chains removes holes from the CuO_2 planes resulting in an insulating phase [57, 59]. Also, the intrinsic relaxation of $YBa_2Cu_3O_{7-\delta}$ at internal interfaces and surfaces likely yields a high density of localized states that are electron like. These facts give evidence that there is an insulating (I) layer at an oxygen deficient and disordered grain-boundary interface, that has a thickness of up to more than 1 nm.

Up to now, several models [49, 60–64] have been developed to describe the electrical transport characteristics of $YBa_2Cu_3O_{7-\delta}$ GBJs. The models can be classified into tunneling (*SIS*) and non-tunneling (*SNS*)-type models. However, as will be discussed below, in some cases the differences between these models are quite small. In order to explain the observed transport properties of GBJs, in *SNS*-type models a normal metal with low conductivity close to the metal–insulator transition is assumed [63]. On the other hand, in *SIS*-type models an insulator with a high density of localized defect states, that is, an insulator close to the insulator–metal transition, is assumed [49]. Up to now, *SNS*-type models [60–64] were favored, since they naturally explain the low values of the I_cR_N products of GBJs and the temperature dependence of the critical current close to T_c. Recently, however, a *SIS*-type model [49] has also been proposed that can explain the observed transport and, moreover, the noise characteristics of $YBa_2Cu_3O_{7-\delta}$ GBJs. In this model the insulator is assumed to contain a high density of localized defect states allowing resonant and intermediate state tunneling [65].

6.6.1 *SIS-Type Models*

There are well-established theories describing the superconducting transport characteristics of ideal *SIS*-type Josephson junctions, which are formed by two ideal superconducting banks separated by a perfect insulator [35]. Using *SIS* models to model high-T_c GBJs two important facts have to be considered. First, it is well-known that a characteristic feature of the high-T_c superconductors is their short coherence length. The implications of the short coherence length have to be taken into account in *SIS*-type junction models. Second, the chemistry of $YBa_2Cu_3O_{7-\delta}$ implies that an insulating layer at grain boundaries will most likely contain a high density of defect states. The influence of such defect states on the electrical transport characteristics has to be included in *SIS*-type junction models.

6.6.1.1 Implications of the Short Coherence Length

Deutscher and Müller [66] pointed out very early that the short coherence length ξ_0 of the high-T_c materials would modify the order parameter at a superconductor–insulator interface. In general, on the Ginzburg–Landau scale (lengths of the order $\xi_S(T) \gg \xi_0$) the effect of the boundary between a superconducting and a non-superconducting material (see Figure 6.11) on the order parameter at the boundary $\Delta_i = \Delta(x = 0)$ can simply be described by means of the boundary condition [67]

$$\left(\frac{d\Delta}{dx} - \frac{2ie}{\hbar c} A_x \Delta\right)_{x=0} = \frac{(\Delta)_{x=0}}{b}. \tag{10}$$

Here A is the vector potential and the so-called extrapolation length b is given by [67]

$$\frac{1}{b} = \frac{2}{L} \int_{-\infty}^{\infty} \frac{\Delta(x)}{\Delta_0} \left(1 - \frac{N(x)}{N(0)}\right) dx, \tag{11}$$

where $L \simeq \xi_0^2$ and Δ_0 and $N(0)$ are the bulk values of the order parameter and the density of states, respectively.

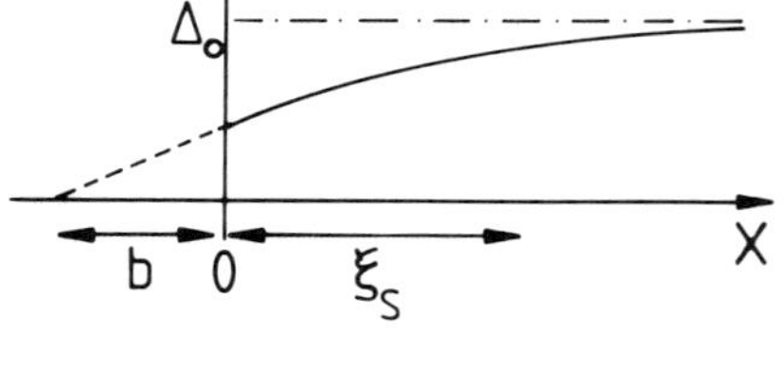

FIGURE 6.11. Spatial variation of the order parameter near a superconductor/non-superconductor interface.

For an S/I interface, $\Delta(x)/\Delta_0$ approaches zero in the insulating region and is about 1 in the superconductor. Similarly, $N(x)/N(0)$ passes from 1 to 0 within a few atomic distances going from the superconducting to the insulating material. Therefore, the integral is non-vanishing only within the width of the order of the interatomic distance a. The extrapolation length is then obtained to $b \simeq L/a \simeq \xi_0^2/a$. By solving the Ginzburg–Landau equations, the order parameter inside the superconducting material is obtained to [67]

$$\Delta(x) = \Delta_0 \tanh\left(\frac{x - x_0}{\sqrt{2}\xi_S(T)}\right), \tag{12}$$

where the value of x_0 is determined by the above boundary conditions (Eqs. (10) and (11)). For large coherence length superconductors we have $\xi_0 \gg a$ resulting in $b \gg \xi_S(T)$ and $\Delta_i/\Delta_0 \simeq 1$. Thus, a reduced value of Δ_i is obtained only very close to T_c due to the divergence of $\xi_S(T)$. In contrast, for small coherence length superconductors such as the high-T_c cuprates we have $\xi_0 \simeq a$ resulting in $b < \xi_S(T)$ and $\Delta_i/\Delta_0 \simeq b/\xi_S(T) < 1$ in a wide temperature regime.

The reduction of Δ_i results in an unusual temperature dependence of the critical current density of SIS-type Josephson junctions close to T_c. According to the Ambegaokar–Baratoff expression [39] the critical current density is proportional to the product of the order parameter on both sides of the junction, $J_c \propto \Delta_i^+ \Delta_i^-$. That is, one obtains $J_c \propto [\Delta_0(T)(b/\xi_S(T))]^2 \propto (1 - T/T_c)^2$ instead of the usual result: $J_c \propto [\Delta_0(T)]^2 \propto (1 - T/T_c)$. This is in good agreement with the dependence observed for GBJs. Furthermore, the reduction of Δ_i results in a reduced value of the characteristic voltage $V_c = I_c R_N \simeq \Delta_i/e$ as observed for GBJs. However, at lower temperatures we have $b/\xi_S(T) \simeq b/\xi_0 \simeq 1$ and much of the strength of the order parameter at the interface is restored. This should be the case especially for the [001] tilt GBJs where the longer ab-plane coherence length ξ_{ab} is the important length scale. However, for all $YBa_2Cu_3O_{7-\delta}$ GBJs low $I_c R_N$ products are also observed at low temperatures. These small $I_c R_N$ values at low temperatures cannot be explained solely by a reduction of the order parameter at the grain boundary due to the small coherence length. Furthermore, the scaling behavior of the $I_c R_N$ product is difficult to be explained within this model. However, the $J_c(T)$ dependence of bicrystal GBJs, which is proportional to $(1 - T/T_c)^2$ close to T_c and Ambegaokar–Baratoff like dependence at low temperatures, is in reasonable agreement with the prediction of the SIS-type model.

6.6.1.2 Resonant and Intermediate State Tunneling

We have considered above the case that the I layer which separates the superconducting electrodes is formed by a perfect insulator. We will now discuss a SIS-type model [49] where the insulator contains a high density of defect states as schematically shown in Figure 6.12. Note that the Ambegaokar–Baratoff expression [Eq. (7)] for the tunneling current of superconducting

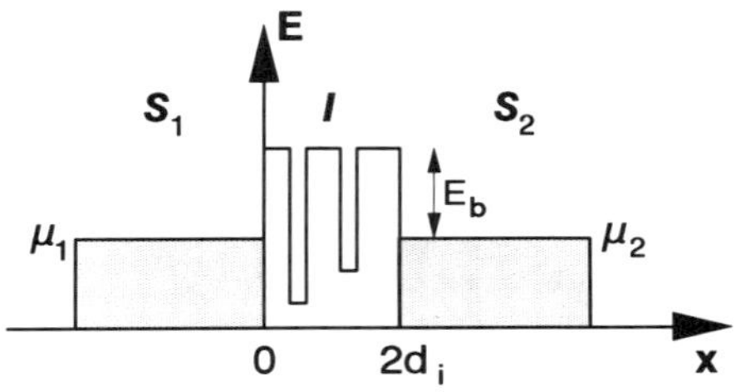

FIGURE 6.12. Sketch of a superconductor–insulator–superconductor junction with a barrier layer containing a high density of localized defect states.

SIS tunnel junctions is derived under the assumption that both the paired and unpaired charge carriers have to tunnel through the same barrier. When localized states are present in a tunneling barrier, two different elastic tunneling processes (channels) are conceivable. Channel 1 is the usual direct tunneling process, in which an electron tunnels all the way across the barrier without any scattering. Channel 2 represents indirect tunneling, in which an electron tunnels from one electrode onto a localized state and then in turn tunnels out to the other electrode. The simplest indirect tunneling process is resonant (elastic) tunneling. According to simple one-dimensional theories of resonant tunneling, the conductivity due to this process is enhanced at biases for which the Fermi level of the electrode coincides with the energies of the localized states in the barrier.

In real materials the impurity states generally have a small radius, $a \geq$ Å. This causes a strong Coulomb repulsion, $U \geq 0.1$ eV, of two electrons that occupy one level. This repulsion usually does not lead to a substantial change of the normal current, since it is given by the tunneling of individual electrons. In the absence of Coulomb repulsion, the paired electrons may also tunnel through resonant states [68]. However, since the Josephson current is a consequence of the correlated tunneling of pairs of charge carriers it can be suppressed considerably by Coulomb repulsion. The importance of the Coulomb repulsion is given by the ratio of two time scales. The first time scale, τ_1, is determined by the decay time of a localized state into the conduction electron state and is inversely proportional to the width Γ of the resonant impurity state. The second, τ_2, is determined by the correlation time of the electrons in a Cooper pair and is inversely proportional to the critical temperature T_c. If the decay time of the localized state is long ($\tau_1/\tau_2 > 1$), the resonant Josephson current would be dominated by processes for which the two electrons of a pair are simultaneously at a center. Usually, Coulomb repulsion prohibits such processes resulting in a strong suppression of the Josephson current. In this case we have two different tunneling channels for the paired (mostly direct tunneling) and unpaired (mostly indirect tunneling) electrons and the Ambegaokar–Baratoff expression can no longer be applied. Due to the suppression of the Josephson current one expects an I_cR_N product which is much smaller than that given by Eq. (7). On the other hand, for $\tau_1/\tau_2 \ll 1$ the tunneling processes of the two electrons of a pair are separated in time. In this case, Coulomb repulsion is not important and the Josephson current is determined by the tunneling amplitudes of the single electrons [69], that is, paired and unpaired electrons tunnel through the same channel.

The decay time of a localized state increases as $\tau_1 \propto \exp(\kappa x)$ with increasing distance x from the superconducting electrode. Thus the importance of Coulomb repulsion will increase strongly with increasing distance of the localized states from the electrodes. Here, $\kappa = \sqrt{2mE_b/\hbar^2}$ is the attenuation constant of the conduction electrons in the insulator. Tunneling experiments indicate a tunnel barrier height $E_b > 0.5$ eV. Therefore, the attenuation is strong and $1/\kappa \approx 0.1$–0.2 nm can be estimated. That is, the resonant or intermediate state tunneling of paired electrons via localized states is expected to be small even if the localized states are quite close to the superconducting electrodes [68]. From this we conclude that the supercurrent through the insulating layer is determined by the direct tunneling current. Assuming that the localized states are in the center of the I layer with thickness $2d_i$ we obtain

$$J_c \propto N(0)^2 \exp(-2\kappa d_i). \tag{13}$$

In contrast, the normal current will be dominated by resonant or intermediate tunneling current supported by the localized states with density N_l [70]:

$$J_N \propto 1/\rho_N \propto N_l \exp(-\kappa d_i), \tag{14}$$

if N_l or d_i is large. As discussed above, the presence of a large density of localized defect states is very likely at grain boundaries in $YBa_2Cu_3O_{7-\delta}$. A density $N_l \approx 10^{18}\ \mathrm{cm}^{-3}$ which seems to be reasonable for a disordered cuprate material [72], would be sufficient to make resonant tunneling dominant. Combining Eqs. (13) and (14) at once yields [49]

$$J_c\rho_N \propto \exp(-\kappa d_i) \propto 1/\rho_N \qquad \text{or equivalently} \qquad J_c\rho_N \propto \sqrt{J_c}. \tag{15}$$

This result is close to the experimentally observed scaling behavior shown in Figure 6.5. Within the proposed model the scaling behavior of the $J_c\rho_N$ product is simply caused by different tunneling channels of the paired and unpaired charge carriers. The strongly reduced values of the $J_c\rho_N$ products result from an increased barrier width for the supercurrent as compared to the normal current. Note that $\exp(-\kappa d_i)$ should be temperature independent. Therefore, with $\rho_N = $ const., similar $J_c(T)$ dependences are expected for junctions with different absolute $J_c\rho_N$ values in agreement with the experimental observation.

According to Eq. (14), the $J_c\rho_N$ product decreases proportional to $\exp(-\kappa d_i)$ with increasing barrier thickness. That is, GBJs with a large intrinsic insulator thickness should have small $J_c\rho_N$ products. The observed decrease of $J_c\rho_N$ with increasing misorientation angle (see Figure 6.6) then suggests that large-angle grain boundaries have a thicker insulating barrier. This is reasonable, since disorder and strain and, hence, the thickness of the barrier layer is expected to increase with increasing misorientation angle. Furthermore, the

$J_c\rho_N$ products of the GBJs can be increased by ozone annealing. Annealing and oxygen supply is expected to improve order at the grain-boundary interface thereby reducing the thickness of the barrier layer and, hence, increasing the $J_c\rho_N$ product.

It is known that localized states in oxides adjacent to metals hybridize with conduction electron states forming interface states [70]. By overlap with the Cooper pairs of the bulk electrodes these surface states have a superconducting gap, $\Delta_S(x) \propto \Delta_0 \exp(-\tau_1(x)/\tau_2)$, which decays rapidly with increasing distance form the bulk electrode due to the increase of the on-site Coulomb repulsion counteracting the superconducting attraction [71]. It is likely that $\Delta_S \leq 5$ meV already for $x \geq 0.2$ nm [72]. Therefore, the supercurrent mediated by the surface states is expected to decrease very fast with increasing x. Qualitatively, the surface states can be considered as to mediate a continuous transition between the metal electrode and the insulating barrier layer (space-wise metal–insulator transition). By this, a narrow part of the insulating layer becomes metal-like and in turn superconducting by proximity effect. This results in a so-called SNS-like temperature dependence of the critical current $J_c \propto (1 - T/T_c)^2$ for temperatures close to T_c. For $T \ll T_c$, a $J_c(T)$ dependence close to the Ambegaokar–Baratoff dependence is expected. Of course, the modification of the order parameter at the S/I interface due to the small coherence length of the high-T_c superconductor has to also be taken into account within the present model. Because of the less abrupt change of $N(x)/N(0)$ and $\Delta(x)/\Delta_0$ at the interface due to the space-wise metal–insulator transition, the extrapolation length b is expected to be smaller. This in turn results in a stronger reduction of the order parameter at the boundary than in the case of an ideal S/I interface discussed above.

The proposed SIS-type model which includes resonant and intermediate tunneling via localized states can explain most characteristics observed for $YBa_2Cu_3O_{7-\delta}$ GBJs. Beyond the $J_c(T)$ dependence and the reduced I_cR_N products it also explains the excess low-frequency noise in $YBa_2Cu_3O_{7-\delta}$ GBJs. The resonant tunneling via localized states depends on the local potential of the localized states in the insulator. This potential and thus the tunneling current can easily be modulated by nearby charging. This usually yields a large amount of low-frequency noise as found for $YBa_2Cu_3O_{7-\delta}$ GBJs and SQUIDs [8, 9, 12–17, 25, 49]. Ozone annealing is likely to decrease the density of defect states and, hence, the amount of low-frequency noise. Moreover, within this model the scaling behavior and the excess low-frequency noise have the same physical origin, namely the tunneling via localized states. This is in perfect agreement with the experimental observation that both the transport (J_c and ρ_N) and noise (δI_c and δR_N) characteristics of $YBa_2Cu_3O_{7-\delta}$ GBJs show the same scaling behavior. Furthermore, a white distribution of localized states explains the linear increase of conductance with increasing bias voltage, $dI(V)/dV \propto |V|$, as found for some bicrystal GBJs [40]. Finally, the model explains that similar $J_c(T)$ dependences are observed for GBJs with different I_cR_N products.

6.6.2 *SNS-Type Models*

We now consider an *SNS*-type junction formed by two superconducting banks, *S*, with normal state resistivity ρ_S separated by a thin normal metal layer, *N*, with resistivity ρ_n and thickness d_n. The analysis of BCS based models for *S*/*N* interfaces shows that the proximity effect at these interfaces can be characterized by two dimensionless parameters. In the dirty limit these parameters are given by [60–62]

$$\gamma_S = \rho_S \xi_S / \rho_n \xi_n, \qquad \gamma_B = \rho_B / \rho_n \xi_n. \tag{16}$$

Here, ξ_n is the decay length of the superconducting order parameter in the normal metal and ρ_B is the boundary resistance of the *S*/*N* interface.

We will first consider the case of negligible boundary resistance, that is, $\gamma_B = 0$. This case was discussed recently by Deutscher and Chaudhari [63] under the assumption that the *N* layer consists of a low conductivity metal. The variation of the order parameter at the *S*/*N* interface can be treated in close analogy to the case of an *S*/*I* interface discussed above. The type of boundary condition that applies to the order parameter at the *S*/*N* interface is given by [63, 74]

$$b = \xi_n \coth\left(\frac{d_n}{\xi_n}\right). \tag{17}$$

The structural analysis of the grain boundaries shows that d_n should be small. Therefore we can assume that $d_n < \xi_n$. This yields $b \propto \xi_n^2/d_n$ in good approximation. Furthermore, since $I_c R_N \ll \Delta_0/e$, we can assume that b is small compared to ξ_S. In this limit the ratio of the order parameter at the grain boundary Δ_i to the bulk order parameter Δ_0 is given by b/ξ_S (see Figure 6.11). With these assumptions we obtain

$$\Delta_i \propto \xi_n^2/d_n. \tag{18}$$

In the dirty limit, $\xi_n^2 \propto 1/\rho_n$. Then, using $I_c R_N \simeq \Delta_i/e$ at small reduced temperatures and $\rho_n d_n = \rho_N = R_N A$ results in

$$J_c \rho_N = I_c R_N \propto 1/\rho_N \qquad \text{or equivalently} \qquad J_c \rho_N = I_c R_N \propto \sqrt{J_c}. \tag{19}$$

This is in close agreement with the experimentally observed scaling behavior.

The assumption that the grain-boundary layer is composed of a low conductivity metal is not essential. It also might be composed of an insulator close to the metal-insulator transition having a localization length $\xi_l > d_n$. By extension of the dirty limit to this case one obtains $\xi_n^2 \propto D(d_n)$ [67] and the above result [Eq. (19)] is recovered. Here, $D(d_n)$ is the diffusion coefficient on scale d_n. Recently, it was shown that the decay length near the metal–insulator transition can be considerably larger than interatomic distances [73]. Therefore, the basic assumption $\xi_n > d_n$ of the above model should remain valid.

Kupriyanov [61, 62] pointed out recently that the condition $\gamma_B \gg \gamma_S \geq 1$ is inherent to high-T_c superconductor–normal metal interfaces. Therefore, the case discussed above ($\gamma_B = 0$) should not apply to such interfaces. In the case of finite γ_B the order parameter undergoes a jump at the S/N interface. The jump is increasing with increasing ρ_B. For $\gamma_B \gg 1$ one can neglect the suppression of the order parameter in the superconducting electrodes due to the proximity effect. The detailed analysis shows that for $\xi_n < d_n$ one obtains $I_c R_N \propto 1/\rho_B$. With $\rho_N \simeq \rho_B$ for $\gamma_B \gg 1$, the same scaling behavior, $I_c R_N \propto 1/\rho_N$ is obtained. That is, the result $I_c R_N \propto 1/\rho_N$ is obtained quite generally no matter whether the normal resistance of the grain boundary is determined by the normal metal layer ($\gamma_B \ll 1; \rho_N \simeq \rho_n d_n$) or the boundary resistance ($\gamma_B \gg 1; \rho_N \simeq \rho_B$).

The SNS-type models can explain both the scaling law $I_c R_N \propto 1/\rho_N$ and the temperature dependence of the critical current near T_c. However, the observation that GBJs with different $I_c R_N$ products show very similar $I_c(T)$ dependences is difficult to explain within these models. Furthermore, these models cannot explain the relationship between the transport and noise properties of $YBa_2Cu_3O_{7-\delta}$ GBJs.

Acknowledgments. The author would like to thank A. Braginski, P. Chaudhari, D. Dimos, A. Gupta, J. Halbritter, R. P. Huebener, M. Kawasaki, M. Kupriyanov, J. Mannhart, B. Mayer, M. Siegel, F. Schmidl, and D. Winkler for valuable discussions. Financial support of this work by the Bundesminister für Forschung und Technologie (Project No. 13N5843) is gratefully acknowledged.

References

1. P. Chaudhari, J. Mannhart, D. Dimos, C.C. Tsuei, C.C. Chi, M.M. Oprysko, and M. Scheuermann, Phys. Rev. Lett. **60**, 1653 (1988).
2. D. Dimos, P. Chaudhari, J. Mannhart, and F.K. LeGoues, Phys. Rev. Lett. **61**, 219 (1988).
3. J. Mannhart, P. Chaudhari, D. Dimos, C.C. Tsuei, and T.R. McGuire, Phys. Rev. Lett. **61**, 2476 (1988).
4. D. Dimos, P. Chaudhari, and J. Mannhart, Phys. Rev. B **41**, 4038 (1990).
5. J. Mannhart, R. Gross, K. Hipler, R.P. Huebener, C.C. Tsuei, D. Dimos, and P. Chaudhari, Science **245**, 839 (1989).
6. R. Gross, P. Chaudhari, D. Dimos, A. Gupta, and G. Koren, Phys. Rev. Lett. **64**, 228 (1990).
7. R. Gross, P. Chaudhari, M. Kawasaki, and A. Gupta, Phys. Rev. B **42**, 10,735 (1990).
8. R. Gross, P. Chaudhari, M. Kawasaki, and A. Gupta, IEEE Trans. Magn. **MAG-27**, 3227 (1991).
9. R. Gross, P. Chaudhari, M. Kawasaki, and A. Gupta, Superconduct. Sci. Technol. **4**, S253 (1991); M. Kawasaki, P. Chaudhari, R. Gross, and A. Gupta (unpublished).
10. N. Tomita, Y. Takahashi, and Y. Ishida, Jpn. J. Appl. Phys. **29**, L30 (1990).

11. M. Kawasaki, P. Chaudhari, A. Gupta, and W. Lee, Appl. Phys. Lett. **62**, 417 (1993).
12. R. Gross, and P. Chaudhari, in *Principles and Applications of Superconducting Quantum Interference Devices*, edited by A. Barone (World Scientific, Singapore, 1991).
13. R. Gross, P. Chaudhari, M. Kawasaki, M.B. Ketchen, and A. Gupta, Appl. Phys. Lett. **57**, 727 (1990).
14. R. Gross, P. Chaudhari, M. Kawasaki, M.B. Ketchen, and A. Gupta, Physica C **170**, 315 (1990).
15. R. Gross, P. Chaudhari, M. Kawasaki, M.B. Ketchen, and A. Gupta, IEEE Trans. Magn. **MAG-27**, 2565 (1991).
16. M. Kawasaki, P. Chaudhari, T. Newman, and A. Gupta, Appl. Phys. Lett. **58**, 2555 (1991).
17. M. Kawasaki, P. Chaudhari, and A. Gupta, Phys. Rev. Lett. **68**, 1065 (1992).
18. Z.G. Ivanov, P.A. Nilsson, D. Winkler, J.A. Alarco, T. Claeson, E.A. Stepantsov, and A.Y. Tzalenchuk, Appl. Phys. Lett. **23**, 3030 (1991).
19. K. Char, M.S. Coclough, S.M. Garrison, N. Newman, and G. Zaharchuk, Appl. Phys. Lett. **59**, 733 (1991).
20. K. Char, M.S. Coclough, L.P. Lee, and G. Zaharchuk, Appl. Phys. Lett. (in press); see also Proceedings of SQUID'91 Conference, Berlin, Germany (1991).
21. L.P. Lee, K. Char, M.S. Coclough, and G. Zaharchuk, Appl. Phys. Lett. **59**, 2177 (1991).
22. R. Simon, J.B. Bulman, J.F. Burch, S.B. Coons, K.P. Daly, W.D. Dozier, R. Hu, A.E. Lee, J.A. Luine, C.E. Platt, and M.J. Zani, IEEE Trans. Magn. **27**, 3209 (1991); Appl. Phys. Lett. **58**, 543 (1991).
23. M. Siegel, F. Schmidl, K. Zach, E. Heinz, J. Borck, W. Michalke, and P. Seidel, Physica C (in print).
24. G. Cui, Y. Zhang, K. Hermann, C. Buchal, J. Schubert, W. Zander, A.I. Braginski, and C. Heiden, in *Nonlinear Electronics and Josephson Devices*, edited by N.F. Pedersen, M. Russo, A. Davidson, G. Costabile, and S. Pagano (Plenum, New York, 1991); see also Supercond. Sci. Techol. **4**, S130 (1991); Physica C **175**, 545 (1991).
25. K. Herrmann, Y. Zhang, H.M. Mück, J. Schubert, W. Zander, and A.I. Braginski, Superconduct. Sci. Techol. **4**, 583 (1991).
26. A.H. Micklich, J.J. Kingston, F.C. Wellstood, J. Clarke, K. Char, M.S. Coclough, and G. Zaharchuk, Appl. Phys. Lett. **59**, 988 (1991).
27. S.E. Russek, D.K. Lathrop, B.H. Moeckly, R.A. Buhrmann, and D.H. Shin, Appl. Phys. Lett. **57**, 1155 (1990).
28. B.H. Moeckly, S.E. Russek, D.K. Lathrop, R.A. Buhrmann, J. Li, and J.W. Mayer, Appl. Phys. Lett. **57**, 1687 (1990); Appl. Phys. Lett. **57**, 2951 (1990).
29. R.A. Camps et al., Nature **329**, 229 (1987).
30. S. Nakahara et al., J. Cryst. Growth **85**, 639 (1987).
31. S.E. Babcok, and D.C. Larbalestier, Appl. Phys. Lett **55**, 393 (1989); see also Physics Today **44**, 74 (1991).
32. D.M. Kroeger et al., J. Appl. Phys. **64**, 331 (1988).
33. M.S. Chisholm, and S.J. Penneycook, Nature **351**, 47 (1991).
34. M.F. Chisholm, and D.A. Smith, Phil. Mag. A **59**, 181 (1989).
35. A. Barone, and G. Paterno, *Physics and Applications of the Josephson Effect* (Wiley, New York, 1982).

36. W.C. Stewart, Appl. Phys. Lett. **12**, 277 (1968); D.E. McCumber, J. Appl. Phys. **39**, 3113 (1968).
37. V. Ambegaokar and B. Halperin, Phys. Rev. Lett. **22**, 1364 (1969).
38. J.R. Waldram, A.B. Pippard, and J. Clarke, Philos. Trans. Roy. Soc. London A **268**, 265 (1970).
39. V. Ambegaokar, and A. Baratoff, Phys. Rev. Lett. **10**, 486 (1963).
40. J. Mannhart, R. Gross, R.P. Huebener, P. Chaudhari, D. Dimos, and C.C. Tsuei, Cryogenics **30**, 397 (1990).
41. M. Gurvitch, J.M. Valles, A.M. Cucolo, R.C. Dynes, J.P. Garno, and L.F. Schneemayer, Phys. Rev. Lett. **63**, 1008 (1989).
42. S.J. Bending, and M.R. Beasley, Phys. Rev. Lett. **55**, 324 (1985).
43. R.H. Koch, W.J. Gallagher, B. Bumble, and W.Y. Lee, Appl. Phys. Lett. **54**, 951 (1989).
44. V. Foglietti, R.H. Koch, W.J. Gallagher, B. Oh, B. Bumble, and W.Y. Lee, Appl. Phys. Lett. **54**, 2259 (1989).
45. Y. Zhang, H.M. Mück, K. Herrmann, A. Braginski, and C. Heiden, Appl. Phys. Lett. **60**, 645 (1992).
46. M. Siegel et al., Proceedings of the 4th German–Soviet Bilateral Seminar on High-Temperature Superconductivity, St. Petersburg, UDSSR (1991).
47. G. Friedl, G. Daalmans, and H.E. Hoenig, Proceedings of the 4th German–Soviet Bilateral Seminar on High-Temperature Superconductivity, St. Petersburg, UDSSR (1991).
48. C.D. Tesche, and J. Clarke, J. Low Temp. Phys. **29**, 301 (1977).
49. R. Gross, and B. Mayer, Physica C **180**, 235 (1991).
50. C.T. Rogers, and A. Buhrman, Phys. Rev. Lett. **53**, 1272 (1984); C.T. Rogers, A. Buhrman, H. Kroger, and L.N. Smith, Appl. Phys. Lett. **49**, 1107 (1986).
51. R.T. Wakai, and D.J. Van Harlingen, Appl. Phys. Lett. **49**, 593 (1986).
52. P.J. Restle, R.J. Hamilton, M.B. Weissmann, and M.S. Love, Phys. Rev. B **31**, 2254 (1985).
53. P. Chaudhari, Proceedings of the M^2HTSC Conference, Kanazawa, Japan (1991) (to be published in Physica B).
54. K. Jagannadham, and J. Narayan, Mater. Sci. Eng. B **8**, 5 (1991).
55. S.E. Babcock, X.Y. Cai, D.L. Kaiser, and D.C. Larbalestier, Nature **347**, 178 (1990).
56. C.B. Eom, A.F. Marshall, Y. Suzuki, B. Boyer, R.W. Pease, and T.H. Geballe, Nature **353**, 544 (1991).
57. H.C. Hass, in *Solid State Physics*, edited by H. Ehrenreich, and D. Thurnbull (Academic, New York, 1989), Vol. 42, p. 213.
58. J.H. Muller, and R. Gruehn, Physica C **159**, 527 (1989).
59. J. Halbritter, Int. J. Mod. Phys. B **3**, 719 (1989).
60. M.Y. Kupriyanov, and K.K. Likharev, Sov. Phys. Uspekhi **160**, 49 (1990).
61. M.Y. Kupriyanov, and K.K. Likharev, IEEE Trans. Magn. **MAG 27**, N2 (1991).
62. M.Y. Kupriyanov, Sov. Phys. JETP **96**, 1420 (1989); see also Extended Abstracts ISEC'89, Tokyo (1989), p. 534; Proceedings of SQUID'91 Conference, Berlin, Germany (1991).
63. G. Deutscher, and P. Chaudhari, Phys. Rev. B **44**, 4664 (1991).
64. J. Mannhart, and P. Martinoli, Appl. Phys. Lett. **58**, 643 (1991).
65. E.L. Wolf, *Principles of Electron Tunneling Spectroscopy* (Oxford University Press, New York, 1985).

66. G. Deutscher, and K.A. Müller, Phys. Rev. Lett. **59**, 1745 (1987).
67. P.G. de Gennes, *Superconductivity in Metals and Alloys* (Benjamin, New York, 1966); see also Rev. Mod. Phys. **36**, 225 (1964).
68. A.G. Aslamasov, and M.V. Fistul, Sov. Phys. JETP **55**, 681 (1982).
69. L.I. Glazman, and K.A. Mateevul, JETP Lett. **49**, 659 (1989).
70. J. Halbritter, Surf. Sci. **122**, 80 (1982); J. Appl. Phys. **58**, 1320 (1985); IEEE Trans. Mag. **MAG-19**, 799 (1983).
71. J. Halbritter, Solid State Commun. **18**, 1447 (1976); see also Solid State Commun. **34**, 675 (1980).
72. J. Halbritter, Phys. Rev. B **46**, 11,238 (1992).
73. J.W.P. Hsu, G. Deutscher, and A. Kapitulnik, Phys. Rev. Lett. (submitted for publication).
74. N.R. Werthamer, Phys. Rev. **132**, 2440 (1963).

7 Overlayer Formation on High-Temperature Superconductors

J.H. WEAVER

7.1 Introduction

Studies of the surface properties and interface properties of the high-temperature superconductors (HTS) have gone hand in hand with other developments in HTS science and technology. The relevance of such surface/interface properties is clear: If these superconductors are to be integrated with other materials, then materials compatibility issues must be thoroughly explored and, ultimately, optimized. In the last few years, the techniques of surface science have been used to advantage to explore such materials compatabilities, and a great deal has been learned [1, 2].

This chapter deals with HTS interfaces, the processes by which they are formed, and the correlation of properties with processing. In 1990, we reviewed [1] the HTS surface and interface literature, and we refer the reader to that review and to the one on the same subject by Lindberg et al. [2]. Here, we will emphasize new developments and insights gained by growth at low temperature and growth with clusters. Photoemission will be used to examine chemical changes at the surface and scanning tunneling microscopy (STM) will be used to explore structural properties, notably clustering on the surface. The plan is to briefly examine overlayer formation by atom deposition at 300 K, emphasizing recent studies of Ti growth on $YBa_2Cu_3O_7$, Y_2BaCuO_5, and CuO and pointing to the common characteristics of surface reactions. (For brevity, we will denote $YBa_2Cu_3O_7$ as YBCO, giving subscripts only when there might be confusion and recognizing the fact that the oxygen content should be written $7 - x$.) Overlayer formation with nonreactive elements, namely Ag and Au, will then be explored for YBCO and $Bi_2Sr_2CaCu_2O_8(001)$, abbreviated BSCCO, and it will be shown that sig-

Department of Materials Science and Chemical Engineering, University of Minnesota, Minneapolis, MN 55455.
* Supported by the Office of Naval Research and the Defense Advanced Research Projects Agency.

nificant surface modifications occur, despite the limited tendency of Ag to form a surface oxide. The mechanism of this modification will be elucidated through STM studies of Ag atom deposition on BSCCO(001) at 300 K. Low-temperature growth will then be explored through an examination of Ti and Cr interactions with YBCO(001) and BSCCO(001), and the discussion will emphasize the role of kinetics in inhibiting reaction. Finally, we will explore growth of overlayers using clusters of atoms rather than individual atoms. This process will be shown to produce a much less reacted layer because of the detailed interactions at the surface. Indeed, cluster assembly produces interfaces by following a different route than conventional atom deposition, and the stability of the clusters in contact with the surface accounts for the much reduced chemical activity of the interface.

The first HTS surface and interface studies focused on $La_{2-x}Sr_xCuO_4$, abbreviated LSCO. As discussed in Refs. [1] and [2], those studies dealt with sintered pellets, and many of them reported spectra that were not intrinsic to the HTS material. In particular, it was quickly demonstrated that air exposure led to severe contamination because of surface reactions. Low-density samples that were fractured and studied in vacuum were also troubled because internal grain boundaries and oxides were exposed and the grain boundaries were repositories for other phases or non-intrinsic species. It was only when high-density samples became available that spectroscopic studies, with their sensitivity to surface conditions, became more reliable. This lesson was learned again with YBCO where the problems were worse because of difficulties in preparing good samples. Many of the resulting XPS spectra showed O 1*s* core level emission dominated by features that are now recognized as being related to contamination [1, 2]. There has been less controversy regarding the intrinsic electronic states for BSCCO, in part because the BSCCO compounds have greater chemical stability. Much less is known about the Tl-based HTS compounds.

Studies of overlayer growth have emphasized photoemission or inverse photoemission investigations of the chemical character of the surface during adatom condensation. Most photoemission studies have been done with Al K_α radiation (XPS, $h\nu = 1486.6$ eV), although studies using synchrotron radiation have also been reported [1, 2]. The advantage of XPS is that the core level signatures can be readily detected and changes in them reflect bonding changes. Moreover, since the electron mean free path increases once the kinetic energy of the electron exceeds about 50 eV, the probe depth in XPS is enhanced relative to lower energy studies. The advantage of lower energy studies is that the valence band cross sections are greater and changes near the Fermi level are more readily determined. In inverse photoemission studies (IPES), the advantage is that the empty states can be probed, and changes in them offer additional insight into chemical modification of the surface. In IPES studies, the surface sensitivity is comparable to that of typical synchrotron radiation studies because the electron energies are below about 40 eV.

7.2 Room Temperature Overlayer Formation with Atoms

It has been well documented that chemical changes can be induced at HTS surfaces by adatom condensation at 300 K [1, 2]. The degree of disruption is greatest for elements with high heats of oxide formation, such as Ti, Cr, and Al, and reaction can be understood in terms of the thermodynamic tendency to form new oxide species at the expense of Cu–O bonding. Disruption is much less following Ag and Au atom deposition, where the heats of oxide formation are small, but detailed studies of Ag and Au growth point to the inherent tendency of the unstable surface layers to restructure, forming new surface oxide layers. In all cases, such changes are influenced by temperature, i.e., the ability to transport atoms to the surface region where reaction will occur.

7.2.1 *Ti Interfaces with $YBa_2Cu_3O_7$, Y_2BaCuO_5, and CuO: Prototypes of Reactive Overlayers*

Figures 7.1 and 7.2 summarize XPS results for Ti overlayer formation on $YBa_2Cu_3O_7$, Y_2BaCuO_5, and CuO, emphasizing changes in the substrate

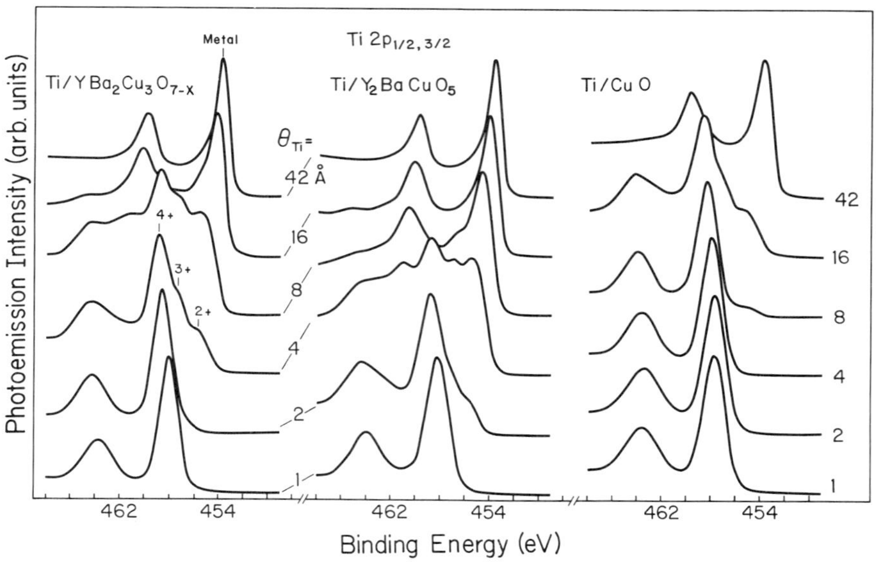

FIGURE 7.1. Ti $2p_{1/2,3/2}$ core level emission for Ti deposition onto $YBa_2Cu_3O_7$, Y_2BaCu_5, and CuO. For low Ti depositions, the $2p$ spectra show well-resolved doublets with binding energies characteristic of TiO_2. At intermediate coverages the line shape is complex and shows overlapping Ti chemical states. At high coverage the spectra are characteristic of Ti metal. (From [3].)

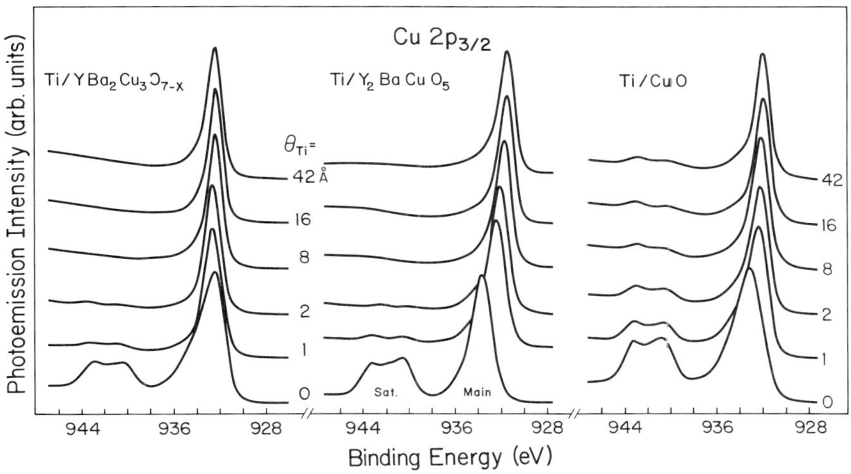

FIGURE 7.2. Cu $2p_{3/2}$ core level emission for Ti deposition onto $YBa_2Cu_3O_7$, Y_2BaCu_5, and CuO. The clean surface spectra show characteristic Cu^{2+} features, namely a broad main line at ~933 eV and satellite emission at higher binding energy. Ti deposition results in the loss of satellite intensity and narrowing of the main line as Cu atoms are reduced during Ti–O interactions. The probe depth of the measurements is ~50 Å, and that region is completely modified for the Y compounds by small amounts of reactive metal. (From [3].)

through the Cu $2p$ emission and in the overlayer through the Ti $2p$ emission. The Ti $2p$ spectra of Figure 7.1 are normalized to constant height to emphasize line shapes and binding energy positions. To obtain these results, Meyer et al. [3] produced clean surfaces under ultrahigh vacuum conditions, first characterizing the surfaces and then exposing them to a Ti atom flux. Comparison of the data for Ti deposition on YBCO [3, 4] to those for Ti deposition onto LSCO [5] demonstrates the close correspondence of surface reactivity for the two Cu–O based materials. Further, comparison to Ti/Y_2BaCuO_5 and Ti/CuO shows a similar pattern for the reaction of Ti atoms on Cu–O based materials [3].

From Figure 7.1, the Ti $2p$ emission for depositions at low coverage is characterized by a doublet with a narrow $2p_{3/2}$ peak, a broader $2p_{1/2}$ peak, and a spin orbit splitting of 5.6 eV. With increasing deposition, the Ti peaks shift rigidly to higher binding energy, showing that the oxidation state is not unique below ~4 Å and that it evolves as the amount of Ti-induced reaction proceeds, that is, as a species that is more like a solid state phase is produced. Comparison with the literature [6] indicates that the fully reacted product is TiO_2, though examination of the $2p_{3/2}$ linewidth shows it to be broader than in bulk TiO_2. This indicates that the reacted region is not homogeneous and it might better be described as aggregates with TiO_2-like bonding than a planar layer.

The results of Figure 7.1 also show shoulders in the Ti 2*p* spectra shifted to lower binding energy. These features appear as the TiO_2-like reaction nears completion and they signal the onset of the growth of Ti-suboxide bonding configurations. The observed chemical shifts correspond to Ti^{3+} and Ti^{2+} oxidation states, as in the suboxides Ti_2O_3 and TiO. They form when the supply of oxygen to the Ti-rich region is limited by kinetics, i.e., when transport to the site of continued reaction is impeded by the reaction products themselves. A metallic component is observed at higher coverage as Ti metal nucleates and grows, though this layer is also not homogeneous [3].

The results of Figure 7.1 suggest that a cross section of the Ti/YBCO interface would reveal buried YBCO, a disrupted layer of Y–Ba–Cu–O of unknown stoichiometry and structure that is deficient in oxygen, a region containing TiO_2 and Ti–O suboxide species, and an overlayer of Ti metal. This interface would be heterogeneous along the surface because of the polycrystalline character of the surface and the kinetics of reaction. The results of Figure 7.1 show that Ti atoms will react with oxygen atoms withdrawn from the substrate in an analogous fashion for all three of these copper-oxide based materials, although the coverage range over which reaction takes place differs.

Figure 7.2 shows the influence of Ti deposition on the Cu 2*p* core level emission. The clean surface spectra are shown at the bottom. Clearly evident is a dramatic loss of Cu 2+ satellite emission and an alteration of the main line shape and position. Since nearly all satellite emission is quenched by 2 Å for Ti/YBCO, the Cu atoms of the substrate have been reduced from a 2+ to a 1+ oxidation state within the sampling depth of the measurements ($\sim$ 50 Å). Similar changes are observed for Y_2BaCuO_5 and CuO, demonstrating that oxygen removal causes changes in the chemical environment of the Cu atoms. The extent of this disruption is large compared to the amount of metal deposited, especially for the HTS compounds. For CuO, the satellite loss is less pronounced, probably because of the more stable three-dimensional Cu–O bonding configuration of CuO.

The photoemission core level and valence band spectra [3] for Ti deposition indicate that interface formation occurs in three steps. First, vapor deposited Ti atoms react with oxygen withdrawn from the substrate to form TiO_2-like species, reducing the Cu oxidation state from 2+ to 1+ and inducing a chemical shift for substrate core level features. The valence bands show loss of emission near E_F as substrate Cu–O hybrid states are modified and the metallic character of the probed region is lost. During this first stage, Ti condensation changes the thermodynamic balance and favors conversion to new configurations. During the second stage, new Ti 2*p* features grow as suboxide species are produced. Finally, Ti metal nucleates and the reacted regions are covered by an inhomogeneous metal layer.

Results that are generally equivalent to those just described have been found for all reactive metals deposited on Cu–O based superconductors [1, 2]. These include Ti, Cr, Fe, Cu, La, Pd, Al, In, Bi as well as Ge and Si [1].

Reactions are limited by the diffusion of O to the surface region where new phases form, driven by thermodynamically favored metal–oxygen reactions (but see below for a discussion of the inherent Bi–O BSCCO surface instability as studied during Ag overlayer growth). When the amount of metal exceeds the amount of available O needed to form a fully oxidized species, other oxide species may form or may evolve into solid solutions as oxygen is trapped and the overlayer thickens. Growth by simple layer-by-layer fashion should not be assumed because clustering on the reacted layer is common [1]. Such interfaces are metastable, and they can be made to react further by increasing the temperature [1].

These general conclusions for room temperature deposition of atoms are important because the control of chemical interaction at interfaces involving HTS materials is crucial if contacts are to be made. Rather than forming simple boundaries, these overlayers on HTS surfaces exhibit complex morphologies parallel and perpendicular to the surface.

7.2.2 *Ag Interfaces with YBCO and BSCCO*(001): *Photoemission and STM Studies*

Ag and Au are elements with very low heats of oxide formation and, logically one might expect that they would exhibit very different interface profiles than those just discussed. Indeed, studies of Ag and Au on HTS surfaces [7–14] have concluded that there is much less disruption for Ag and Au than for any other metals. Nonetheless, results for Ag adatom deposition onto single crystal YBCO and BSCCO demonstrate changes in the Cu $2p_{3/2}$ line shape that indicate Ag-induced surface disruption [7], and this raises questions as to the mechanisms for such modification.

Figure 7.3 summarizes Cu $2p_{3/2}$ spectra for Ag deposition onto YBCO(001) and BSCCO(001). Cu $2p_{3/2}$ line shape changes can be seen as the main line narrows and shifts and the satellite emission decreases. For Ag deposition onto BSCCO(001), modification of the near surface region occurs by 4 Å deposition, with little change thereafter. Comparison shows that the extent of disruption is less for BSCCO than for YBCO.

Evidence for BSCCO(001) surface disruption can be seen by examining the Bi $4f_{5/2,7/2}$ core levels of Figure 7.4 where an adatom-induced Bi component appears at lower binding energy relative to the substrate. It is not easily distinguished below $\sim$4 Å, but it is easily seen by 15 Å. Its presence reflects Bi atoms that are dissociated from the superconductor surface layer [15]. Its appearance and the variation in relative intensity for the substrate and new Bi components give strong evidence for segregation of Bi atoms to the surface and near surface region of the thickening film. Analogous changes in the Bi emission have been seen for other overlayers deposited onto BSCCO, but the effect was smallest for Ag [8, 12, 15].

The Ag growth morphology can be determined with photoemission by examining the intensity of the substrate and overlayer core level emission as

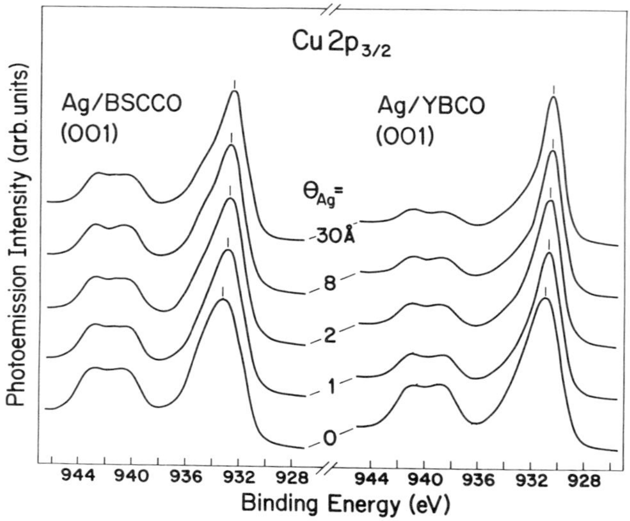

FIGURE 7.3. Cu $2p_{3/2}$ core level spectra for BSCCO(001) and YBCO(001) as a function of Ag coverage showing a reduction of satellite emission and sharpening of the main line. These results demonstrate changes in Cu–O bonding near the surface. (From [7].)

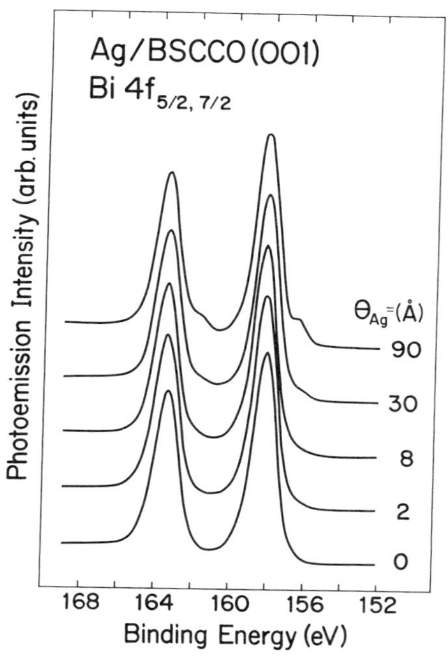

FIGURE 7.4. Bi $4f_{5/2,7/2}$ core level spectra for BSCCO(001) as a function of Ag coverage. The broadening on the low binding energy side is attributed to Bi atoms dissociated from the substrate and segregated to the surface region. (From [7].)

a function of deposition. Those results indicate cluster growth starting from the lowest coverages, a result that is confirmed by the STM results. Since the changes in the Cu $2p_{3/2}$ line shape ceased after ~4 Å on BSCCO(001) but continued until 8 Å on YBCO(001), we conclude that Ag deposition affects the former less than the latter.

For Ag/BSCCO, minimal changes in the O 1*s* energy or line shape were observed. For Ag/YBCO, slight variations were evident and they suggest rearrangements of substrate oxygen atoms following Ag-induced disruption of Cu–O bonds. The changes are too subtle to be caused by new Ag–O bond formation, given the amount of reduction of the Cu satellites. Moreover, the heats of formation of AgO and Ag_2O are −7 and −8 kcal/mol, and one would not expect Ag–O reaction species. The changes in the O 1*s* spectra were accompanied by slight increases for total emission from both Ba and Y, again suggesting atomic rearrangement at the surface.

Scanning tunneling microscopy studies of the Ag/BSCCO(001) interface make it possible to correlate the chemical information obtained with photoemission to the structural changes of the surface region induced by the adatoms [16]. Figure 7.5 shows an 85 × 85 Å^2 STM image of cleaved $Bi_2Sr_2CaCu_2O_8$(001) acquired with a tip bias of 1 V with no correction for thermal drift. As discussed by Luo et al. [16], these experiments were performed in an ultrahigh vacuum STM system equipped with low-energy electron diffraction (LEED) capabilities. $Bi_2Sr_2CaCu_2O_8$ single crystals, having superconducting transition temperatures of 85 K, were cleaved in situ. STM imaging was done on mirror-like surface regions that were a few mm^2 in size.

The bright protrusions of Figure 7.5 form a square lattice with a lattice constant of 3.8 Å, consistent with the fact that the cleavage plane of $Bi_2Sr_2CaCu_2O_8$ is the Bi–O plane [17, 18]. STM images reveal a single protrusion per unit cell [18–20], even though the unit cell is comprised of one Bi atom and one O atom. It is still not clear whether the protrusions are

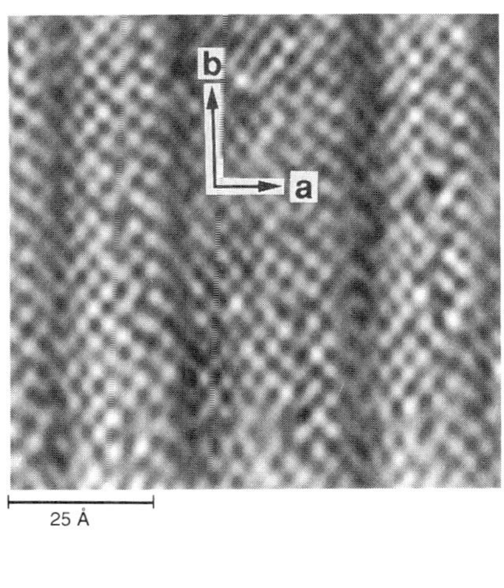

FIGURE 7.5. 85 × 85-Å^2 STM image of cleaved $Bi_2Sr_2CaCu_2O_8$(001) acquired at tip bias of 1 V and tunneling current of 0.18 nA. The stripes parallel to the *b* axis reflect height modulations perpendicular to the *a*–*b* plane. The spacing between stripes is typically 4.5–5 tetragonal unit cells along the *a* axis. The central portion shows an unusually large modulation period. (From [16].)

derived from Bi atoms, O atoms, or an electronic charge distribution consistent with both.

Figure 7.5 also shows bright and dark stripes that run along the diagonal of the square lattice, i.e., the *b* axis, that represent height modulations perpendicular to the *a*–*b* plane. X-ray diffraction results have shown that such modulations also exist in the bulk [21]. It is caused by the insertion of extra O atoms into Bi–O planes [22, 23]. If the distance between adjacent dark stripes is designated as the modulation period, then extra O atoms are accommodated at the center of each period such that there are extra O-atom rows on top of the modulations [23]. Modulation periods for the Bi–O plane are typically 4.5–5 tetragonal unit cells along the *a* axis, giving an extra oxygen atom every 9 or 10 Bi atoms [18, 21]. The image of Figure 7.5 shows an exceptionally large period of ~7 cells at the center. The left and right portions show periods where the modulation is ~4.5 tetragonal unit cells. For the 7 cell period, there is 1 extra O atom for every 14 Bi atoms, indicating a local reduction in the O to Bi atom ratio.

Figures 7.6(a) to 7.6(f) show STM images for Ag deposition on $Bi_2Sr_2CaCu_2O_8(001)$ at 300 K in amounts ranging from 0.1 to 20 Å. They measure 310 × 310 Å^2 in all cases except Figure 7.6(a) where the image is 290 × 290 Å^2. The lines superimposed on Figure 7.6(a) to 7.6(c) are parallel to the *b* axis of the substrate, corresponding to the modulation direction of Figure 7.5. Their spacing is ~25 Å, equal to the prevalent modulation period. Atomic resolution was not possible after Ag had been deposited.

Figure 7.6(a) shows the effect of 0.1-Å Ag deposition. Circular clusters having diameters from ~16 to ~25 Å and an average height of ~6 Å are clearly evident. Most are centered on modulation lines. The total cluster volume per unit area based on the cluster density and the average size was ~9 times greater than expected, based on the volume of Ag deposited. Ag deposition to 0.2 Å, Figure 7.6(b), produced more clusters with more apparent alignment of cluster rows but no significant change in cluster size distribution. For 0.2-Å deposition, the volume of the clusters was ~11 times that of the deposited Ag. Deposition to 0.6 Å produced more obvious alignment, Figure 7.6(c) and a sharper cluster size distribution, ranging from 20 to 26 Å in diameter. The average height was still ~6 Å, and the volume of the clusters was 5 times greater than expected.

The results of Figure 7.6(a) to 7.6(c) demonstrate Ag induced cluster development and the fact that the amount of Ag is insufficient to account for them. The photoemission results showed conversion of Cu^{2+} to Cu^{1+} bonding configurations, depletion of O from the Cu–O layers, and Bi segregation [24] to the surface of the Ag film. Thus, atoms of the terminating Bi–O plane are rearranged as a consequence of Ag condensation, and there is modification of the near surface Cu–O planes. The driving force for these changes cannot be the thermodynamics of AgO formation.

The explanation for these Ag induced modifications lies within the layered

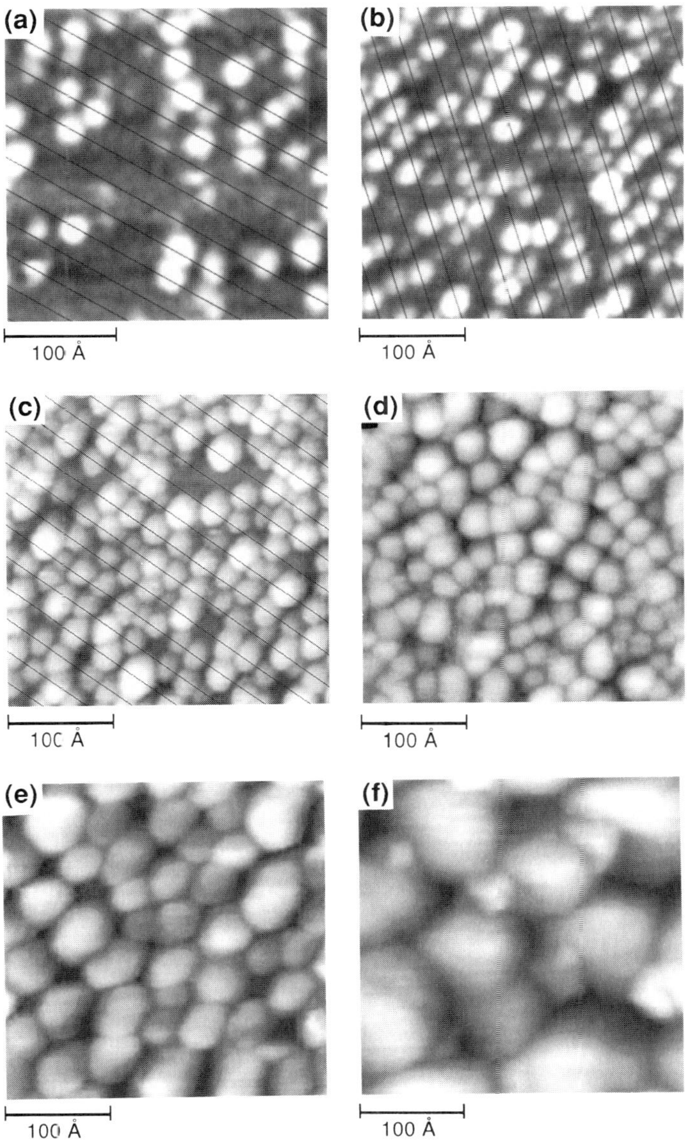

FIGURE 7.6. (a)–(f) STM images taken after Ag depositions of 0.1, 0.2, 0.6, 1, 5, and 20 Å. Tip biases ranged from −2 to −2.34 V and tunneling currents were 0.03 to 0.5 nA. Lines parallel to the *b* axis with spacing of 25 Å are superimposed on (a)–(c) to show cluster alignment. The cluster volumes are many times the amount of Ag deposited because they reflect the conversion of planar Bi–O to Bi_2O_3-like clusters induced by Ag clustering. Ag accumulates over the Bi_2O_3 clusters after conversion of the Bi–O layer is complete, and images (d)–(f) reflect Ag aggregation. (From [16].)

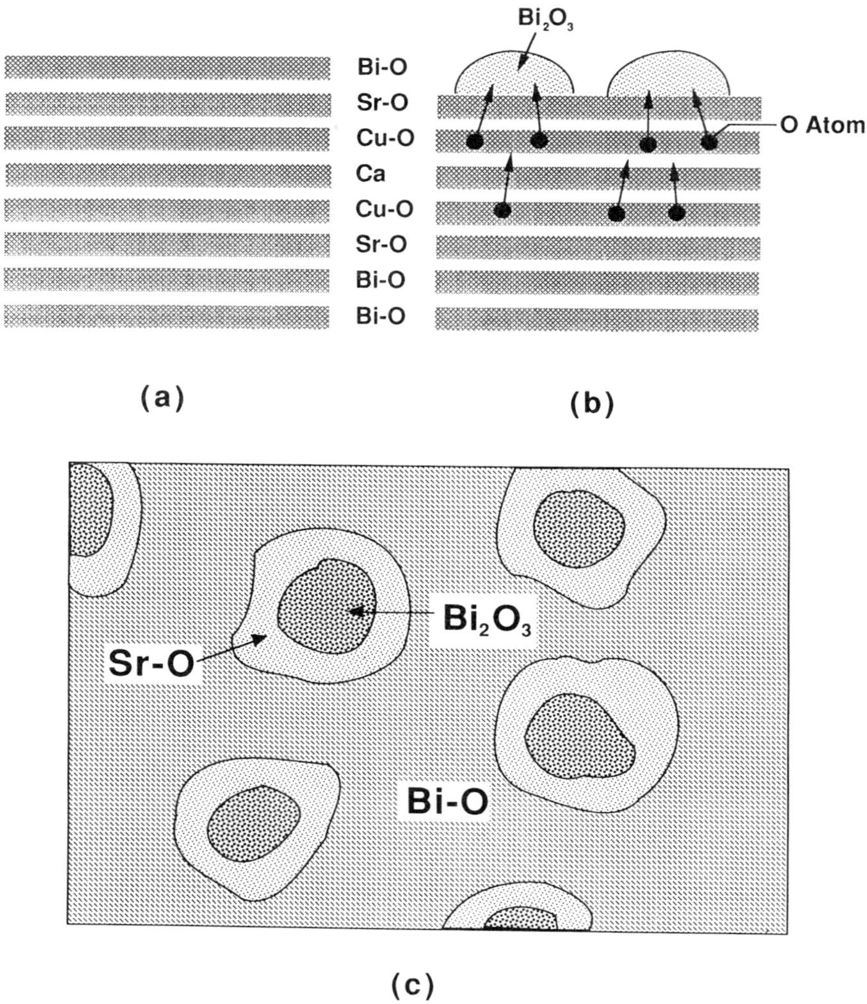

FIGURE 7.7. (a) Schematic cross section of the $Bi_2Sr_2CaCu_2O_8$ layered structure parallel to the c direction before Ag deposition with a Bi–O plane terminating the structure. (b) Schematic showing conversion of the top Bi–O plane into Bi_2O_3-like species and O withdrawal from the Cu–O double layer. Ag clusters account for less than 20% of the surface and are omitted. (c) Schematic top view of the $Bi_2Sr_2CaCu_2O_8$ surface after Ag deposition depicting partial conversion of the Bi–O plane as Bi_2O_3-like clusters are formed. Structural changes involving the Bi–O plane expose the Sr–O plane around the clusters. (From [16].)

$Bi_2Sr_2CaCu_2O_8$ structure itself, depicted in Figure 7.7(a), particularly the surface exposed by cleaving. Although Bi–O planes are stable as part of the bulk structure, cleaving to produce a surface monolayer of Bi–O establishes a less stable structure. The conversion of the exposed Bi–O layer to a more bulk-like Bi_2O_3 cluster configuration is thermodynamically favored. (The bulk heat of formation of Bi_2O_3 is -140 kcal/mol but that of a vacuum-terminated Bi–O monolayer supported by the HTS structure is not known.) The Bi–O structure of the HTS is unique to this class of materials, and the fact that only bilayers are formed in bulk materials suggests that single Bi–O layers are metastable.

There are two ways that the conversion of the Bi–O plane could be accomplished, one involving the addition of oxygen and the other involving separation into Bi and Bi_2O_3 phases. Some separation definitely occurs because the photoemission results show Bi segregation to the surface in an amount that corresponds to 3% of a single Bi–O layer. Hence, there is insufficient oxygen for complete Bi_2O_3 formation. The thermodynamic demand for oxygen atoms then serves as the driving force for withdrawal from Cu–O layers, as depicted in Figure 7.7(b). This is responsible for the changes in Cu bonding states.

The cleaved $Bi_2Sr_2CaCu_2O_8(001)$ surface is stable at 300 K against repeated STM scanning and time in ultrahigh vacuum. The absence of spontaneous decomposition of the surface layer indicates that there is an activation barrier for the conversion. The addition of Ag provides the energy needed to overcome such a barrier. The kinetic energy of the impinging Ag atom is one source of energy but it is relatively small. Moreover, if atom impact were the activating process, then the number of clusters would be much larger. The energy released by Ag–Ag bonding and cluster formation is considerably greater. Such cluster-related disruption has been observed frequently for metal growth on semiconductors [25]. It is likely that the energy released when mobile Ag atoms on the surface form clusters is responsible for activating Bi–O conversion and associated events.

The model depicted in Figure 7.7 can be checked by analyzing the Cu $2p$ photoemission intensity. Luo et al. [16] calculated that the Cu $2p$ satellite-to-main intensity ratio should decrease by 23% to supply enough O to allow complete conversion of the surface Bi–O layer to Bi_2O_3 species. The photoemission results of Meyer et al. [7] show that the Cu $2p$ satellite to main line intensity ratio decreased from 0.44 to 0.33 by 4-Å Ag deposition, in agreement with the calculation. Additional Ag deposition did not induce further change.

The cluster rows of Figure 7.6(a) to 7.6(c) provide further support for the model. If Bi and O atoms occupy their ideal lattice sites, then all Bi sites would be equivalent and adatom nucleation or surface layer disruption would occur randomly. The atomic modulation alters the translational symmetry of the Bi–O plane and gives a unit cell of $\sim 25 \times \sim 5.4$ Å. According to LePage et al. [23], the Bi lattice is slightly expanded around the top of the atomic modulation to allow an extra O atom row to be inserted. Conversely, the Bi lattice is contracted around the bottom of that modulation. These areas should have different stabilities, and the less stable areas would be preferred sites for clustering. While the STM results show a tendency for the clusters to form rows running along the *b* direction with a spacing of the modulation period, the resolution was not sufficient after Ag deposition to determine the particular nucleation sites. Luo et al. postulated that it starts along the extra O rows because conversion there needs less O from deeper layers.

Figure 7.7(c) depicts the surface morphology following Ag deposition, in the context of the model. For simplicity, the Ag clusters have been omitted. Bi_2O_3 agglomeration exposes the Sr–O layer around each cluster. LEED results show the persistence of the 5×1 pattern characteristic of the clean surface or the Sr–O layer [26]. Since this pattern persists, although attenuated, we speculate that the exposed Sr–O layer is structurally stable despite Bi–O conversion. The Sr–O layer is less likely to be preserved beneath the clusters because of changes in the buried Cu–O layer, Figure 7.7(b).

Continued deposition of Ag to 1.6 Å leads to coalescence, and this alters the cluster pattern, as evident in Figure 7.6(d). The STM images of Figures 7.6(e) and 7.6(f) for coverages of 5 and 20 Å demonstrate that smaller clusters merge and rearrange to maintain circular shapes, presumably to minimize surface free energy. By these coverages, the photoemission results show that atom mixing is largely completed and Ag covers the disrupted region [7]. Ag cluster growth is then controlled by coalescence, but it is also affected by the disrupted substrate and the presence of segregated Bi atoms. The surface of the Ag clusters is disordered even though they appear to have a preferred shape. This disorder can be explained by the presence of segregated Bi. Based on the large cohesive energy of Ag, the clusters evident in Figure 7.6(f) are most likely crystalline.

This model emphasizes the instability of the Bi–O layer and the perturbation introduced by adatom clustering. Equivalent disruption should occur for other adatoms if the perturbation is large enough to activate the process. Such effects can be inferred from photoemission experiments for Au deposition, Ne sputtering, and thermal annealing [8, 27]. Indeed, STM studies of Au overlayer growth on BSCCO(001) at 300 K have shown a surface morphology that is almost identical to that for Ag deposition at similar coverage [16]. Again, this type of disruption, where overlayer metal oxide formation is not favored, should be independent of the chemical nature of the (nonreactive) overlayer atoms. When the overlayer atoms are also likely to form oxide species of their own at the expense of the substrate, such as Ti discussed above, then there is a greater demand for oxygen and surface disruption is greater, driven by both thermodynamic processes. STM studies of Cr growth on BSCCO(001) have demonstrated just this effect [16].

7.3 Low-Temperature Overlayer Formation with Atoms

7.3.1 *Ti Interfaces with YBCO and BSCCO*(001)

The results of Sect. 7.2 showed that Ti overlayer growth at 300 K on polycrystalline YBCO and related oxide compounds is characterized by oxygen withdrawal and substrate disruption [3, 4]. Growth at 20 K alters the overlayer profile [28], as summarized by the normalized core level spectra of Figure 7.8. In particular, the Ti $2p_{3/2}$ results reveal TiO_2-like bonding configurations by ~1-Å deposition, and suboxide formation is also apparent because of emission at lower binding energy. The nucleation of Ti metal on this oxidized surface by 4 Å (compared to ~10 Å for 300-K deposition) demonstrates reduced oxygen outdiffusion and oxide formation at low temperature. Moreover, the metallic Ti layer is relatively uniform since emission from the oxides is rapidly attenuated by the growing Ti layer. By 25-Å deposition, the Ti spectral features are indistinguishable from those of pure Ti. This was not the case for 300-K growth because Ti formed clusters on the oxide layer.

Examination of the substrate core level spectra of Figure 7.8 shows that there is much less disruption than for 300-K growth. The lowest Cu and Ba curves for clean YBCO at 20 K are typical of those measured for nearly phase-pure polycrystalline YBCO and YBCO(001) [1, 29]. Again, the Cu 2*p*

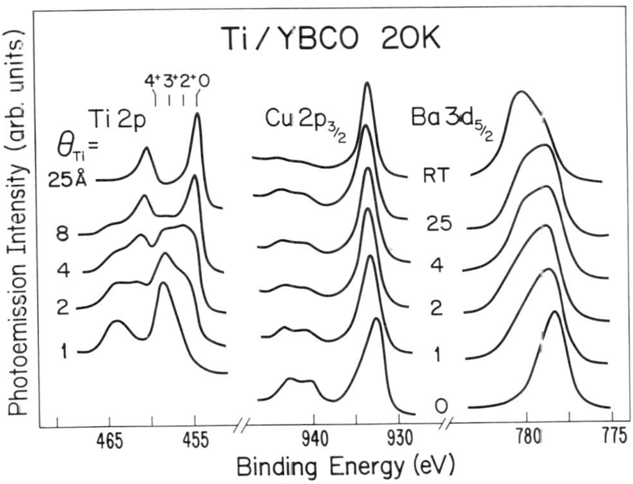

FIGURE 7.8. Core level spectra for Ti deposition on polycrystalline YBCO at 20 K. The Ti 2*p* results show the TiO_2-like layer formation at a low coverage, followed by suboxides and a uniform metallic Ti layer. (The low coverage Ti emission is exaggerated by normalization to constant height.) Changes in the substrate spectra show much reduced reaction compared to 300-K growth. Warming to 300 K (RT) activated slight additional reaction at the buried interface. (From [28].)

changes result from oxygen removal and disruption of the Cu–O planes and chains [1]. Significantly, however, the Cu $2p_{3/2}$ emission remained unchanged for coverages above ~2 Å as reaction stopped and Ti metal formed. The Ba $3d_{5/2}$ emission is also changed as a new feature appears, but there were minimal changes after ~4-Å Ti deposition.

The stability of the Ti/YBCO interface grown at 20 K was tested by warming to 300 K after 25-Å deposition, yielding the curve labeled RT in Figure 7.8. This led to a reduction in the Cu $2p_{3/2}$ satellite-to-main ratio and a narrowing of the main line. Further reaction involving Ba was evident. While this demonstrates enhanced reaction beneath the surface, comparison with results for growth at 300 K shows that the amount of reaction is still very much less (compare to Figures 7.1 and 7.2).

To compare the effects of low-temperature growth for BSCCO and YBCO, we show photoemission spectra in Figure 7.9 for Ti growth on BSCCO(001) at 20 K. In this case, the Ti $2p$ emission overlaps a broad Bi $4d_{3/2}$ feature at ~465 eV, and the latter has been removed mathematically [28]. Deposition of 0.5 Å of Ti results in the formation of Ti–O bonds, and 1-Å Ti on BSCCO results in a broader peak that resembles that formed on YBCO. This suggests more than one Ti–O bonding configuration, with the dominant feature near the binding energy of bulk TiO_2. Further Ti deposition yields increased suboxide formation. The nucleation of small amounts of Ti metal began at 2 Å and dominated at higher coverages while attenuating the oxide emission.

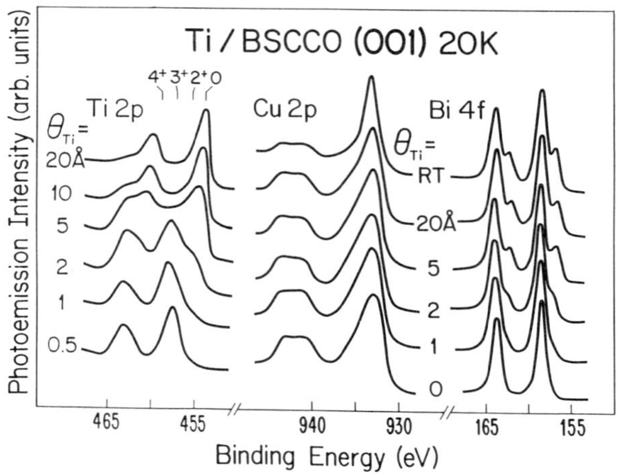

FIGURE 7.9. Core level spectra for Ti/BSCCO(001) formation at 20 K. The Ti results show the formation of overlayer oxides followed by the nucleation of Ti metal, as for YBCO. Overlayer formation at 20 K reduced the Cu $2p_{3/2}$ satellite intensity to 80% of its clean value and warming reduced it further as limited reaction occurred. The appearance of a small reacted component for Bi reflects Bi atoms released from the substrate and kinetically trapped in the thickening overlayer. (From [28].)

Figure 7.9 shows that Ti deposition on BSCCO(001) reduced the Cu $2p_{3/2}$ satellite-to-main line intensity ratio, but there were no changes in shape after 2 Å. Kimachi et al. [28] estimated that the conversion of about half of the first Cu–O layer to a 1+ state would decrease the satellite-to-main ratio to the value observed for growth of Ti/BSCCO(001) at 20 K. The Bi $4f$ results show a new spin orbit split doublet due to Bi atoms released by interface reaction. The fact that the Bi $4f$ feature and the substrate component were attenuated at the same rate after ~5 Å indicates kinetic trapping of Bi by the Ti overlayer.

Warming the 20-Å Ti/BSCCO(001) interface to 300 K produced only slight changes, as shown in Figure 7.9, associated with further modification of the Cu–O layers. Warming also causes an increase in the total Cu $2p_{3/2}$ emission and a decrease in the Ti $2p$ emission due to a roughening of the Ti overlayer, an effect driven by the high surface free energy of Ti. Hence, 20-K deposition produced a metastable interface. Solid state reactions continued when the kinetic constraints were lessened but the final amount of reaction upon warming to 300 K was dramatically less than that for growth by atom deposition at 300 K.

The slight reaction observed upon warming is aided by the instability of the interface between Ti metal and the TiO_x layer. In particular, the Ti/TiO_x interface is not a stable structure, and the heats of formation of the oxide are high. Hence, the metal can be converted locally into suboxide-like bonding structures by oxygen withdrawal from underlying TiO_x. The buried HTS can provide oxygen to sustain the TiO_x layer but at the expense of the HTS structure. This thermodynamically driven process is further enhanced by inhomogeneities in the interface [3, 4]. At the same time, reactions encountered upon warming the already-formed interface are solid state reactions that are thermally activated. Thus, formation at low temperature followed by heating to 300 K represents a different pathway than when a system responds to the deposition of adatoms at 300 K. From the results of Figures 7.8 and 7.9, reactions are more likely for Ti/YBCO than Ti/BSCCO, consistent with the greater stability of BSCCO.

7.3.2 *Cr Interfaces with BSCCO*(001) *and YBCO*(001)

Studies of Cr atom deposition on BSCCO(001) at 20 K offered additional insight into low-temperature growth and oxide-layer stability. In this case, the broad Cr $2p_{3/2}$ emission shown in Figure 7.10 suggests rather poorly defined, oxygen-deficient Cr bonding configurations caused by low atom mobility at 20 K. Further Cr deposition induced a shift to lower binding energy, possibly altering the structure of the oxide locally as Cr was added because a variety of local Cr–O bonding states were stabilized by low-temperature kinetics. There was no evidence of the distinct configurations observed for 300-K growth when the spectra could be deconvolved into Cr_2O_3-like and

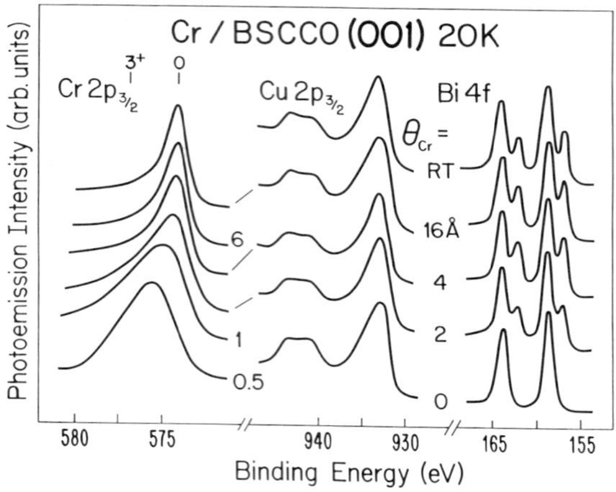

FIGURE 7.10. Core level spectra for Cr/BSCCO(001) formation at 20 K. The Cr results indicate that the Cr–O bonding configurations are different from those produced by 300-K deposition. The continuous shift for Cr toward the bulk metal position indicates that well-defined Cr–O bonding configurations are not present. Significantly, the Cu $2p_{3/2}$ emission is changed very little, indicating only slight modification of the Cu–O planes. Disruption of the outermost Bi–O layer is evident from the appearance of the new Bi 4f doublet. In contrast to the results for Ti/BSCCO, the Cu $2p_{3/2}$ satellite also does not change after warming to 300 K, indicating that the overlayer is stable. (From [28].)

Cr-like states [30]. By 4 Å, the Cr $2p$ line shape was dominated by emission from metallic bonding, with some contributions from buried Cr oxides. By 16 Å, the Cr $2p_{3/2}$ emission was essentially that of Cr metal.

The deposition of ~2 Å of Cr at 20 K led to changes that corresponded to valence conversion of about half of the first Cu–O plane. At higher coverage, the Cu line shape remained almost constant, consistent with inhibited reaction after ~2 Å. Low coverage reaction also produced a new Bi $4f$ doublet due to Bi atoms released from the Bi–O surface layer. Intensity analysis showed that the reacted components reached a relative maximum after 3–4 Å deposition and metallic Cr growth appeared to be nearly layer-by-layer thereafter.

Warming the 16-Å Cr/BSCCO interface to 300 K did not change the Cu $2p_{3/2}$ line shape, as is evident from Figure 7.10. However, warming did cause surface roughening since the Cr $2p_{3/2}$ emission intensity decreased and the substrate intensities almost doubled. Despite these morphology changes, the persistence of the Cu $2p_{3/2}$ satellite (Figure 7.10, curves labeled RT) demonstrates that low-temperature deposition dramatically reduces the amount of interface disruption, even when warmed. In particular, growth at 300 K led to the disappearance of the satellite by 4 Å and much more substrate disruption [1, 30].

Analogous results were obtained for Cr deposition on YBCO(001) at 20 K [28], with the curtailing of reaction by ~2-Å deposition and the formation of metallic Cr on the reacted layer. Kimachi et al. [28] reported that warming the 12-Å Cr/YBCO(001) interface to 300 K had no significant effect on the Cu $2p_{3/2}$ emission.

The absence of continued disruption upon warming Cr/BSCCO and Cr/YBCO interfaces reflects the stability of the Cr/Cr–oxide interface. The binary phase diagram for Cr–O indicates that there are no stable oxides with lower oxygen content than Cr_2O_3, in contrast to Ti–O [31]. Warming Cr/HTS interfaces from 20 to 300 K should then assist in the formation of Cr and Cr_2O_3 from the oxygen-deficient intermediate oxide layer [30]. In turn, Cr_2O_3 has very low diffusion coefficients for Cr and O, an important property for corrosion resistant films, and oxygen diffusion from the HTS to continue reaction with Cr is small. Finally, the activation energy for O diffusion in Cr_2O_3 is about twice that in TiO_2, showing that the Cr oxide will be a better barrier than even single-phase Ti oxides [32]. Much less reaction is observed for warmed interfaces than for those formed at 300 K, in part because of the kinetics of reaction for already-formed solid overlayers.

7.4 Cluster Assembled Overlayers

The above sections have reviewed the consequences of overlayer growth by atom deposition at 300 and 20 K. In each case, there are complex surface and near surface processes associated with impinging atoms, diffusing atoms, and reacting atoms. Such processes are unavoidable, given the procedures of growth, but they produce less than ideal surfaces.

To create more ideal boundary regions, we recently developed a method of joining preformed metal clusters with atomically clean surfaces [33]. The approach has been to isolate the substrate from the impinging atom flux and thereby avoid complexities at the surface due to adatom impact and bonding. In this way, it is possible to prevent adatoms from interacting with the surface until they have agglomerated into clusters. To achieve this, we first condense a thin layer of solid Xe on the clean substrate at temperatures below ~60 K and then deposit atoms onto this buffer layer. Adatom mobility is sufficient to assure the formation of clusters containing tens to thousands of atoms, depending on the amount of material deposited. The clusters come into contact with the clean surface when the Xe buffer is thermally desorbed. One of the advantages of this novel growth process is that it changes the kinetic pathways at the surface enough to produce undisrupted interfaces [33]. In particular, it offers the possibility of examining HTS interfaces with little or no modification of the superconducting surface region, provided the thermodynamics driving forces for cluster–solid reaction are not too severe. In the following, we examine cluster-assembled HTS interfaces, focusing on issues related to surface structure, surface chemistry, and reaction mechanisms. The range of cluster reactivity is explored by considering Cr and Au overlayers, but other systems including Ge have been reported [34, 35].

Ohno et al. [34, 35] prepared their single crystal surfaces by mounting samples of YBCO or BSCCO onto rigid posts and prying off a rod epoxied to the surface. Typical cleaved YBCO surfaces were flat and smooth but they had irregular pits that were visible under an optical microscope. The cleaved BSCCO surfaces had flat, mirror-like regions a few mm in size but they also had rough fractured regions.

For cluster deposition, the measured substrate photoelectron emission was derived from areas under the clusters and from areas that remained exposed. During overlayer growth, the amount of covered surface increased and the core level line shapes increasingly reflected interfacial effects. Information concerning the morphology of the cluster-assembled layers can be obtained by examining the rate at which the substrate emission is attenuated. With cluster assembly, the rate of attenuation is slower than would be observed for the same amount of material distributed in a uniform layer. This deviation reflects clusters with thicknesses greater than the mean free path of the photoelectron [33]. Waddill et al. [33] used a model based on uniformly shaped hemispheres to estimate the cluster diameter as a function of the amount of material deposited. For a deposition of 2 Å, the cluster diameter was ~30 Å so that a representative cluster contained ~500 atoms.

7.4.1 *Cluster Assembly of Au on BSCCO*(001): *Weak Substrate Interactions*

The experiments reported by Ohno et al. [34] involving Au cluster assembly on BSCCO(001) followed the procedures outlined above with cleaving at 20 K, the formation of a buffer layer, the deposition of 2-Å Au onto the buffer, and then warming to ~100 K to remove the buffer. Data acquisition was done after the sample cooled to 20 K. The Au clusters were metallic for 2-Å assembly [33], and the Au emission showed bulk characteristics. The spectra for Au (clusters)/BSCCO are not reproduced here because there were no changes in overlayer and substrate core level spectra at any coverage from 2 to 20 Å. Hence, the Au clusters did not disrupt the substrate. In drawing this conclusion, it is important to note that the clusters were sufficiently thin that the substrate could be seen and chemical changes could have been deleted. Analysis of substrate emission showed nearly exponential attenuation as a function of deposition at low temperature. The simplest model for Au(clusters)/BSCCO would be clusters thinner than the photoelectron attenuation length that are initially well-separated. The clusters appeared to wet the surface and, with increased density, probably formed connected patterns with sintering when clusters made contact. An interface formed in this way by successive depositions, 2 Å at a time, will be composed of a large number of small crystallites. However, since assembly is done at ~100 K, the structure will also reflect kinetic constraints as the Xe layer desorbs.

To investigate the stability of a 20-Å Au(cluster)/BSCCO interface formed in 3-Å increments, Ohno et al. warmed the sample to 300 K in situ. There were no changes in the Au $4f$ line shape or the Cu $2p_{3/2}$ satellite-to-main ratio but the Bi line shape broadened. Although minimal chemical modification of the Cu–O plane occurred, there were changes in the morphology of the overlayer as the Au $4f$ intensity decreased and the Bi and O intensities increased 50%–80%. This suggests delamination of the Au layer from the HTS surface to reduce the total surface area of the Au and possibly a restructuring of the Bi–O surface layer.

From these results, it is evident that Au cluster assembly at low temperature produces an overlayer that interacts weakly with the Bi–O terminated surface. Cluster–substrate van der Waals interactions result in wetting, but kinetic factors trap the system in a metastable configuration at low temperature. Upon warming to 300 K, the clusters tend to delaminate and structures appear that have smaller surface areas. These results can be understood in terms of the competition between equilibrium thermodynamic growth structures and low-temperature growth constraints imposed by limited transport and van der Waals attraction. It is these constraints associated with assembly with clusters that gives novel interface structures, as we shall see in the following for reactive systems.

7.4.2 *Cluster Assembly of Cr on BSCCO*(001): *Suppressed Interface Reactivity*

To investigate Cr cluster assembly on BSCCO(001), Ohno et al. [34, 35] formed clusters in 2-Å increments up to 14 Å, then in increments of 4 and 12 to 30 Å. These clusters were fully metallic at all coverages, based on the Cr $2p_{3/2}$ emission. Representative core level spectra for 2 and 30 Å shown in Figure 7.11 demonstrate that cluster assembly produced limited BSCCO surface changes. Indeed, the Cu $2p_{3/2}$ line shape was unchanged by 30-Å Cr cluster assembly at $T \leq 100$ K. The O $1s$ emission did broaden to higher binding energy and the Bi $4f$ emission showed a new (but small) component, as in Figure 7.4. These results indicate substrate modification that was probably restricted to the terminal Bi–O layer, consistent with the discussion of Sect. 7.2 related to perturbation of the surface.

For Cr cluster assembly on BSCCO(001), the rate of attenuation of the substrate emission was similar to a uniform layer for the first three cluster-assembly cycles, but the Cr clusters apparently coalesced at higher coverage. This is consistent with the relatively large surface-free energy for Cr. Although surface diffusion would be slow at ~100 K, the amount of energy released by the formation of a Cr–Cr grain boundary is appreciable. This very low-temperature sintering in the presence of a substrate probably also

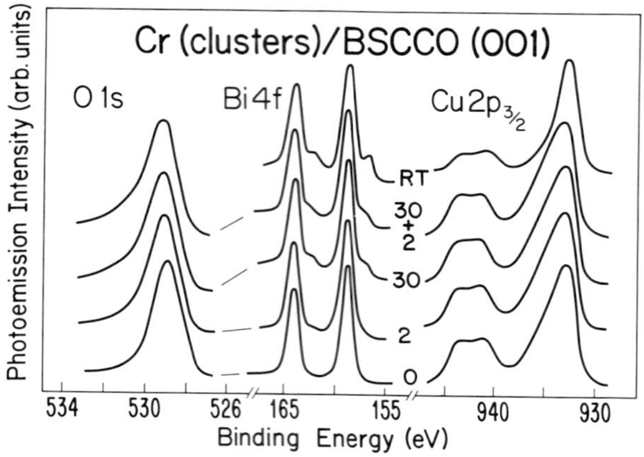

FIGURE 7.11. Core level spectra for Cr(clusters)/BSCCO(001). The Cu $2p_{3/2}$ EDCs have been corrected for small contributions from condensed Xe, and a linear background has been subtracted. The appearance of a shoulder on the Bi $4f$ features indicate slight substrate disruption, and there is a small reduction of the Cu $2p_{3/2}$ satellite. The addition of 2-Å Cr as atoms at 20 K does not change the substrate line shapes, indicating that the surface was covered. Warming to 300 K led to further reaction at the buried interface but the amount was small compared to that seen for atom deposition. (From [34].)

gains from the small volume-surface ratio of the nanoclusters as the energy released is dissipated by a small volume. Despite surface rearrangement of the clusters, the reactivity with the substrate Cu–O planes is still very small and the effect on the Bi–O planes is comparable to that for Ag deposition.

Ohno et al. [34] noted that the coalescence of the Cr clusters introduced some uncertainty as to whether the HTS surface was completely covered. This was tested by atom deposition of 2-Å Cr over the 30-Å cluster-assembled layer. The results shown in Figure 7.11 reveal no substantial changes in the substrate emission. Figure 7.11 also shows the effect of subsequent warming to 300 K, namely an increase in the low binding energy Bi emission, a reduction of the Cu satellite-to-main-line ratio, and a sharpening of the main line. Even with these changes, the modification of the BSCCO surface region is significantly less than for Cr atom deposition at 300 K. These results therefore provide another indication of the barrier formed by cluster deposition and the temperature dependence of the reaction pathway.

7.4.3 *Cluster Assembly of Cr on YBCO*(001)

Cr cluster assembly on YBCO(001) was undertaken to generalize the results for cluster assembly. From atom deposition studies, YBCO is less stable than BSCCO, probably because of self-passivation due to the Bi–O terminal

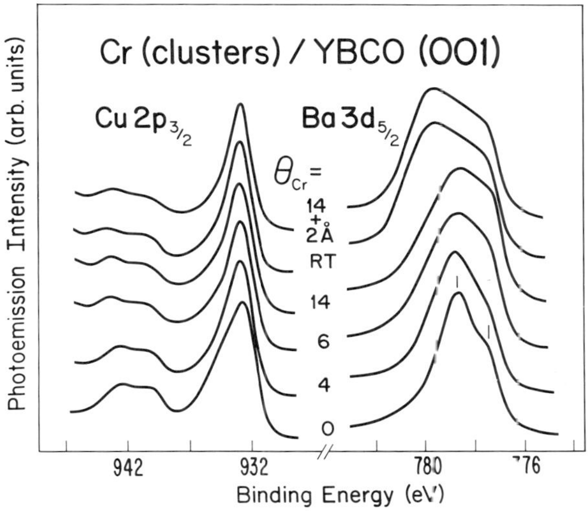

FIGURE 7.12. Cu $2p_{3/2}$ and Ba $3d_{5/2}$ spectra for Cr(clusters)/YBCO(001). Cluster deposition results in partial loss of Cu $2p_{3/2}$ satellite emission and a small shift of the Ba $3d_{5/2}$ centroid. These changes are very small compared to atom deposition onto YBCO. Even this reaction was curtailed by ~6-Å clusters. Warming to 300 K resulted in some additional reaction, as shown by an increase in the reacted Ba $3d_{5/2}$ component at ~780 eV and loss of Cu satellite emission. The addition of 2 Å of Cr atoms at 300 K produced no further changes because the overlayer was complete. (From [34].)

planes. Figure 7.12 shows results for Cr(clusters)/YBCO(001). These clusters were formed in 2-Å increments up to 14 Å. As shown, the Cu $2p_{3/2}$ satellite-to-main-line ratio decreased to 60% of the clean value by 6-Å Cr cluster deposition while the main line sharpened, consistent with 2+ to 1+ conversion. The Ba $3d_{5/2}$ emission broadened as the high binding energy surface doublet was reduced, and additional intensity was added at higher binding energy because of a reacted feature. The O 1*s* emission (not shown) revealed a new component due to Cr–O formation, and this contribution increased until ~6 Å. There were no changes in line shape for the substrate features after 8-ÅCr cluster deposition. This indicates that reaction had been curtailed. The Cr $2p_{3/2}$ emission was characteristic of metallic Cr, although comparison to results for Cr(clusters)/BSCCO(001) suggested oxidation at the buried interface because there was additional intensity at higher binding energy.

Warming to 300 K produced chemical changes at the interface, as is evident from the EDCs of Figure 7.12. In particular there was a shift and broadening of the Ba $3d_{5/2}$ emission because of conversion of the substrate component into the reacted-Ba component at ~780 eV. This reaction product is similar to that formed upon reactive atom deposition, as for Ti/YBCO, but to a lesser extent. Remarkably, the persistence of the Cu $2p_{3/2}$ satellite indicates that substantial amounts of Cu^{2+} remained in the probed area. There were no changes in intensities of the various components that would indicate delamination. To rule out the possibility of contributions from uncovered areas, 2 Å of Cr was deposited as atoms at 300 K onto the cluster-assembled layer. As shown, there were no changes in the substrate emission. The significant amount of remaining Cu $2p_{3/2}$ satellite and the unmodified Ba $3d_{5/2}$ substrate emission demonstrates that Cr cluster deposition provides a method for forming relatively abrupt junctions with YBCO, even for very reactive materials.

7.5 Summary

This chapter has reviewed the formation of overlayers on HTS surfaces, emphasizing YBCO and BSCCO and comparing the reactivities of the two HTS materials. By focusing on different growth procedures and different adatoms, we have been able to examine the extent to which the properties of the HTS surface region can be varied.

In a pictorial sense, Figure 7.13 shows the consequence of atom deposition

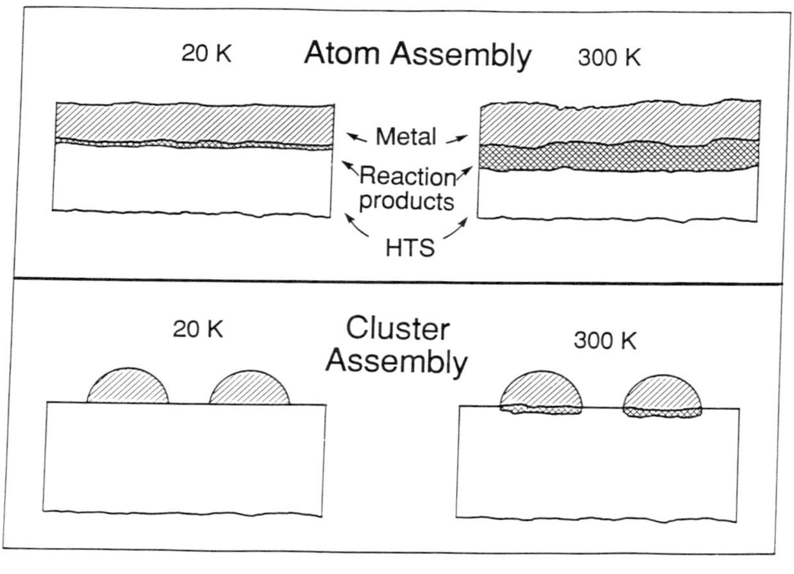

FIGURE 7.13 Pictorial representation of overlayer formation by atom deposition at low temperature and room temperature where the difference in the width of the reacted/intermixed region is related to kinetic effects. For cluster deposition, intermixing is again restricted by the relative stability of the preformed cluster.

at 300 K where reaction products form, derived from the constituents of the HTS near surface region and the adatoms. In this region, the interface is disordered and heterogeneous parallel and perpendicular to the surface. Even very small amounts of reactive metal or non-metal atoms deposited on the surface give rise to disruption of at least 50 Å of the HTS material because of reactions with substrate oxygen. Changes at the surface have been observed following condensation of even the most non-reactive of adatoms, Ag, on BSCCO because of the intrinsic instability of the Bi–O terminating surface layer. In this case, the surface changes are driven by the thermodynamic gain in converting from a Bi–O monolayer to Bi_2O_3-like configurations. Undoubtedly, such a conversion is also important for the reactive atoms, as demonstrated by Luo et al. [16] in their STM studies of Cr deposition on BSCCO(001).

Reductions in reaction have been accomplished by low-temperature atom deposition. As depicted in Figure 7.13, this produces a less disrupted HTS surface region than that at 300 K because of kinetic factors related to oxygen transport. Reactions with the adatoms will still drive by thermodynamics, of course, and annealing will enhance it. Reductions in reaction also have been accomplished by cluster assembly because of the relative stability of the preformed clusters relative to individual adatoms. This stability is depicted on the bottom of Figure 7.13, where the boundary layer is modified only minimally at 20 K. Such an interface is metastable, of course, because warming will allow thermally activated processes to produce more stable bonding states.

While the stability of a solid couple will be determined ultimately by thermodynamics, we have described ways of reaching states that are metastable and potentially important. In all cases, we have discussed the processes related to surface modification when reaction occurs or when there are surface perturbations for the HTS materials.

Acknowledgments. It is a pleasure to acknowledge the contributions of H.M. Meyer III, T.J. Wagener, D.M. Hill, T.R. Ohno, Y.-N. Yang, Y.S. Luo, G.H. Kroll, J.C. Patrin, Y. Kimachi, Y. Hidaka, D.W. Capone II, C.F.Gallo, K.C. Goretta, and B.K. Flandermeyer. This work was supported by the Office of Naval Research and the Defense Advanced Research Projects Agency.

References

1. H.M. Meyer III and J.H. Weaver, in *Physical Properties of High-Temperature Superconductors II*, edited by D.M. Ginsberg (World Scientific, Singapore, 1990), Chap. 6, and extensive references therein.
2. P.A.P. Lindberg, Z.-X. Shen, W.E. Spicer, and I. Lindau, Sur. Sci. Rep. **11**, 1 (1990), and extensive references therein.
3. H.M. Meyer III, J.H. Weaver, and K.C. Goretta, J. Appl. Phys. **67**, 1995 (1990).

4. H.M. Meyer III, D.M. Hill, T.J. Wagener, Y. Gao, J.H. Weaver, D.W. Capone II, and K.C. Goretta, Phys. Rev. B **38**, 6500 (1988).
5. H.M. Meyer III, D.M. Hill, S.G. Anderson, J.H. Weaver, and D.W. Capone II, Appl. Phys. Lett. **51**, 1750 (1987).
6. R. Rocker and W. Gopel, Surf. Sci. **181**, 530 (1987).
7. H.M. Meyer III, D.M. Hill, T.J. Wagener, J.H. Weaver, C.F. Gallo, and K.C. Goretta, J. Appl. Phys. **65**, 3130 (1989).
8. P.A.P. Lindberg, P. Soukiassian, Z.-X. Shen, S.I. Shah, C.B. Eom, I. Lindau, W.E. Spicer, and T.H. Geballe, Appl. Phys. Lett. **53**, 1970 (1988).
9. H.M. Meyer III, T.J. Wagener, D.M. Hill, Y. Gao, S.G. Anderson, S.D. Krahn, J.H. Weaver, B. Flandermeyer, and D.W. Capone II, Appl. Phys. Lett. **51**, 1118 (1987).
10. T.J. Wagener, Y. Gao, I.M. Vitomirov, J.J. Joyce, C. Capasso, J.H. Weaver, and D.W. Capone II, Phys. Rev. B **37**, 232 (1988).
11. C. Laubschat, M. Domke, M. Prietsch, T. Mandel, M. Bodenbach, G. Kaindl, H.J. Eickenbucsh, R. Schoellhorn, R. Miranda, E. Moran, F. Garcia, and M.A. Alario, Europhys. Lett. **6**, 555 (1988).
12. E. Weschke, C. Laubschat, M. Domke, M. Bodenbach, G. Kaindl, J.E. Ortega, and R. Miranda, Z. Phys. B **74**, 191 (1989).
13. M.T. Schmidt, Q.Y. Ma, X. Wu, and E.S. Yang, *High T_c Superconducting Thin Films: Processing, Characterization, and Applications*, edited by R. Stockbauer and R. Kurtz (AIP, New York, 1989).
14. B.D. Hunt, M.C. Foote, and R.P Vasquez, *High T_c Superconducting Thin Films: Processing, Characterization, and Applications*, edited by R. Stockbauer and R. Kurtz (AIP, New York, 1989).
15. D.M. Hill, H.M. Meyer III, J.H. Weaver, C.F. Gallo, and K.C. Goretta, Phys. Rev. B **38**, 11,331 (1988).
16. Y.S. Luo, Y.-N. Yang, and J.H. Weaver, Phys. Rev. B **46**, 1114 (1992).
17. M.D. Kirk, C.B. Eom, B. Oh, S.R. Spielman, M.R. Beasley, A. Kapitulnik, T.H. Geballe, and C.F. Quate, Appl. Phys. Lett. **52**, 2071 (1988); P.A.P. Lindberg, Z.-X. Shen, B.O. Wells, D.S. Dessau, D.B. Mitzi, I. Lindau, W.E. Spicer, and A. Kapitulnik, Phys. Rev. B **39**, 2890, (1989).
18. C.K. Shih, R.M. Feenstra, J.R. Kirtly, and G.V. Chandrashekhar, Phys. Rev. B **40**, 2682 (1989).
19. M.D. Kirk, J. Nogami, A.A. Baski, D.B. Mitzi, A. Kapitulnik, T.H. Geballe, and C.F. Quate, Science **242**, 1674 (1988).
20. C.K. Shih, R.M. Feenstra, and G.V. Chandrashekhar, Phys. Rev. B **43**, 7913 (1991).
21. Y. Gao, P. Lee, P. Coppens, M.A. Subramanian, and A.W. Sleight, Science **241**, 954 (1988).
22. H.W. Zandbergen, W.A. Groen, F.C. Mijlhoff, G. van Tendeloo, and S. Amelinckx, Physica C **156**, 325 (1988).
23. Y. LePage, W.R. Mckinnon, J.M. Tarason, and P. Barboux, Phys. Rev. B **40**, 6810 (1989).
24. J.H. Weaver, Z. Lin, and F. Xu, in *Surface Segregation Phenomena*, edited by P.A. Dowben and A. Miller (CRC Press, Boca Raton, 1990) Chap. 10, pp. 259–289, and references therein.
25. J.H. Weaver, in *Electronic Materials: A New Era of Materials Science*, edited by J.R. Chelikowsky and A. Franciosi (Springer-Verlag, Berlin, 1991), Chap. 8, pp. 135–214, and references therein.

26. P.A.P. Lindberg, Z.-X. Shen, B.O. Wells, D.B. Mitzi, I. Lindau, W.E. Spicer, and A. Kapitulnik, Appl. Phys. Lett. **53**, 2563 (1988).
27. P.A.P. Lindberg, Z.-X. Shen, I. Lindau, W.E. Spicer, C.B. Eom, and T.H. Geballe, Appl. Phys. Lett. **53**, 529 (1988); D.S. Dessau, Z.-X. Shen, B.O. Wells, W.E. Spicer, R.S. List, A.J. Arko, R.J. Bartlett, Z. Fisk, S.-W. Cheong, D.B. Mitzi, A. Kapitulnik, and J.E. Schirber, Appl. Phys. Lett. **57**, 307 (1990).
28. Y. Kimachi, Y. Hidaka, T.R. Ohno, G.H. Kroll, and J.H. Weaver, J. Appl. Phys. **69**, 3176 (1991).
29. D.E. Fowler, C.E. Brundle, J. Lerczak, and F. Holtzberg, J. Electron Spectrosc. Relat. Phenom. **52**, 323 (1990).
30. T.R. Ohno, J.C. Patrin, H.M. Meyer III, J.H. Weaver, Y. Kimachi, and Y. Hidaka, Phys. Rev. B **41**, 11,677 (1990).
31. *Binary Alloy Phase Diagrams*, edited by T.B. Massalski (American Society for Metals, Metals Park, OH, 1986) Vol. 1, pp. 845–846, and Vol 2, pp. 1793–1994.
32. *The Oxide Handbook*, edited by G.V. Samsonov (IFI/Plenum, New York, 1982), p. 133.
33. G.D. Waddill, I.M. Vitomirov, C.M. Aldao, S.G. Anderson, C. Capasso, J.H. Weaver, and Z. Liliental-Weber, Phys. Rev. B **41**, 5293 (1990), and references therein. See also J.H. Weaver and G.D. Waddill, Science **251**, 1444 (1991).
34. T.R. Ohno, Y.-N. Yang, G.H. Kroll, K. Krause, L.D. Schmidt, J.H. Weaver, Y. Kimachi, Y. Hidaka, S.H. Pan, and A.L. de Lozanne, Phys. Rev. B **43**, 7980 (1991).
35. T.R. Ohno, J.C. Patrin, H.M. Meyer III, J.H. Weaver, Y. Kimachi, and Y. Hidaka, Phys. Rev. B **41**, 11,677 (1990).

Index